Biomedical Sensors for Functional Mapping: Techniques, Methods, Experimental and Medical Applications

Biomedical Sensors for Functional Mapping: Techniques, Methods, Experimental and Medical Applications

Editors

Alfonso Mastropietro
Alessandro Scano
Massimo W. Rivolta

MDPI

Basel • Beijing • Wuhan • Barcelona • Belgrade • Novi Sad • Cluj • Manchester

Editors

Alfonso Mastropietro
Consiglio Nazionale delle Ricerche
Milano, Italy

Alessandro Scano
Consiglio Nazionale delle Ricerche
Milano, Italy

Massimo W. Rivolta
Università degli Studi di Milano
Milan, Italy

Editorial Office
MDPI
St. Alban-Anlage 66
4052 Basel, Switzerland

This is a reprint of articles from the Special Issue published online in the open access journal *Sensors* (ISSN 1424-8220) (available at: https://www.mdpi.com/journal/sensors/special_issues/bm-sensors).

For citation purposes, cite each article independently as indicated on the article page online and as indicated below:

Lastname, A.A.; Lastname, B.B. Article Title. *Journal Name* **Year**, *Volume Number*, Page Range.

ISBN 978-3-0365-9823-9 (Hbk)
ISBN 978-3-0365-9824-6 (PDF)
doi.org/10.3390/books978-3-0365-9824-6

Contents

sensors

MDPI

Editorial

Biomedical Sensors for Functional Mapping: Techniques, Methods, Experimental and Medical Applications

Alfonso Mastropietro [1,*], Massimo Walter Rivolta [2] and Alessandro Scano [3]

1 Istituto di Tecnologie Biomediche, Consiglio Nazionale delle Ricerche, 20054 Segrate, Italy
2 Dipartimento di Informatica, Università degli Studi di Milano, 20133 Milan, Italy; massimo.rivolta@unimi.it
3 Istituto di Sistemi e Tecnologie Industriali Intelligenti per il Manifatturiero Avanzato, Consiglio Nazionale delle Ricerche, 20133 Milan, Italy; alessandro.scano@stiima.cnr.it
* Correspondence: alfonso.mastropietro@itb.cnr.it

1. Introduction

The rapid advancement of biomedical sensor technology has revolutionized the field of functional mapping in medicine, offering novel and powerful tools for diagnosis, clinical assessment, and rehabilitation. The ability to collect and analyze various physiological signals, even in real-time, has provided unprecedented insights into the "hidden" functioning of the human body. Biomedical sensors have not only enhanced our understanding of human physiology but have also significantly impacted clinical decision-making, patient management, and the development of personalized medical interventions.

This Special Issue presents a collection of 14 papers that showcase the diverse applications of biomedical sensors in the context of functional mapping. The papers can be grouped into three sections, highlighting their contributions to (i) medical diagnosis, detection and prediction; (ii) neurological and rehabilitation assessment; and (iii) medical applications and monitoring. Together, these papers shed light on the transformative role of biomedical sensors in understanding physiological mechanisms and enhancing healthcare practices.

2. Biomedical Sensors for Diagnosis, Detection and Prediction

This section focuses on the application of biomedical sensors for medical diagnosis, detection and prediction. The papers included in this section have a specific focus on the detection of conditions such as COVID-19 and hand osteoarthritis and the prediction of emotions by biosignals. Furthermore, novel approaches based on artificial intelligence and cutting-edge technologies are described.

The paper "COVID-19 Detection Using Photoplethysmography and Neural Networks" [1] presents a groundbreaking approach that utilizes deep learning and raw photoplethysmography signals acquired from a pulse oximeter to identify COVID-19 patients. Achieving an impressive 83.86% accuracy and 84.30% sensitivity in identifying COVID-19 patients, this non-invasive and cost-effective method holds promise for early detection and management of the COVID-19 pandemic, particularly in resource-limited healthcare settings.

In the paper "Toward Early and Objective Hand Osteoarthritis Detection by Using EMG during Grasps" [2], researchers explore the potential of electromyography (EMG) in detecting hand osteoarthritis at an early stage. By studying EMG characteristics during hand grasping tasks, the study provides valuable insights into identifying hand osteoarthritis patients before joint degeneration occurs, enabling timely intervention and improved patient outcomes.

The third paper "Applications of Laser-Induced Fluorescence in Medicine" [3] explores the various medical applications of laser-induced fluorescence (LIF). This highly sensitive spectroscopic method proves valuable in diagnosing and monitoring conditions such as cancer, dental diseases, and fungal infections, offering a versatile tool for medical diagnostics.

Citation: Mastropietro, A.; Rivolta, M.W.; Scano, A. Biomedical Sensors for Functional Mapping: Techniques, Methods, Experiments and Medical Applications. *Sensors* **2023**, *23*, 7063. https://doi.org/10.3390/s23167063

Received: 27 July 2023
Accepted: 7 August 2023
Published: 10 August 2023

Additionally, the paper "Predicting Emotion with Biosignals: A Comparison of Classification and Regression Models for Estimating Valence and Arousal Level Using Wearable Sensors" [4] delves into predicting emotions using biosignals collected via wrist-worn sensors. By comparing different prediction models, the study highlights the effectiveness of regression models, particularly LSTM-based, in estimating emotional valence and arousal levels, enhancing our understanding of human emotions and their applications in healthcare.

3. Biomedical Sensors for Neurological and Rehabilitation Assessment

This section explores the applications of biomedical sensors in neurological assessment and rehabilitation. The papers collected in this section describe interesting advancements in the analysis of electroencephalography (EEG), EMG and Near-Infrared Spectroscopy (NIRS) signals for the extraction of biomarkers to characterize individual status in neuromuscular applications that can have a potential impact, for example, on the assessment of rehabilitation effectiveness.

The paper "Reliability of Mental Workload Index Assessed by EEG with Different Electrode Configurations and Signal Pre-Processing Pipelines" [5] evaluates the reproducibility and sensitivity of mental workload assessment from EEG signals using different electrode configurations and pre-processing pipelines. The findings provide valuable insights into developing reliable methods for assessing cognitive tasks, crucial for enhancing human performance in various domains.

In the paper "A Novel Approach for Segment-Length Selection Based on Stationarity to Perform Effective Connectivity Analysis Applied to Resting-State EEG Signals" [6], researchers proposed a novel approach for selecting appropriate segment lengths in EEG-based effective connectivity analysis. By addressing the critical issue of segment-length selection, this method offers valuable insights into studying brain network interactions during resting-state, improving our understanding of brain function and connectivity.

The paper "Reliable Fast (20 Hz) Acquisition Rate by a TD fNIRS Device: Brain Resting-State Oscillation Studies" [7] introduces a high-power setup for multichannel time-domain functional NIRS measurements. This high-speed acquisition method holds potential applications in studying brain resting-state oscillations, providing valuable information for neuroscientific and clinical research.

The paper "Combined Use of EMG and EEG Techniques for Neuromotor Assessment in Rehabilitative Applications: A Systematic Review" [8] presents a systematic review of combined EEG and EMG techniques in neuromotor assessment during rehabilitation. The review highlights the potential of cortico-muscular interactions for improving rehabilitation approaches in patients with impaired locomotor functions, paving the way for innovative rehabilitation strategies.

Next, the paper "Whole-Body Adaptive Functional Electrical Stimulation Kinesitherapy Can Promote the Restoring of Physiological Muscle Synergies for Neurological Patients" [9] introduces a novel treatment approach, Adaptive Functional Electrical Stimulation Kinesitherapy (AFESK™), for neurological patients using whole-body adaptive functional electrical stimulation kinesitherapy. This treatment shows promise in restoring physiological muscle synergies, enhancing motor functionality, and improving rehabilitation outcomes.

Finally, the paper, "Technology Acceptance Model for Exoskeletons for Rehabilitation of the Upper Limbs from Therapists' Perspectives" [10] addresses the challenges of integrating exoskeleton technology into clinical practice for upper limb rehabilitation. By investigating therapists' perspectives on exoskeleton acceptability, this study reveals factors influencing their willingness to adopt the technology. The findings suggest that integrating exoskeletons with multi-sensor feedback systems may improve acceptance and facilitate better patient outcomes.

4. Biomedical Sensors for Medical Applications and Monitoring

This last section emphasizes the role of biomedical sensors in medical applications and monitoring, describing novel technologies and tools to improve health monitoring in different medical scenarios.

The paper "Towards a Practical Implementation of a Single-Beam All-Optical Non-Zero-Field Magnetic Sensor for Magnetoencephalographic Complexes" [11] introduces a single-beam all-optical two-channel magnetic sensor scheme developed for non-zero-field magnetoencephalography and magnetocardiography applications. This innovative sensor scheme utilizes a single laser beam with time-modulated linear polarization to detect magnetic resonance, providing valuable insights for neurological assessments and diagnostic applications.

The paper "Experimental Assessment of Cuff Pressures on the Walls of a Trachea-Like Model Using Force Sensing Resistors: Insights for Patient Management in Intensive Care Unit Settings" [12] investigates the pressures exerted by endotracheal tube cuffs on the walls of a test bench mimicking the laryngotracheal tract. The study provides valuable insights for patient management in intensive care unit settings, highlighting the need for periodic checks of cuff pressure to prevent pressure-related complications.

Continuing in the realm of prosthetic control, the paper "Questioning Domain Adaptation in Myoelectric Hand Prostheses Control: An Inter- and Intra-Subject Study" [13] delves into the challenges of domain adaptation techniques in myoelectric hand prosthesis control. The results question the conventional approach based on transfer learning and suggest the need for further exploration in this area.

The last paper in this section, "Multi-Scale Evaluation of Sleep Quality Based on Motion Signal from Unobtrusive Device" [14], introduces a multi-scale method for evaluating sleep behavior using motion signals obtained from a pressure bed sensor. The algorithm provides a good correlation between sleep quality measures obtained with polysomnography and pressure bed sensors, offering potential applications for home monitoring of sleep and improving subjects' awareness of potential sleep disorders.

5. Conclusions

The Special Issue *"Biomedical Sensors for Functional Mapping: Techniques, Methods, Experimental and Medical Applications"* presents a comprehensive collection of cutting-edge research in the field of biomedical sensors. The papers cover a wide range of applications, including medical diagnosis and detection, neurological assessment and rehabilitation, and medical monitoring. These advancements pave the way for improved healthcare practices, patient outcomes, and personalized medicine. As biomedical sensor technology continues to evolve, the findings from these research studies hold significant promise in revolutionizing medical practices and addressing complex health challenges, ultimately leading to better human health and well-being.

Author Contributions: Conceptualization, A.M., M.W.R. and A.S.; writing—original draft preparation, A.M., M.W.R. and A.S.; writing—review and editing, A.M., M.W.R. and A.S. All authors have read and agreed to the published version of the manuscript.

Funding: This work was partially funded by Fondazione Cariplo and Regione Lombardia, Project "Active3—Everyone, Everywhere, Everyday (ref. 2021-0612)".

Acknowledgments: We would like to extend our heartfelt gratitude and appreciation to all the contributors to this Special Issue. The success of this collection of research papers is a testament to the dedication, expertise, and hard work of each and every author. Their valuable insights, innovative methodologies, and groundbreaking findings have enriched the field of medical diagnostics and detection using biomedical sensors. We are also deeply thankful to the peer reviewers for their diligent and meticulous evaluations, which have ensured the high quality and rigor of the published papers. Additionally, we extend our thanks to the editorial and technical teams for their unwavering support throughout the process, ensuring a smooth and efficient publication journey. Finally, we express our gratitude to the readers and the broader scientific community for their interest and engagement in this

Special Issue, as it is their enthusiasm that drives the advancement of knowledge and fosters progress in healthcare technologies. Together, we have achieved a remarkable and significant contribution to the field, and we look forward to continued collaboration and further advancements in biomedical sensor applications.

Conflicts of Interest: The authors declare no conflict of interest. The funders had no role in the design of the study; in the collection, analyses, or interpretation of data; in the writing of the manuscript; or in the decision to publish the results.

References

1. Lombardi, S.; Francia, P.; Deodati, R.; Calamai, I.; Luchini, M.; Spina, R.; Bocchi, L. COVID-19 Detection Using Photoplethysmography and Neural Networks. *Sensors* **2023**, *23*, 2561. [CrossRef] [PubMed]
2. Jarque-Bou, N.J.; Gracia-Ibáñez, V.; Roda-Sales, A.; Bayarri-Porcar, V.; Sancho-Bru, J.L.; Vergara, M. Toward early and objective hand osteoarthritis detection by using EMG during Grasps. *Sensors* **2023**, *23*, 2413. [CrossRef] [PubMed]
3. Kwaśny, M.; Bombalska, A. Applications of laser-induced fluorescence in medicine. *Sensors* **2022**, *22*, 2956. [CrossRef] [PubMed]
4. Siirtola, P.; Tamminen, S.; Chandra, G.; Ihalapathirana, A.; Röning, J. Predicting Emotion with Biosignals: A Comparison of Classification and Regression Models for Estimating Valence and Arousal Level Using Wearable Sensors. *Sensors* **2023**, *23*, 1598. [CrossRef] [PubMed]
5. Mastropietro, A.; Pirovano, I.; Marciano, A.; Porcelli, S.; Rizzo, G. Reliability of Mental Workload Index Assessed by EEG with Different Electrode Configurations and Signal Pre-Processing Pipelines. *Sensors* **2023**, *23*, 1367. [CrossRef] [PubMed]
6. Góngora, L.; Paglialonga, A.; Mastropietro, A.; Rizzo, G.; Barbieri, R. A Novel Approach for Segment-Length Selection Based on Stationarity to Perform Effective Connectivity Analysis Applied to Resting-State EEG Signals. *Sensors* **2022**, *22*, 4747. [CrossRef] [PubMed]
7. Re, R.; Pirovano, I.; Contini, D.; Amendola, C.; Contini, L.; Frabasile, L.; Levoni, P.; Torricelli, A.; Spinelli, L. Reliable Fast (20 Hz) Acquisition Rate by a TD fNIRS Device: Brain Resting-State Oscillation Studies. *Sensors* **2022**, *23*, 196. [CrossRef] [PubMed]
8. Brambilla, C.; Pirovano, I.; Mira, R.M.; Rizzo, G.; Scano, A.; Mastropietro, A. Combined use of EMG and EEG techniques for neuromotor assessment in rehabilitative applications: A systematic review. *Sensors* **2021**, *21*, 7014. [CrossRef] [PubMed]
9. Scano, A.; Mira, R.M.; Gabbrielli, G.; Molteni, F.; Terekhov, V. Whole-Body Adaptive Functional Electrical Stimulation Kinesitherapy Can Promote the Restoring of Physiological Muscle Synergies for Neurological Patients. *Sensors* **2022**, *22*, 1443. [CrossRef] [PubMed]
10. Luciani, B.; Braghin, F.; Pedrocchi, A.L.G.; Gandolla, M. Technology Acceptance Model for Exoskeletons for Rehabilitation of the Upper Limbs from Therapists' Perspectives. *Sensors* **2023**, *23*, 1721. [CrossRef] [PubMed]
11. Petrenko, M.; Vershovskii, A. Towards a Practical Implementation of a Single-Beam All-Optical Non-Zero-Field Magnetic Sensor for Magnetoencephalographic Complexes. *Sensors* **2022**, *22*, 9862. [CrossRef] [PubMed]
12. Crivello, A.; Milazzo, M.; La Rosa, D.; Fiacchini, G.; Danti, S.; Guarracino, F.; Berrettini, S.; Bruschini, L. Experimental assessment of cuff pressures on the walls of a trachea-like model using force sensing resistors: Insights for patient management in intensive care unit settings. *Sensors* **2022**, *22*, 697. [PubMed]
13. Marano, G.; Brambilla, C.; Mira, R.M.; Scano, A.; Müller, H.; Atzori, M. Questioning domain adaptation in myoelectric hand prostheses control: An inter-and intra-subject study. *Sensors* **2021**, *21*, 7500. [PubMed]
14. Coluzzi, D.; Baselli, G.; Bianchi, A.M.; Guerrero-Mora, G.; Kortelainen, J.M.; Tenhunen, M.L.; Mendez, M.O. Multi-Scale Evaluation of Sleep Quality Based on Motion Signal from Unobtrusive Device. *Sensors* **2022**, *22*, 5295. [CrossRef] [PubMed]

sensors

MDPI

Systematic Review

Combined Use of EMG and EEG Techniques for Neuromotor Assessment in Rehabilitative Applications: A Systematic Review

Cristina Brambilla [1,†], Ileana Pirovano [2,†], Robert Mihai Mira [1], Giovanna Rizzo [2,*], Alessandro Scano [1,‡] and Alfonso Mastropietro [2,‡]

1 Istituto di Sistemi e Tecnologie Industriali Intelligenti per il Manifatturiero Avanzato (STIIMA), Consiglio Nazionale delle Ricerche (CNR), Via Previati 1/E, 23900 Lecco, Italy; cristina.brambilla@stiima.cnr.it (C.B.); robertmihai.mira@stiima.cnr.it (R.M.M.); alessandro.scano@stiima.cnr.it (A.S.)

2 Istituto di Tecnologie Biomediche (ITB), Consiglio Nazionale delle Ricerche (CNR), via Fratelli Cervi 93, 20054 Segrate, Italy; ileana.pirovano@itb.cnr.it (I.P.); alfonso.mastropietro@itb.cnr.it (A.M.)

* Correspondence: giovanna.rizzo@itb.cnr.it; Tel.: +39-02-2642-2210

† These authors contributed equally to the work.

‡ These authors contributed equally to the work.

Abstract: Electroencephalography (EEG) and electromyography (EMG) are widespread and well-known quantitative techniques used for gathering biological signals at cortical and muscular levels, respectively. Indeed, they provide relevant insights for increasing knowledge in different domains, such as physical and cognitive, and research fields, including neuromotor rehabilitation. So far, EEG and EMG techniques have been independently exploited to guide or assess the outcome of the rehabilitation, preferring one technique over the other according to the aim of the investigation. More recently, the combination of EEG and EMG started to be considered as a potential breakthrough approach to improve rehabilitation effectiveness. However, since it is a relatively recent research field, we observed that no comprehensive reviews available nor standard procedures and setups for simultaneous acquisitions and processing have been identified. Consequently, this paper presents a systematic review of EEG and EMG applications specifically aimed at evaluating and assessing neuromotor performance, focusing on cortico-muscular interactions in the rehabilitation field. A total of 213 articles were identified from scientific databases, and, following rigorous scrutiny, 55 were analyzed in detail in this review. Most of the applications are focused on the study of stroke patients, and the rehabilitation target is usually on the upper or lower limbs. Regarding the methodological approaches used to acquire and process data, our results show that a simultaneous EEG and EMG acquisition is quite common in the field, but it is mostly performed with EMG as a support technique for more specific EEG approaches. Non-specific processing methods such as EEG-EMG coherence are used to provide combined EEG/EMG signal analysis, but rarely both signals are analyzed using state-of-the-art techniques that are gold-standard in each of the two domains. Future directions may be oriented toward multi-domain approaches able to exploit the full potential of combined EEG and EMG, for example targeting a wider range of pathologies and implementing more structured clinical trials to confirm the results of the current pilot studies.

Keywords: EMG; EEG; rehabilitation; neuromotor; evaluation; assessment; review

Citation: Brambilla, C.; Pirovano, I.; Mira, R.M.; Rizzo, G.; Scano, A.; Mastropietro, A. Combined Use of EMG and EEG Techniques for Neuromotor Assessment in Rehabilitative Applications: A Systematic Review. *Sensors* **2021**, *21*, 7014. https://doi.org/10.3390/s21217014

Academic Editor: Yvonne Tran

Received: 23 August 2021
Accepted: 20 October 2021
Published: 22 October 2021

1. Introduction

Neuromotor disorders are developmental or acquired conditions usually caused by neurological diseases affecting the central nervous system that typically impair movement, gross and fine motor ability, and posture. It was recently reported that neurological disorders are the third most common cause of disability and premature death in the European Union [1], and their prevalence will increase with the progressive aging of the population. Therefore, neuromotor disorders associated with neurological diseases currently represent a burden for patients in all age ranges, health systems, and caregivers (formal and informal). A detailed comprehension of the processes underlying motor impairment, with direct involvement of the central nervous/peripheral nervous system, is at the basis of motor recovery understanding [2]. Moreover, the use of quantitative and instrumental neuromotor assessments can foster the design of effective therapeutic interventions and promote the development of personalized therapies to maximize motor recovery [3].

Different instruments and techniques have been developed to offer clinically reliable assessments of the neuromotor performances of patients. Two of the most valuable techniques used for analyzing, evaluating, and assessing motor performance employed in the rehabilitation field are EEG and surface EMG (sEMG, hereby only EMG). They record the electrical potentials that originated at cortical and muscular levels, respectively. EEG is a noninvasive and versatile technique that measures electrical activity related to neuron pools at the cortical level and is suitable for clinical, experimental, and real-life scenarios [4], whereas EMG measures the train of motor unit action potentials, generated by muscular contraction, through surface electrodes placed on the skin overlying muscle fibers [5]. They thus provide insights into neuromotor integrity/impairment by monitoring cortical activation and its motor correlates. Both techniques have been employed for neuromotor assessments, especially in research studies on rehabilitative applications.

In particular, EEG can explore the brain activity at the cortical and subcortical level and allows neuronal brain dynamics to be monitored with a high temporal resolution to explore whole-brain neuronal networks organization [6]. In the reference literature, EEG signal analysis was historically used in different applications to assess the activation pathways and to understand basic mechanisms underlying motor functions [7]. This led to more recent specific studies focused on neuromotor rehabilitation for investigating how activity patterns change depending on the location of the cortical lesions and on different rehabilitation treatments in different diseases such as stroke [8], Parkinson's disease [9], and others. The evolution of the EEG signal analysis during the last decades shifted the focus from the time domain to the frequency domain analysis [10] with the more recent use of functional and effective connectivity approaches [11] to better understand neural network changes occurring in physiological and pathological conditions. Another relevant application is related to the use of EEG signal for interactively guiding the rehabilitation session using brain-computer interfaces (BCI) or biofeedback methods to control a rehabilitation robot [12] or a lower limb exoskeleton [13], to monitor the status of recovery [14] and to evaluate the patient's engagement in traditional motor rehabilitation [15] and in virtual reality environments [16].

Analogously, in the rehabilitation field, EMG analysis has been used for a variety of assessments. Current applications of EMG are mainly related to the physiological investigation, monitoring of neurological disorders, and planning of treatments [17]. The study of muscle activity and coordination patterns is a useful tool for the identification of motor disorders and the evaluation of motor recovery after rehabilitation. Muscle activation patterns were identified in both upper [18] and lower limbs [19]; furthermore, EMG was employed for studying abnormal muscular activity, such as spasticity [20], and effect such as muscle fatigue [21]. Moreover, the factorization of the EMG signal is also at the basis of motor control theories, such as muscle synergies, which provide insight into the control mechanisms for motor planning [22,23]. Applications of this theory have been oriented toward quantifying motor control abnormalities [24] and changes in the muscular activation patterns [25]. EMG signals are also used to control exoskeletons for improving motor rehabilitation and to support daily life activities [26]. Interesting applications were also found in prosthetic control for amputees using residual EMG near the amputated region [27].

However, the employment of EEG and EMG signals has been only partially explored so far, especially in combined applications, although their coupling seems natural and effective [28]. It is indeed clear from existing literature that these techniques carry critical and complementary information regarding several aspects related to neuromotor assessment. In fact, it has been shown that these techniques allow a better understanding of pathologies involving the central nervous systems causing motor deficits, especially from the neuromotor point of view. EEG and EMG combined usage also contributes with detailed insights to the customization and tailoring of therapies by supporting the clinicians with relevant data on motor organization. Another potential impact provided with EEG and EMG is the outcome prediction. This issue has been explored with EEG [29] and EMG [30] in separate studies and acquires variability in a scenario that is evolving toward rationalization of the resources, containment of the costs, and rehabilitation efficiency [31].

Interestingly, we noticed that, despite their potential, EEG and EMG have been considered simultaneously in applications with assessment aims only occasionally. They might help in profiling the level of disability with multi-parameters approaches [32] and can constitute solid bases for novel approaches based on detailed multimodal assessments [33,34]. We also noticed a lack of comprehensive reviews describing which scenarios have been explored, what applications, setups, methods of analysis, and potential developments can be foreseen for such techniques, whereas most of the works where EEG and EMG are coupled focuses on brain-computer and multimodal interfaces for feedback and control [35]. Indeed, BCI and biofeedback are the first research fields for which the combination of the two signals has been successfully employed, and the literature of the past few years focused on these applications, providing an overview of the possibilities offered by the techniques until now.

Following the previous considerations, this systematic literature analysis aims to cover a field that has been less exhaustively described, reviewing all available studies in which EMG and EEG were combined for clinical practice, targeting applications of the two combined techniques not only for guiding rehabilitation but mostly for the evaluation and the assessment of physio-pathological motor function in both healthy subjects and patients. This review also provides critical comments on the current state-of-the-art approaches and future trends and directions.

2. Materials and Methods

This review attempts to answer the main research question (RQ 0): "How have EMG and EEG been combined in clinical practice for assessment of people in rehabilitation?". RQ 0 is furtherly split into the following research questions:

RQ (1) Which type of experimental study design was employed?

RQ (2) Which groups of subjects, pathologies, and anatomical segments were targeted with the combined EEG-EMG approach in rehabilitation?

RQ (3) What setups were used for rehabilitation and signal acquisition?

RQ (4) What analysis techniques have been employed and what results were achieved?

We thus considered papers that applied EMG and EEG simultaneously provided an overview of which scenarios were considered for applications and which setups were used for rehabilitation and acquisitions, explored the data analysis techniques and the achieved results. The international guidelines established by PRISMA (Preferred Reporting Items for Systematic Reviews and Meta-Analyses) [36] were used.

2.1. Criteria for Papers Classification

Our review of the previous literature was organized to summarize the state-of-the-art of the field by detailing the following categories:

- TYPE OF STUDY: lists the papers based on the study type described in the text (e.g., Observational, Pilot, Randomized Controlled Trial, and Methodological).

This section answers the question RQ 1.

- SUBJECTS AND ANATOMICAL TARGETS: describes to which clinical scenarios EEG and EMG were applied together. This section answers the question RQ 2.

This section was further specialized into:

Cohort of Subjects: aiming to summarize what kind of subjects were enrolled for combined EEG-EMG studies (e.g., post-stroke subjects and healthy controls) and what sample size of patients/subjects was enrolled in the experimental studies.

Anatomical Targets: aiming to review which anatomical segments were assessed and/or rehabilitated in concurrent EEG-EMG studies (e.g., upper limb).

- EXPERIMENTAL SETUP and PROTOCOLS: describes which experimental setups were employed for rehabilitation and data acquisitions. This section answers the question RQ 3.

Experimental setup and protocols were further divided into:

Setup for tests/rehabilitation: aiming to describe which setup was used for rehabilitation (e.g., robotic assistive device)

Setup for signal acquisition: aiming to review what setups were used for data collection (e.g., 16 channel s-EMG).

- DATA ANALYSIS: describes which techniques and findings were used for data analysis. This section answers the question RQ 4.

Data analysis was further divided into:

Analysis Techniques: aiming to describe which techniques have been employed and which domains and features were considered in the analysis (e.g., time/frequency).

Benefits of combined EEG-EMG applications: aiming to describe which were the main findings that were achieved using the combined EEG-EMG acquisition and analysis.

2.2. Bibliographic Research Criteria

With the above-mentioned aims, the following procedure was employed for the literature screening. A collection of articles was obtained by screening PubMed, Scopus, and Web of Science (WOS), using a query based on the keywords: "EEG", "EMG", "MUSCL*", "MOTOR*", "MOVEMENT", "MOTION", "REHABILITATION" and excluding the keyword "BCI". Articles strictly concerning BCI and biofeedback implementation with electrical biological signals were excluded since their main aim is commonly not focused on combined EEG/EMG functional assessment.

The formal logical query was (EEG) AND (EMG OR MUSCL*) AND (MOTOR* OR MOVEMENT OR MOTION) AND (REHABILITATION) AND NOT (BCI).

2.3. Eligibility Criteria

In the eligibility phase, we distinguished the papers relevant to the aim of this review. For being eligible, screened papers had to satisfy all the following criteria:

(A) To include the specified query in the abstract and/or title and/or in the keywords
(B) To involve the simultaneous use of EMG and EEG
(C) EMG and EEG had to be used for neuromotor assessment
(D) To target rehabilitation scenarios
(E) To be indexed in at least one of the screened databases
(F) To be a full article (at least 4 pages)
(G) To be available in English

The papers were screened, one by one, for inclusion by two different groups (composed of subgroups of the authors of the paper) independently. The main inclusion criteria had to include: "criteria A AND B AND C AND D AND E AND F AND G". Each paper was screened by two different reviewers who blindly classified it as eligible or non-eligible. Any disagreement in the classification was settled by discussion between the two groups, and a consensus was reached in all cases.

3. Results

3.1. Selected Papers

As a result of the screening, 174 papers were found on Scopus, 58 on PubMed, and 144 on WOS. The total number of articles was 385. Out of all these articles, 163 were duplicates across the 3 databases. The number of studies eligible for the detailed screening was 213.

After the screening phase, the number of papers identified as eligible, meeting all the selection criteria, and included in the review was 55. In the next sections, the results of our research are presented. The PRISMA flow chart summarizing all the steps for screening and inclusion is presented in Figure 1.

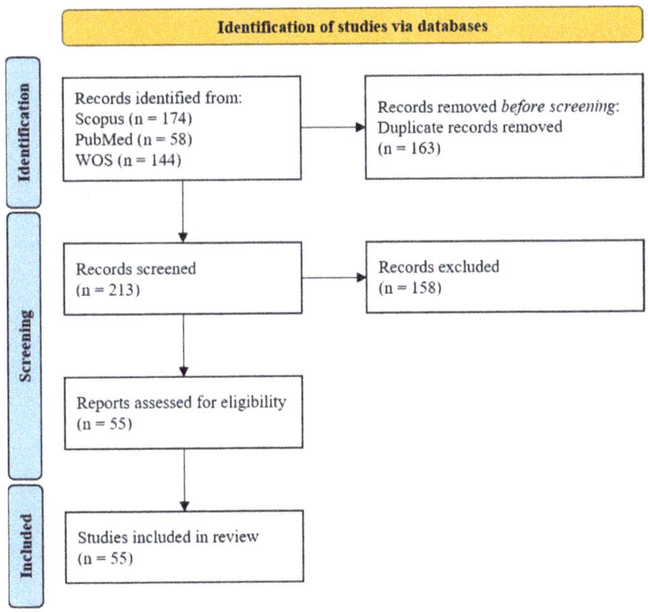

Figure 1. The PRISMA flow chart for the proposed literature review [36].

As shown in Figure 2, most of the papers describing concomitant applications of EEG and EMG in the assessment of neuromotor skills in rehabilitation were recently published. Indeed, more than 50% of the papers included in this review were produced in the last 5 years.

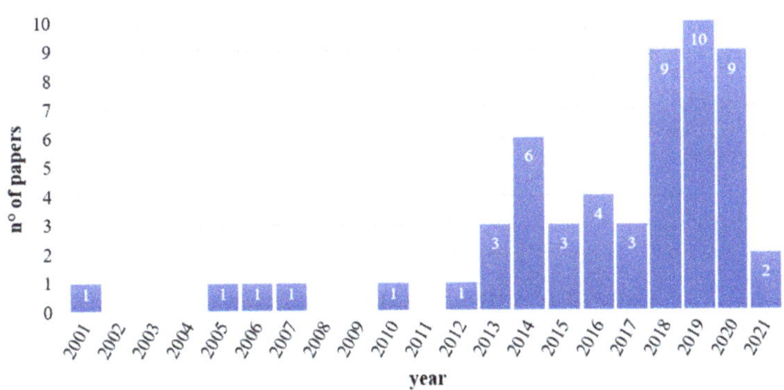

Figure 2. Temporal distribution (number of papers published per year) of selected papers.

An exploratory analysis of the 50 most cited words within the selected papers has highlighted that they appear overall 34,540 times in the text, and they represent more than 10% of all the words composing the whole documents. Among them, the most cited word is motor (1900 citations), followed by EEG (1845) and stroke (1805). Among the first 10 most cited words, we also found: EMG, patients, study, data, movement, muscle, and coherence. A pictorial representation is shown in Figure 3.

Figure 3. Word-cloud representing the 50 most cited words (excluding all the words not representing nouns and not relevant acronyms) included in the text of the papers selected in this review. The higher the size, the higher the number of citations inside the papers.

3.2. *Type of Study*

In this section, the papers were subdivided according to the study design proposed by the experimenters. In Table 1, we grouped the works into four categories: observational study, pilot study, randomized controlled trial, and methodological study. For each paper, we also detailed the aim of the study. The distribution of papers in the categories is shown in a pie chart in Figure 4.

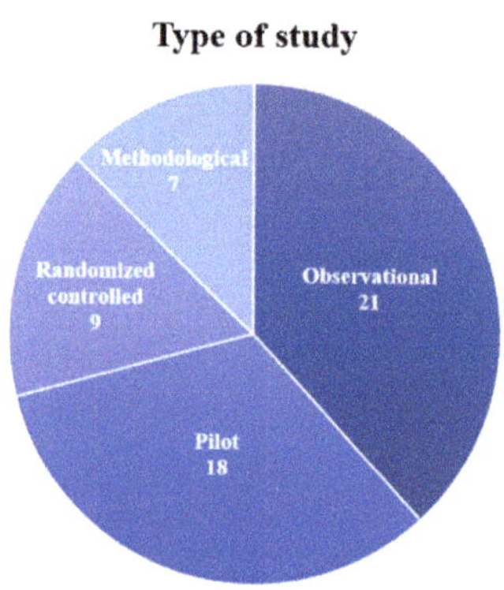

Figure 4. Pie chart portraying the number of selected studies for each study design (Observational, Pilot, Randomized Controlled, Methodological).

Table 1. Type of study and aim.

Type of Study	Aim
Observational study	Study of the cortico-muscular coupling (1) during motor tasks [37–41] and (2) with electrical stimulation [42–46] Investigation of the effects of exoskeleton on functional connectivity [47–49] Investigation of the effects of visual feedback [50,51] Detection of movement intention [52,53] Study of the interhemispheric interaction with TMS [54] Study of a neurophysiological marker of stress [55] Study of slow cortical potentials in stroke [56] Study of correlation between lower back pain and altered postural stabilization [57]
Pilot study	Test new rehabilitation paradigm [34,58–62] Investigation of the efficacy of BMI [63] and EEG feedback [64] Study of movement classification combining EEG and EMG [65] Investigation of movement intention [66,67] and motor imagery detection [68] Study of the cortico-muscular coupling [69–71] Study of the neuroplasticity with electrical stimulation [72] Investigation of engagement in game rehabilitation [73] Study of the effects of the use of VR in facial rehabilitation [74]
Randomized controlled trial	Study the effects of transcranial [75,76] and peripheral electrical stimulation [77–79] Investigation of the efficacy of lower limb exoskeleton rehabilitation [80] Assessment of a novel gait training paradigm [81] Investigation of the efficacy of neurologic therapy based on music [82] Investigation of the effects of biofeedback [83]
Methodological study	Presentation of a multivariate approach for motor assessment [32,33] Presentation of a method for compressing EEG-EMG signal [84] Presentation of algorithms for motion detection [85,86] and motion classification [87,88]

A total of 21 papers out of 55 (37%) presented observational studies in which functional parameters or effects of treatments were investigated on healthy subjects and patients. An aim commonly found in these works was the assessment of the cortico-muscular coupling during movements [37–41] as a method to better understand motor control mechanisms for improving the rehabilitation design. Cortico-muscular coherence was also tested as a tool for investigating the effects of functional electrical stimulation [42–46]. Some studies analyzed the effects of treatments based on exoskeletons on neuromotor outcomes [47–49]. The efficacy of visual feedback was assessed to explore novel rehabilitation paradigms [50,51]. Two studies only [52,53] were interested in detecting movement intention coupled with EEG and EMG recordings. Other works investigated specific parameters typical of each study: Palmer et al. [54] studied the interhemispheric interaction using transcranial stimulation, Vladimirov et al. [55] searched neurophysiological markers of stress, Yilmaz et al. [56] investigated slow cortical potentials in stroke patients, and Jacobs et al. [57] studied the correlation between low back pain and postural stabilization.

Papers that tested a novel experimental setup or concept design on a limited number of subjects were classified as pilot studies. A total of 18 out of 55 papers (33%) were classified as pilot studies. Many of these studies presented novel rehabilitation paradigms based on robotics [58–60] or exoskeletons [34,61,62]. Donati et al. [63] tested a multi-stage brain-machine interface (BMI), while Hashimoto et al. [64] used the EEG feedback for improving rehabilitation. Some studies presented preliminary results for methods of movement classification [65], detection of movement intention [66,67], and motor imagery [68]. Three studies investigated cortico-muscular coupling [69–71] as a novel method to evaluate the motor recovery of post-stroke patients. Neuroplastic changes induced by TMS were studied by Dutta et al. [72] to improve rehabilitation technologies. Moreover, pilot studies were conducted for the evaluation of the level of engagement during game rehabilitation [73] and of the effects of virtual reality on facial rehabilitation [74].

A total of 9 studies out of 55 (17%) presented randomized controlled trials that subdivided the enrolled cohorts into treatment groups compared to the control groups to test the validity of an experimental setup. Bao et al. [75] and Benninger et al. [76] studied the efficacy of employing transcranial stimulation on stroke and parkinsonian patients, respectively, while three studies applied peripheral electrical stimulation on stroke patients [77–79]. Rehabilitation for stroke patients was investigated by Calabrò et al. [80] using an exoskeleton and by Chen et al. [81] with a novel treadmill. Furthermore, the efficacy of sensorial feedback based on music [82] or EEG/EMG biofeedback [83] in stroke patients was assessed during rehabilitation to improve motor recovery.

Finally, seven papers (13%) were classified as methodological studies: they presented novel methods and algorithms for analyzing together EEG and EMG signals. Cisotto et al. [84] provided a method for compressing EEG and EMG signals. Other studies developed algorithms for detecting motion [85,86] and classifying it [87,88]. Belfatto et al. [32] and Pierella et al. [33], instead, showed a methodology for a multivariate motor assessment aiming at proposing a novel methodology for evaluating rehabilitation.

3.3. Subjects and Anatomical Targets

This paragraph describes the properties of the cohorts of subjects involved in the experimental sessions considering the clinical status of the subjects (healthy vs. pathologic), age range, and the sample size. Furthermore, the anatomical targets for the functional assessment and/or rehabilitation are described. A summary of the most relevant results described in this section is reported in Table 2.

Table 2. Subjects and anatomical targets.

Categories		Details and References
Type of subjects	Healthy (Age range)	Target: 20–30 [61], n.a. [84], 24–36 [72], 60–62 [62], 18–35 [51], 57.8 ± 4.7 [66], 24 ± 2.32 [47,48], 23–27 [67], 26.86 ± 3.39 [52], 25.0 ± 1.7 [60], 22–28 [41], 22 [86], 22.8 ± 3.3 [87], 24.9 ± 5.4 [88], 21.2 ± 1.1 [44], 26.5 ± 6.5 [49], 23–27 [45,46] Control group: 20–39 [42], 53–62 [58], 27 ± 4 [38], 50.08 ± 15.8 [39], 35.4 ± 5.25 [57], 24 ± 1.5 [43], 27 ± 4 [65], 53 ± 14 [54], 58 ± 16 [33], 24–27 [40], 20 ≥ 60 [85], 33.5 ± 7.9 [53], 55.1 ± 2.1 [55], 35.9 ± 7.7 [68], n.a. [69,70], 42 ± 13 [71]
	Pathologic (Age range)	Chronic stroke: 53–72 [37], 37–72 [42], 35–63 [75], 61 ± 11 [32], 55–77 [80], 52–63 [58], 52.5 ± 9.7 [81], 56.5 ± 9.5 [39], 43–79 [43], 69.9 ± 10.5 [77], 46–81 [54], 56.5 ± 9.5 [78], 68 ± 18 [33], 52.7 ± 8.4 [74], 45–51 [40], 49.9 ± 10.9 [79], 56 [68], n.a. [70], 46–60 [71], 51.4 ± 11.1 [56]; Subacute stroke: 32–79 [50]; Subacute and chronic stroke: 52–64 [69], 42–92 [82] Parkinson disease: 40–80 [76]; Cerebral palsy: 5–19 [34]; Spinal cord injury: 32.5 ± 6.2 [38], 26–38 [63], 32 ± 6 [65], 43.5 ± 12.4 [53]; Writer's cramp: 67 [64]; Low back pain: 39.2 ± 6.33 [57]; Mixed injuries and diseases: 33–54 [73], 60–80 [59]; Facial palsy: 23 [74]; MCI: n.a. [85]; Cardiovascular diseases: 56.3 ± 1.0 [55]
Nr of subjects	≤10	6 [37], 5 [32], 6 [34], 3 + 4 [58], 6 [61], 1 [84], 8 [63], 10 [72], 2 [62], 1 [64], 7 [66], 5 [47], 8 [67], 7 [52], 10 [33], 1 [74], 5 + 5 [40], 10 [41], 1 [86], 4 + 4 [53], 4 + 4 [71], 6 [49], 10 [45,46]
	10 < n ≤ 20	11 [75], 14 [50], 9 + 9 [81], 10 + 8 [38], 10 + 10 [57], 16 [51], 7 + 4 [69], 10 + 8 [65], 15 [77], 12 [78], 6 + 6 [33], 18 [87], 13 [44], 10 + 1 [68], 7 + 7 [70], 20 [56]
	>20	16 + 12 [42], 26 [76], 20 + 20 [80], 14 + 10 [39], 15 + 15 [43], 30 [47,48], 23 [59], 12 + 30 [82], 19 + 14 [54], 30 [83], 14 + 14 [79], 28 + 7 [85], 32 [88], 14 + 14 [55]

Table 2. *Cont.*

Categories		Details and References
Rehabilitation target	Distal Upper Limb	Hand and wrist: [50,59,61,69,84]; Wrist: [37,40,44,75,82]; Hand: [39,43,45,46,51,54,56,62,64,67,73,76,78,83]
	Proximal Upper Limb	Shoulder and elbow: [32,33,58,60,70,71,82,84]; Elbow: [37–40,62,65,76,87,88]; Arm: [52,85,86]; Shoulder: [55]
	Distal Lower Limb	Ankle: [42,47–49,51,63,80]; Leg: [41,72,77,79,81]; Foot: [81]
	Proximal Lower Limb	Knee: [34,42,47–49,80]; Hip and knee: [63]
	Other	Torso: [57]; Face: [74]; Neck: [68]

3.3.1. Cohorts of Subjects

In the papers analyzed in this review, the cohorts of subjects involved during the experimental sessions could be divided into two macro-categories: (i) healthy subjects and (ii) patients affected by different diseases and pathological conditions affecting the neuromotor system. In particular, 36 out of 55 (64%) studies have enrolled healthy volunteers as either target groups (19 out of 36–53%) or control groups (17 out of 36–47%). Conversely, 36 out of 55 (65%) papers have enrolled patients.

Most of the studies involving patients were focused on stroke (23 out of 36–64%) at different stages. The chronic stage was studied in 20 out of 23 papers (87%), the subacute phase was described in 1 paper (4%), whereas a longitudinal analysis (subacute and chronic stage) was performed in 2 out of 23 papers (9%).

A total of 13 out of 36 papers considered other types of pathological conditions such as: spinal cord injury (4 out of 13–31%), mixed injuries and diseases (2 out of 13–15%), Parkinson's disease, cerebral palsy, writer's cramp, low back pain, facial palsy, mild cognitive impairment and cardiovascular diseases (1 document each). See Figure 5 for a schematic representation of the results.

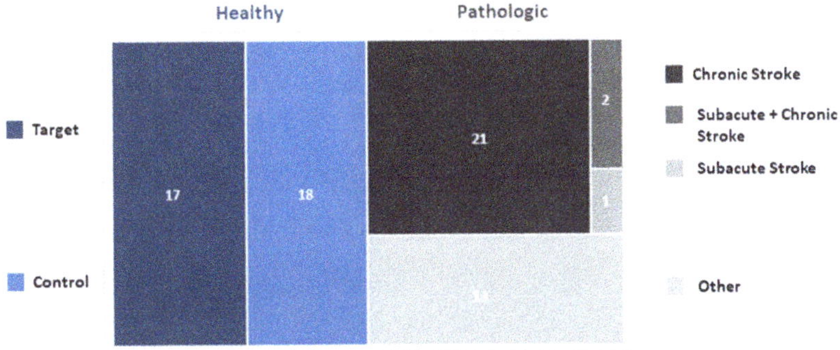

Figure 5. Graph representing the distribution of selected papers based on the cohort of subjects enrolled.

The age of the subjects involved in the studies ranged from 5 to 92. Most of the healthy subjects' cohorts were composed of young adults (up to 40) (21 out of 36 documents—58%), whereas the patients' cohorts were mostly composed of adults (> 40) or older adults. (>65) (29 out of 36 papers—80%).

Considering the sample size, 24 out of 55 (44%) papers involved at most 10 subjects, 16 out of 55 papers (29%) enrolled up to 20 subjects, whereas 15 out of 55 papers (27%) enrolled more than 20 subjects. It is worth noticing that four papers describe results based on a single subject analysis, whereas the highest number of subjects involved was 42. See Figure 6 for a schematic representation of the results.

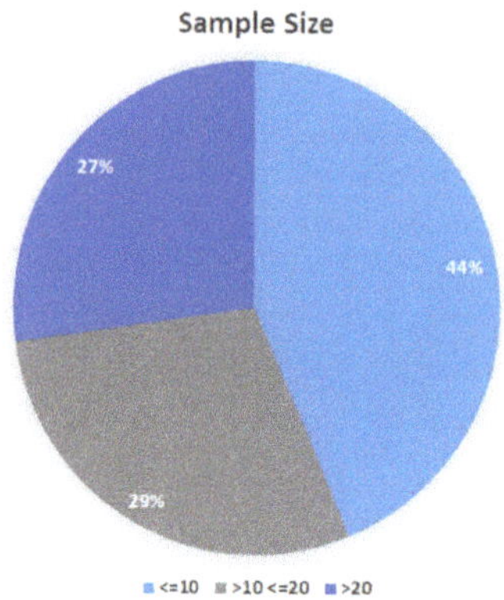

Figure 6. Pie chart portraying the percentage of selected studies for the sample size.

3.3.2. Anatomical Targets

Regarding the anatomical regions that were objects of study, the distal upper limb was predominantly considered in 24 out of 55 papers (44%), the proximal upper limb was evaluated in 21 out of 55 documents (38%), the distal lower limb is included in 13 out of 55 papers (24%) whereas the proximal lower limb in 8 out of 55 papers (15%). It is worth noticing that in 3 out of 55 documents (5%), the focus was put on other regions (torso, face, neck).

Specifically, considering the distal upper limb, most of the applications were focused on the assessment/rehabilitation of hand movements (14 out of 24 papers—58%); 5 out of 24 studies (21%) considered the wrist, whereas 5 out of 24 (21%) documents described the concurrent analysis of wrist and hand. As to the proximal upper limb, 9 out of 21 (43%) papers were focused on the elbow, 3 out of 21 (14%) generically on the arm, 1 out of 21 (5%) on the shoulder, whereas 8 out of 21 (38%) studies were focused on the combined analysis of shoulder and elbow.

As to the lower limb, in the distal part, 7 out of 13 (54%) papers focused on the ankle, 5 out of 13 (38%) documents on the leg, and 1 out of 13 (8%) studies on the foot movements. Finally, considering the proximal lower limb, 6 out of 8 (88%) papers focused on the knee whereas 1 out of 8 (12%) on the simultaneous analysis of hip and knee. A schematic representation of the results is shown in Figure 7.

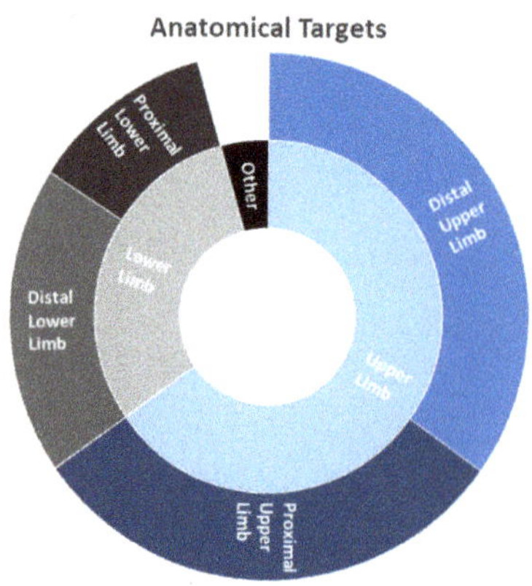

Figure 7. Hierarchical representation of anatomical targets considered in the analyzed documents.

3.4. Experimental Setups and Protocols

The screened papers were subdivided according to the setup employed for the rehabilitation and/or assessment. The main categories reported in Table 3 were identified as miscellaneous techniques for free movement and rehabilitation, robotic assistance, peripheral electrical stimulation, transcranial electrical stimulation, and assisted rehabilitation. Papers presenting techniques possibly ascribable to multiple categories were assigned to the best fitting one.

Among the selected studies, we grouped studies addressing miscellaneous techniques in the group "Miscellaneous techniques for free movement rehabilitation". In this category, we selected papers that did not include aids, robots, and supports or that used devices that are very typical of a specific study or do not belong to a specific category. Some studies assessed the motor function only with simple movements performed by subjects, such as wrist [69,84], elbow [38], arm [37,57,70,71,87,88], hand [40,67,83,86] and leg tasks [53] and respiratory movements [55]. Additional sensorial feedbacks were employed for evaluating the effects of visual [65,73], auditory [82], and audiovisual feedback [56]. In one paper, the additional neurofeedback allowed the improvement of functional recovery of the hand in dystonic patients [64]. Moreover, in two studies [53,85], movements were compared to motor imagery. Bartur and colleagues [50] employed mirror visual feedback in the rehabilitation setup for hemiparetic stroke patients. The effectiveness of a novel balance handle was assessed with arm movements [52].

Table 3. Setup for tests/rehabilitation.

Setup	Details and References
Miscellaneous techniques for free movements and rehabilitation	Mirror visual feedback [50] Wrist movements [69,84]; wrist movements + visual feedback [73] Elbow movements [38]; elbow movements + visual feedback [65] Arm movements [37,57,70,71,87,88]; arm movements + auditory feedback [82]; arm movements + balance handle [52] Hand movements [40,67,83,86]; hand movements + motor imagery + auditory/smelling stimulus [85]; hand movements + audiovisual feedback [56]; hand movements + biofeedback [64] Leg movements [53] Motor imagery swallow and tongue protrusion [68] Respiratory movements [55] Oculus rift [74]
Robotic assistance	Exoskeleton lower limb + virtual reality [34]; exoskeleton lower limb + virtual reality + BMI [63]; exoskeleton lower limb + walking [47,48,80]; exoskeleton lower limb + treadmill [49] Robotic end effector upper limb [32,33,58,59] Robotic mirror therapy upper limb [60] Hand mobilizer exoskeleton [39,61,62]; hand mobilizer exoskeleton + motor imagery [66] Ankle mobilizer [41]
Peripheral electrical stimulation	FES + hand movements [43,78]; FES + walking [79]; FES + hand movements + motor imagery [45,46] ePAS + ankle movements [77] NMES + wrist movements [44]
Transcranial electrical stimulation	HD-tDCs + wrist contractions [75]; anode tDCs + ankle dorsiflexion [72] rTMS [76]; TMS + wrist contractions [54]
Assisted rehabilitation	Pedaling system + NMES [42] Treadmill [81]; Treadmill + visual feedback [51]

Robotic solutions and exoskeletons were employed in 16 papers for motor rehabilitation. Exoskeletons for lower limbs were present in six studies and were tested in walking tasks [47,48,80], on a treadmill [49], and in virtual reality environments [34,63]. One study only [41] presented a mobilizer specific for the ankle. In the selected papers, exoskeletons for upper limbs were designed for hand mobilization and finger movements [39,61,62,66]. Four papers [32,33,58,59] presented a robotic end effector for the upper limb, and Park et al. [60] developed a robotic mirror therapy for the arm.

For improving motor recovery, peripheral stimulation, in which the stimulus is applied to the nerve to induce the contraction of the muscles, was employed in different studies. Functional electrical stimulation (FES) is a popular technique, and it was used for stimulating hand muscles [43,45,46,78] and ankle muscles [79]. Other techniques for peripheral stimulation are endogenous paired associative stimulation (ePAS), used by Olsen et al. [77] for the ankle joint muscles, and neuromuscular electrical stimulation (NMES), employed by Xu et al. [44] for wrist muscles. In transcranial electrical stimulation, instead, the electrical stimulus is delivered at the cortical level through electrical current, as in transcranial direct current stimulation (tDCS), or magnetic field, as in transcranial magnetic stimulation (TMS). Bao et al. [75] presented a high-density tDCS associated with wrist contractions for stroke rehabilitation, while Dutta et al. [72] used an anodal tDCS for assessing neuroplastic changes. Palmer et al. [54] employed TMS for studying the cortico-muscular coherence in stroke patients, while Benninger et al. [76] assessed the safety of using repetitive TMS for treating parkinsonian symptoms.

In the assisted rehabilitation group, we collected experimental setups in which devices that helped the movements performed during rehabilitation were included. In Bao et al. [42], a pedaling system coupled with NMES was employed for motor rehabilitation in stroke patients. A novel turning-based treadmill was presented by Chen et al. [81], while Jensen et al. [51] added the visual feedback to rehabilitation on a motorized treadmill.

Very few papers included virtual reality (VR) environments in rehabilitation: VR environment alone was present in only one study [74], while Bulea et al. [34] and Donati et al. [63] used virtual reality in combination with exoskeletons.

3.5. Setups for Signal Acquisition

We analyzed the setup used for the acquisition of EEG and EMG signals according to the type of system, the number, and the positioning of the electrodes. The electrode positioning of EEG was related to the area of the brain from which the signal was recorded and was identified in the motor area only, sensorimotor area, and whole cortex; for EMG signal, we subdivided the positioning based on the number of joints and limbs involved: therefore, we sorted papers in single-joint, multi-joint and multi-limb. The details are reported in Table 4, divided into EEG and EMG acquisition setups. Unfortunately, not all the papers declared the systems used in detail; therefore, we reported only the studies in which the recording systems were clearly specified.

Table 4. Setup for signal acquisition.

EEG ACQUISITION		
Setup		**Details and References**
Number of electrodes	N > 100	163 [70,71]; 160 [58]; 128 [42,59];
	100 < N < 30	64 [32–34,38,44,50,65,66,75]; 62 [47,48]; 56 [37]; 40 [68,77]; 35 [67]; 32 [40,52–54,81,84];
	30 < N < 10	21 [39,80]; 20 [69,82]; 16 [43,49,56,63,78]; 15 [85]; 14 [62,73]; 10 [57,61,64];
	N < 10	8 [83]; 5 [72]; 3 [60,86–88]; 2 [45,46]; 1 [41,51,55,79];
Electrodes positioning	Motor area only [41,61,67] Sensorimotor area [39,43,60,64,78] Whole cortex [32,34,37,38,42,47,48,50,54,58,62,63,66,70,71,75,77]	

Table 4. *Cont.*

EMG ACQUISITION		
Setup		**Details and references**
Number of electrodes		16 [53]; 15 [33,58]; 13 [71]; 10 [37,62]; 8 [32,56,70,73,76,80]; 6 [40,49]; 5 [57,63,84]; 4 [38,39,42,47,48,51,59,65,68,69,75,81]; 3 [52,60,61,64,67]; 2 [34,44,54,87,88]; 1 [41,43,55,66,72,77–79,82,83,85,86];
Electrodes positioning	Single-joint	Wrist [43,44,50,54,59,61,62,64,66,67,73,75,78,81,83,85,86] Arm [38,55,60,65,82,87,88] Leg [34,72] Ankle [41,51,77,79] Face [74]
	Multi-joint	Lower limb [42,47–49,53,63,80] Upper limb [32,33,37,39,40,52,56,58,70,71,76,84] Trunk-upper limb [57]
	Multi-limb	Upper limb [37,70,71,76]

For the EEG signal, different types of systems were employed for the signal acquisition, and all the instruments were commercial systems, reported in Table 4. Two papers only used customized amplifiers [51,86]. We noticed that no system was preferred with respect to the others, but a variety of instruments were employed in the literature. In addition, the set sampling frequency and the impedance were different among the studies.

The number of EEG channels changed depending on the study design: a higher number of electrodes was used for a more comprehensive mapping of the whole cortical area, while a lower number gives details on the activity of specific areas. Five papers employed more than 100 electrodes for recording EEG signals: 163 electrodes were used for studying cortico-muscular coupling [70,71], 160 for mapping brain activity [58], and 128 for neurophysiological assessments [42,59].

Many papers employed the standard number of electrodes that are usually provided with EEG caps: nine studies used 64 electrodes [32–34,38,44,50,65,66,75], six used 32 electrodes [40,52–54,81,84], five used 16 electrodes [43,49,56,63,78] and one applied 8 electrodes [83]. No standard number of electrodes was found in the other papers. Li et al. in two studies [47,48] employed 62 electrodes, Mima et al. [37] used 56 electrodes, Olsen et. al. [77] and Yang et al. [68] applied 40 electrodes and Lou et al. [67] used 35 channels. Other studies recorded the EEG signal with a lower number of electrodes: 21 electrodes were used in two papers [39,80] and 20 in other two [69,82]; six studies employed between 15 and 10 electrodes [57,61,62,64,73,85].

Finally, some papers applied very few electrodes: Dutta et al. [72] used five channels, four studies used three electrodes [60,86–88], and Zhai et al. in two studies [45,46] employed only two electrodes. Four studies [41,51,55,79] applied only one channel for recording EEG signal: the electrode was positioned in Cz of the 10–20 electrode placement system in all these studies.

Papers that employed a high number of electrodes recorded the whole cortical activity with a high density of probes, but other papers acquired the activity of the whole cortex also with a lower number of probes [62,63]. Usually, few electrodes were employed for recording the activity of the sensorimotor area [39,43,60,64,78] or the motor area only [41,61,67].

As for the EEG acquisition setups, the types of systems employed for EMG signal were different among all the studies, and two papers only customized the setups [42,86], while all the other used commercial systems, reported in Table 4. Moreover, the sampling rate and the impedance used for the recording were different among the studies.

Many studies applied few electrodes for recording the activity of a single muscle or a pair of agonist and antagonist muscles: 4 electrodes were used in 12 studies, 3 in 5 studies, 2 electrodes were employed in 5 papers, and a single probe was used in 12 papers. Using a few electrodes, the activity was recorded from muscles controlling only one joint. The employment of more probes allows recording the activity of some muscles that underlie multi-joint coordination: three papers [57,63,84] used five electrodes, two papers [40,49] six electrodes and six studies applied eight electrodes [32,56,70,73,76,80]. Finally, the EMG activity was recorded from more than 10 electrodes in 6 papers [33,37,53,58,62,71].

The EMG electrode positioning was classified based on the number of joints that are controlled with the muscles recorded. The upper limbs were investigated more than the lower limbs in both single and multiple joint categories. The wrist joint was studied the most: 17 papers analyzed muscles of the forearm moving the hand. Seven studies, instead, recorded muscles that move the whole upper arm. Two papers [34,72] placed the EMG probes on knee muscles, while muscles moving the ankle joints were recorded in four papers [41,51,77,79]. One study [74] acquired the activity of facial muscles. Studies that employed a higher number of EMG electrodes recorded the activity of muscles that regarded more than one joint: 7 papers analyzed the muscular activity of the lower limb, 12 studied the upper limb, and Jacobs et al. [57] included the trunk analysis to support the upper limb one. Investigating more joints allows the analysis of muscle activation patterns and muscle synergies, giving an insight into motor control [23]. Only four studies [37,70,71,76] involved both upper limbs in the EMG acquisitions.

3.6. Data Analysis

3.6.1. Analysis Techniques

In literature, various techniques are employed to analyze EEG and EMG signals accordingly to the different aims of the specific studies. In Table 5, we identified five different macro-categories of analysis most frequently used for EEG and EMG independently. Three further categories of combined EEG-EMG analysis were identified. For each macro-category, a further detailed sub-classification was made, based on specific approaches implemented. Frequently, more than one technique was employed in the same study. The list of papers reported in Table 5 was sorted accordingly as follows.

Table 5. Signal analysis techniques.

	EEG Analysis Techniques
Time domain analysis	ERP: [56,57,65,76,77] Cortical waves amplitude/slope: [53,74,87,88]
Frequency domain analysis	Average PSD: [34,41,42,47,49,52,59,65,72,75,79,83,86–88] Quantitative index calculation: [55,60,62,73]
Time-frequency analysis	ERD/ERS: [32,33,38,44,50,57,64,68,81] ERSP: [42,47,63]
Connectivity	Functional: [81,85] Effective: [48,54,80]
Sources reconstruction	ICA: [34,42,63] LORETA: [58,71,80]
	EMG Analysis Techniques
Time domain analysis	Amplitude/RMS: [34,55,58–60,63,73,74,77,79,80,83,86] Additional time features: [39,64,65,71,87,88]
Frequency domain analysis	Average PSD: [41,47,49] Median frequency: [44,60,79]
Time-frequency analysis	Channels coherence: [51] Time-frequency decomposition: [47]
Thresholding/landmark identification	EMG-EEG temporal synchronization: [54,57,62,72,81] Identification of events: [48,56,71,85,86]
Synergies	Non-negative matrix factorization algorithm: [32,33,53]

Table 5. *Cont.*

EEG-EMG Combination	
Cortico-muscular coherence	EEG-EMG coherence: [37,39,41,43–45,49,61,64,66,67,69,70,75,78,82,84] EEG-EMG PDC/gPDC: [42,46]
Time-frequency connectivity	Wavelet cross-spectrum: [38,51] Cross-mutual information: [66,81] Pearson's correlation: [47] Copula Granger's causality: [40]
Feature fusion for classification	Linear discriminant analysis: [65] Support vector machine: [87,88]

Among the 55 papers selected, we found that 72.7% of them analyzed the EEG signal individually, while 61.8% performed an analysis on the EMG signal alone. Only 49.1% of the papers extrapolated features combining EEG and EMG signals. In Figure 8, a representation of the different categories of found data analysis is provided, divided by the macro-categories identified in Table 5. For what concern the EEG analysis, we noticed that most of the papers (47.5%) exploited a frequency domain approach, while the time domain one is the most represented approach for the analysis of EMG signal (55.9% of EMG alone papers). Finally, the extraction of the cortico-muscular coherence (CMC) was the most widespread metric employed (70.4%) among papers that considered the EEG and EMG signals in combination.

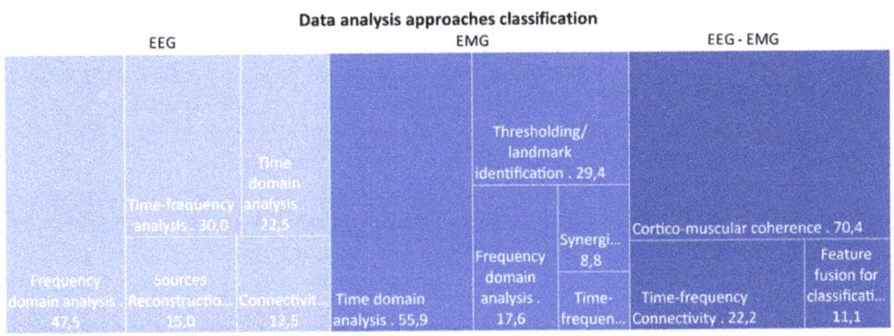

Figure 8. Graphical representation of data analysis approaches categories, divided according to the signal considered: only EEG, only EMG, or EEG and EMG combined. Percentage of papers using features belonging to each specific category are reported in the graph.

Hereafter, we focus on the description of the identified macro-categories and on the features proposed in the literature.

In the first instance, the signals analysis can be classified based on the domain of feature extraction, i.e., in time, frequency, or time-frequency domain. In time domain approaches, EEG and EMG temporal series are directly analyzed after a pre-processing step to remove artifacts.

Only nine of the selected papers focused on the extraction of EEG features in the time domain: in five works, movement event-related potentials (ERP) were analyzed with respect to an external stimulus or a voluntary movement. In [53,74,87,88], specific features of the cortical electric potentials were considered, such as amplitude, slope, fractal dimension, and Hjorth parameters of the cortical response [89].

In contrast with EEG, time domain features are frequently extrapolated from the envelope of the EMG signal. A total of 19 papers implemented this type of approach (see Table 5). Specifically, 13 of them focused on the information extrapolated by the amplitude or the root mean square (RMS) of the envelope to quantify the muscles fibers' activity, while in 6 works, additional time features were calculated. For example, Guo and colleagues [39], as well as Hashimoto and colleagues [64], considered the envelope integral. Tryon and colleagues [87,88] fitted the EMG experimental signal with an auto-regressive model, also calculating the mean absolute value, the mean absolute slope, the waveform length, and zero crossings. Moreover, Yao et al. [70] were able to define a muscle selection index from the temporal series of EMG electrodes.

Even though the time series can provide usual information on the biological processing underling the recorded signals, a complementary analysis can be performed in the frequency domain. The Fourier transforms of the temporal signals are calculated, and the spectral content at specific frequency bands is usually evaluated.

For what concerns the EEG analysis, 19 papers among the ones selected (second raw Table 5) employed this type of approach. Typically, in EEG, the power spectral density (PSD) averaged over epochs of the entire signal is calculated, and five spectral bands of interest are identified, i.e., delta (δ: 0.5–4 Hz), theta (θ: 4–8 Hz), alpha (α: 8–13 Hz), beta (β: 13–30 Hz) and gamma (γ: 30–150 Hz). The power amount in each of these bands and their ratio provides information on a particular mental state and cognitive involvement. Only in 4 of the 19 papers identified, specific quantitative bands power-based indexes were calculated, e.g., a relative amplitude value [60], an engagement index ($\beta/(\alpha + \theta)$) [62,73] and the θ/β ratio [55].

Only six papers exploited the frequency approach to investigate the EMG signal. As for EEG, in the work of [41,47,49], the EMG signal was segmented and the average PSD calculated, identifying responses at specific frequencies or spectral correlation between different muscles signals. In [44,60,79], a specific index, i.e., the median frequency, was calculated to investigate the occurrence of muscles fatigue during exercise.

Especially in the assessment of rehabilitation, it is important to evaluate the neuromotor response related to specific movement or intervention in time. The time-frequency domain analysis allows combining the spectral information retrieved from the EEG and EMG signals as they vary during time. This type of analysis has been proposed in EEG studies to evaluate the rise of cerebral waves at a different frequency. In particular, the event-related desynchronization/synchronization (ERD/ERS) is usually calculated as the percentage power decrease or increase at specific frequency bands following a movement onset. The ERD/ERS represents the synchronization or desynchronization of neuron populations in response to a voluntary muscle activation [7]. Among the selected, nine papers exploited this type of time-frequency analysis to explore the frequency-specific brain response over time. In three further studies [42,63,73], this approach has been extended in the evaluation of active/passive muscle stimulation during cycling or walking and brain-computer interface application, thus using the more general term event-related spectral perturbation (ERSP) to indicate the type of outcome obtained.

Also, for EMG, it is possible to exploit the conjunction between spectral and temporal information, even though this approach is not often used in the applications that we considered in this review work. In fact, only two papers exploited this analysis. Jensen in 2018 [51] performed a time-frequency-based analysis of the coherence between five EMG channels during a visually guided walking task. Li and colleagues [47] evaluated the correlation of EMG PSD of four channels in six frequency bands during the gait cycle.

As explained, an accurate time-frequency analysis of both EEG and EMG signals requires the identification and synchronization of the biological signals time series with the movement, or more in general to the experimental event of interest. Many studies reported in literature exploit the EMG signal for the exact timing of the onset of movement/experiment. In Table 5, we found 11 works in which EMG thresholding algorithms were employed, primarily to identify the onsets, thus leading the following analysis on both EEG and EMG signals.

The analysis approaches described until now usually considered the signal registered from each electrode (EEG or EMG) independently. However, a second-level analysis can be performed, taking into consideration not only the temporal or frequency information but also their spatial distribution and connection. For the cerebral signal, this approach includes connectivity analysis among brain areas, while for EMG, muscle synergies represent spatial patterns involving the recruitment of multiple muscles. Synergies are coordinated activations of groups of muscles as a consequence of a common control signal from the central nervous system [90]. We identified three papers [32,33,53] that implemented this type of investigation in rehabilitation assessment, exploiting the non-negative matrix factorization algorithm [91].

Brain connectivity analysis aims to identify those areas that are synchronously active both at rest and during a specific task. Two types of approaches can be distinguished: functional and effective analysis. In functional analysis, the functional network organization is investigated; in effective analysis, also the directionality and the causal influence between structures are evaluated [92]. We found two works exploring the EEG functional brain connectivity [81,85] and three papers quantifying the effective connectivity [48,54,80] to investigate the effect of rehabilitation or intervention on a patient's cortex connections reorganization.

The EEG analyses can be performed either by directly analyzing data on the electric potential difference registered at each electrode or, as an alternative approach, an intermediate step of reconstruction of the cortical sources can be added to retrieve the temporal and spectral series of the generators of the brain electric field. In our review, we found that most studies were conducted exploiting the electrode signals directly, and only in six papers (see table) the reconstruction of sources was performed. Among these, two different main approaches were used: the first one based on independent component analysis (ICA) of the electrodes signal and fitting of the dipole model [93]; and the second one based on the low-resolution brain electromagnetic tomography (LORETA) [94].

Even though all the papers selected in this literature review combine the acquisition of EMG and EEG, authors often conclude with a separate analysis of the two signals and a combined observation of the results. Only in 28 over 55 works a quantitative combination metric of EEG and EMG was considered. Mostly, the cortico-muscular coherence (CMC), defined as the coherence function between the EEG and EMG signals, was quantified [95] in 16 works. In [42,46], the extended concept of partial directed coherence (PDC) [96] and generalized PDC (gPDC) [97] was applied to identify also causal information in CMC. Six studies also explored the application of time-frequency connectivity methods for the investigation of the relation between muscular and cerebral electrical signals. Cremoux et al. [38], as well as Jensen et al. [51], exploited a wavelet cross-spectrum-based approach, while Chen et al. [81] and Kim and colleagues [66] employed the cross-mutual information metric. In [47], Pearson's correlation coefficients between EEG and EMG channels were calculated. In [40], an effective connectivity method was implemented based on Copula Granger's causality.

Finally, the studies by Leerskov [65] and Tryon and colleagues [87,88] must be mentioned since they pointed out two different classification approaches for the classification of motion and control of robotic rehabilitation devices through the fusion of features derived from EEG and EMG.

3.6.2. Benefits of Combined EEG-EMG Applications

In some papers included in this review, EMG has an ancillary role with respect to EEG since it was used to synchronize EEG signals with respect to relevant functional events composing the experimental/rehabilitation protocol (e.g., target movements, cognitive stimuli, electrical muscle stimulations, etc.). More interestingly, there are papers that combine the EEG/EMG signals to extract new relevant combined features. For example, the use of CMC can help to detect voluntary movements in spastic subjects or can be used to evaluate changes in cortico-muscular phase coherence to assess the effectiveness of rehabilitation strategies (i.e., passive vs. active, with or without exoskeleton or different level of engagement) and to serve as a biomarker for motor recovery in different pathologies [44,61,66,67,69,82]. In particular, the effect on CMC in post-stroke patients is mainly investigated [37,39,50,67,70,75,78]. However, the combination of EEG/EMG has demonstrated to efficiently evaluate the residual integrity of the neuromuscular system also in the spinal cord injury-affected subjects [38,63,65], in identifying low back pain-affected rehabilitation strategies [57] or to study the sensorimotor cortex in cerebral palsy-affected children [34]. As an example of stroke recovery evaluation, Chen et al. [81] demonstrated as a novel turning-based treadmill training was effective for enhancing brain functional reorganization underlying cortico-cortical and cortico-muscular mechanisms and thus might result in gait improvement in people with chronic stroke. Lai et al. [43] compared the outcome of functional electrical stimulation on 15 healthy subjects and 15 post-stroke patients and demonstrated that EEG-EMG coherence can detect electrical stimulation-induced changes in the neuromuscular system.

The literature describes other interesting applications based on other techniques of concurrent analysis. In Pierella et al. 2020 [33], the combined EEG/EMG analysis by PCA has shown the potential role to extract significant biomarkers for patient stratification as well as for the design of more effective rehabilitation protocols. Furthermore, the parameters extracted by the combined analysis of EEG and EMG signals can also be used to improve the classification of motor tasks in robotic rehabilitation if used to feed artificial intelligence approaches [68,87,88]. It is worth noticing that more advanced approaches, using Granger causality and PDC, were used to explore cortico-muscular connectivity and developed to detect complex functional coupling between cortical oscillations and muscle activities and provide a potential quantitative analysis measure for motion control and rehabilitation evaluation [40,42,45,46].

4. Discussion

In this systematic review, we analyzed papers in which EEG and EMG signals were simultaneously recorded and analyzed for evaluating or assessing motor performance in clinical rehabilitation scenarios.

From the distribution of selected papers over the years, the combined application of EEG and EMG signals to assess rehabilitation-related studies is in a growing trend. Indeed, most studies were published in the last decade with a remarkable increase in the last three years. This multi-domain approach was promoted by the improvement of technology and the availability of integrated low-cost commercial solutions aimed at combining EEG and EMG sensors. It arises from the literature that a multi-parametric analysis of EEG/EMG signals allows a more comprehensive investigation of complex neuromotor mechanisms with respect to exploiting a single technique. In fact, applying the two techniques independently cannot provide insight into mechanisms such as the functional connection between the central control system, the brain, and the actuators of movement, the muscles. Considering the EMG signal alone can provide information about muscles activation strategies, but no information can be inferred about the role of the cerebral control in it. Similarly, EEG signal alone during movement can provide information about the cerebral activation without control on the muscle-effective activation. These aspects can explain why the use of combined EEG/EMG analysis is rapidly becoming an emerging topic. All this finding supports the drafting of this review to summarize the results achieved so far.

In our sample of articles, we found that many studies were observational studies, mainly focused on the assessment of cortico-muscular coupling and the evaluation of the effects of treatment administration or rehabilitation methods, and pilot studies in which novel experimental setups or concept design were tested on a limited number of subjects. Only a few papers presented randomized clinical trials that evaluated the efficacy of rehabilitation paradigms by comparing the effects of rehabilitation interventions or treatments on a target group to a control one. The predominance of the pilot and observational studies indicated that the concomitant use of the techniques was found mainly in studies that aim at exploring novel research purposes rather than standard clinical practice. While this is perfectly understandable due to costs, invasive setups, time-consuming procedures, we conclude that, so far, the application and applicability of combined EEG-EMG setup is very limited in clinical practice.

It followed that these techniques were mostly employed in preliminary studies for evaluating the effects and efficacy of novel rehabilitation platforms and for proposing novel methodologies assessing motor performance. The exploratory design of the studies is also confirmed by the fact that in 45% of the analyzed papers, less than 10 subjects were enrolled. This shows their intrinsic nature as pilot studies. Future directions should foresee more structured studies; clinical trials could be developed starting from pilot studies already available so that more reliable conclusions on concomitant EEG-EMG applications can be drawn.

Moreover, EEG and EMG combined analysis was widely used for assessing functional connectivity and comparing cortico-muscular coherence in patients with respect to healthy subjects. Rarely was it used for evaluating clinical outcomes of motor rehabilitation, although it could provide a detailed assessment of patients' status.

Indeed, many studies enrolled healthy subjects as a target for investigating physiological parameters concerning treatments or rehabilitation paradigms. The results obtained in healthy subjects can help to better understand the physiological mechanism underlying cortico-muscular activity and can be used as a benchmark for the pathological changes occurring in neuromotor disorders.

The pathologic subjects involved in the studies were mainly post-stroke patients, probably because stroke is one of the most diffuse cerebrovascular diseases affecting motor control. Other types of pathological conditions were investigated in a few or single articles. Future directions could be the extension of EEG-EMG combined analysis to neuromuscular pathologies that received little or less attention, such as neuromuscular diseases.

The integration of EEG and EMG signals can be useful for the evaluation of motor impairment and recovery, allowing the investigation of the motor system in its complexity. For example, the investigation of deficits of motor control exploits at best the potential of both EEG and EMG domains. Many studies exploit the cortico-muscular coherence, coupling the two domains in a single analysis [98]. Although this metric is the simplest approach to quantify the interactions between all motor system actors during movement, a more complex analysis could be of interest. There are very few studies that tend to exploit the full potential of each of the domains to unify data after assessment. In a recent paper [32], authors used domain-specific measurements (such as ERD-ERS for EEG and muscle synergies for EMG) and tried to interpret critically the detailed findings achieved in each domain. While some interpretations were still debated and not always fully agreed, authors could find a general agreement with other more easily interpreted data such as clinical scales and kinematics. A suggestion for future applications is thus to promote the use of advanced techniques for each of the domains under analysis. In fact, we also noticed that in many applications, EMG is seen as a supporting outcome measure for interpreting EEG data or even to simply allow signal synchronization through thresholding algorithms for detecting movement onset. Of course, while these approaches are perfectly scientifically sound, they might reduce the potential of advanced EMG analysis that allows refined measurements of motor control such as muscle synergies [90]. We also found that synthetic approaches suggested multimodal analysis as a tool for creating novel protocols and metrics based on the coupling of EEG and EMG (and other domains) [32,33]. The EEG/EMG bi-modal analysis, including also further domains, is an unexplored field that might help to shed light on the mechanisms of motor recovery.

For motor rehabilitation and assessment, a variety of experimental setups were explored in the analyzed papers. Many of them assessed motor function using only free movements coupled with sensorial feedbacks, motor imagery, or simple instruments. Robotics and exoskeletons are widely employed in motor rehabilitation to support and guide the movements of the patients preventing injuries and improving recovery using different loads and modes. Exoskeletons for assisting either upper limbs or lower limbs were quite diffused in literature, while robotic devices were found only for upper limb rehabilitation. Interestingly, while they have been used mainly for BCI setups [99,100], EEG and EMG may find wide application in the evaluation of human-robot interaction under a biomechanical perspective for assessment and evaluation.

Furthermore, particular attention was given to the rehabilitation of the hand because the upper limbs are the most affected in neuromotor disorders, and hand movements are involved in many daily life activities. Therefore, hand impairments limit heavily the ability to perform these activities [101,102], and motor recovery becomes essential for the patients' quality of life. Different studies employed electrical stimulation, applying the stimulus either at the peripheral or at the cortical level. This application is useful for studying cortico-muscular connections in both healthy and pathological subjects. Moreover, functional electrical stimulation is demonstrated to be effective in improving muscle strength and motor coordination in patients [103,104].

Among all the papers included, different techniques were employed for the analysis of EEG and EMG signals, according to the aim of each study. EMG signal was mainly analyzed in the time domain, extracting indexes from muscle envelopes, while frequency analysis was used principally for evaluating muscle fatigue from the median frequency. Frequency and time-frequency analysis were predominant in EEG signal processing because the power spectrum can provide information about the mental and cognitive involvement of the subject [105]. Spatial distribution of the signals, as brain connectivity and muscle synergies, was assessed in only a few papers. However, studying the functional and effective connectivity of the brain can provide insight into the cortical reorganization and functional recovery of patients. Moreover, muscle synergies can be used to evaluate motor control and movement coordination that are affected in neuromuscular disorders. Therefore, including the analysis of the spatial distribution of the signals can provide further information about neuromotor impairment and motor improvement in rehabilitation. In this way, the potentiality of the instruments can be deeply exploited.

5. Conclusions

The evaluation of the complementary contribution of EEG and EMG signals to the assessment of cortico-muscular interactions in clinical rehabilitation of neuromotor diseases is a promising topic, and an increased number of applications and scenarios is foreseen in the next future. The combined analysis of EEG and EMG can be boosted by the development of consolidated pipelines, which warranties results in robustness and direct comparison among different studies, putting a special focus on the signal interactions in terms of functional and effective connectivity. Currently, the use of bi-modal EEG/EMG analysis helps to elucidate physiological and pathological mechanisms to assess the rehabilitation treatments and to evaluate their effectiveness. However, prospectively, multi-domain approaches should be developed to exploit the full potential of EEG and EMG, and more pathologies should be targeted with more structured clinical trials to improve the scientific evidence.

Author Contributions: Conceptualization, A.S. and A.M.; methodology, A.M., A.S. and G.R.; formal analysis, C.B., I.P., R.M.M., A.S. and A.M.; investigation, C.B., I.P., R.M.M., G.R., A.S. and A.M.; resources, A.M., A.S. and G.R.; writing—original draft preparation, C.B., I.P., R.M.M., G.R., A.S. and A.M.; writing—review and editing, C.B., I.P., R.M.M., G.R., A.S. and A.M.; visualization, C.B., I.P., R.M.M., A.S. and A.M.; supervision, A.S., A.M. and G.R.; funding acquisition, G.R. All authors have read and agreed to the published version of the manuscript.

Funding: This work was partially funded by Fondazione Cariplo and Regione Lombardia, Grant/Award Number: Progetto Empatia@Lecco, ref 2016-1428.

Institutional Review Board Statement: Not applicable.

Informed Consent Statement: Not applicable.

Data Availability Statement: Not applicable.

Conflicts of Interest: The authors declare no conflict of interest.

References

1. Deuschl, G.; Beghi, E.; Fazekas, F.; Varga, T.; Christoforidi, K.A.; Sipido, E.; Bassetti, C.L.; Vos, T.; Feigin, V.L. The burden of neurological diseases in Europe: An analysis for the Global Burden of Disease Study 2017. *Lancet Public Health* **2020**, *5*, e551–e567. [CrossRef]
2. Jones, T.A. Motor compensation and its effects on neural reorganization after stroke. *Nat. Rev. Neurosci.* **2017**, *18*, 267–280. [CrossRef]
3. Maggioni, S.; Melendez-Calderon, A.; Van Asseldonk, E.; Klamroth-Marganska, V.; Lünenburger, L.; Riener, R.; Van Der Kooij, H. Robot-aided assessment of lower extremity functions: A review. *J. Neuroeng. Rehabil.* **2016**, *13*, 1–25. [CrossRef]
4. Debener, S.; Minow, F.; Emkes, R.; Gandras, K.; de Vos, M. How about taking a low-cost, small, and wireless EEG for a walk? *Psychophysiology* **2012**, *49*, 1617–1621. [CrossRef]
5. Drost, G.; Stegeman, D.F.; van Engelen, B.G.M.; Zwarts, M.J. Clinical applications of high-density surface EMG: A systematic review. *J. Electromyogr. Kinesiol.* **2006**, *16*, 586–602. [CrossRef]
6. Seeber, M.; Cantonas, L.M.; Hoevels, M.; Sesia, T.; Visser-Vandewalle, V.; Michel, C.M. Subcortical electrophysiological activity is detectable with high-density EEG source imaging. *Nat. Commun.* **2019**, *10*, 1–7. [CrossRef]
7. Pfurtscheller, G.; Aranibar, A. Evaluation of event-related desynchronization (ERD) preceding and following voluntary self-paced movement. *Electroencephalogr. Clin. Neurophysiol.* **1979**, *46*, 138–146. [CrossRef]
8. Park, W.; Kwon, G.H.; Kim, Y.H.; Lee, J.H.; Kim, L. EEG response varies with lesion location in patients with chronic stroke. *J. Neuroeng. Rehabil.* **2016**, *13*, 21. [CrossRef] [PubMed]
9. Wang, Q.; Meng, L.; Pang, J.; Zhu, X.; Ming, D. Characterization of EEG data revealing relationships with cognitive and motor symptoms in Parkinson's disease: A systematic review. *Front. Aging Neurosci.* **2020**, *12*, 373. [CrossRef]
10. Pfurtscheller, G.; Lopes Da Silva, F.H. Event-related EEG/MEG synchronization and desynchronization: Basic principles. *Clin. Neurophysiol.* **1999**, *110*, 1842–1857. [CrossRef]
11. Sakkalis, V. Review of advanced techniques for the estimation of brain connectivity measured with EEG/MEG. *Comput. Biol. Med.* **2011**, *41*, 1110–1117. [CrossRef] [PubMed]
12. Xu, B.; Peng, S.; Song, A.; Yang, R.; Pan, L. Robot-aided upper-limb rehabilitation based on motor imagery EEG. *Int. J. Adv. Robot. Syst.* **2011**, *8*, 88–97. [CrossRef]
13. Yu, G.; Wang, J.; Chen, W.; Zhang, J. EEG-based brain-controlled lower extremity exoskeleton rehabilitation robot. In Proceedings of the International Conference on Cybernetics and Intelligent Systems (CIS) and IEEE Conference on Robotics, Automation and Mechatronics (RAM), Ningbo, China, 19–21 November 2017; pp. 763–767.
14. Kanna, S.; Heng, J. Quantitative EEG parameters for monitoring and biofeedback during rehabilitation after stroke. In Proceedings of the International Conference on Advanced Intelligent Mechatronic, Singapore, 14–17 July 2009; pp. 1689–1694.
15. Bartur, G.; Joubran, K.; Peleg-Shani, S.; Vatine, J.J.; Shahaf, G. An EEG Tool for Monitoring Patient Engagement during Stroke Rehabilitation: A Feasibility Study. *Biomed. Res. Int.* **2017**, *2017*, 9071568. [CrossRef]
16. Rogers, J.M.; Jensen, J.; Valderrama, J.T.; Johnstone, S.J.; Wilson, P.H. Single-channel EEG measurement of engagement in virtual rehabilitation: A validation study. *Virtual Real.* **2021**, *25*, 357–366. [CrossRef]
17. Campanini, I.; Disselhorst-Klug, C.; Rymer, W.Z.; Merletti, R. Surface EMG in Clinical Assessment and Neurorehabilitation: Barriers Limiting Its Use. *Front. Neurol.* **2020**, *11*, 934. [CrossRef]
18. Lucas, K.R.; Rich, P.A.; Polus, B.I. Muscle activation patterns in the scapular positioning muscles during loaded scapular plane elevation: The effects of Latent Myofascial Trigger Points. *Clin. Biomech.* **2010**, *25*, 765–770. [CrossRef]
19. Ivanenko, Y.P.; Poppele, R.E.; Lacquaniti, F. Five basic muscle activation patterns account for muscle activity during human locomotion. *J. Physiol.* **2004**, *556*, 267–282. [CrossRef]
20. Kim, D.Y.; Park, C.I.; Chon, J.S.; Ohn, S.H.; Park, T.H.; Bang, I.K. Biomechanical assessment with electromyography of post-stroke ankle plantar flexor spasticity. *Yonsei Med. J.* **2005**, *46*, 546–554. [CrossRef]
21. McDonald, A.C.; Mulla, D.M.; Keir, P.J. Using EMG Amplitude and Frequency to Calculate a Multimuscle Fatigue Score and Evaluate Global Shoulder Fatigue. *Hum. Factors* **2019**, *61*, 526–536. [CrossRef]
22. Lee, W.A. Neuromotor synergies as a basis for coordinated intentional action. *J. Mot. Behav.* **1984**, *16*, 135–170. [CrossRef]
23. Cheung, V.C.K.; D'Avella, A.; Bizzi, E. Adjustments of motor pattern for load compensation via modulated activations of muscle synergies during natural behaviors. *J. Neurophysiol.* **2009**, *101*, 1235–1257. [CrossRef] [PubMed]
24. Bowden, M.G.; Clark, D.J.; Kautz, S.A. Evaluation of abnormal synergy patterns poststroke: Relationship of the fugl-meyer assessment to hemiparetic locomotion. *Neurorehabil. Neural Repair* **2010**, *24*, 328–337. [CrossRef]
25. Cheung, V.C.K.; Turolla, A.; Agostini, M.; Silvoni, S.; Bennis, C.; Kasi, P.; Paganoni, S.; Bonato, P.; Bizzi, E. Muscle synergy patterns as physiological markers of motor cortical damage. *Proc. Natl. Acad. Sci. USA* **2012**, *109*, 14652–14656. [CrossRef]
26. Kiguchi, K.; Hayashi, Y. An EMG-based control for an upper-limb power-assist exoskeleton robot. *IEEE Trans. Syst. Man Cybern. Part. B Cybern.* **2012**, *42*, 1064–1071. [CrossRef]
27. Castellini, C.; Van Der Smagt, P. Surface EMG in advanced hand prosthetics. *Biol. Cybern.* **2009**, *100*, 35–47. [CrossRef]
28. Iandolo, R.; Marini, F.; Semprini, M.; Laffranchi, M.; Mugnosso, M.; Cherif, A.; De Michieli, L.; Chiappalone, M.; Zenzeri, J. Perspectives and challenges in robotic neurorehabilitation. *Appl. Sci.* **2019**, *9*, 3183. [CrossRef]

29. Trujillo, P.; Mastropietro, A.; Scano, A.; Chiavenna, A.; Mrakic-Sposta, S.; Caimmi, M.; Molteni, F.; Rizzo, G. Quantitative EEG for predicting upper limb motor recovery in chronic stroke robot-Assisted rehabilitation. *IEEE Trans. Neural Syst. Rehabil. Eng.* **2017**, *27*, 1058–1067. [CrossRef]
30. Curt, A.; Dietz, V. Electrophysiological recordings in patients with spinal cord injury: Significance for predicting outcome. *Spinal Cord* **1999**, *37*, 157–165. [CrossRef]
31. Stinear, C.M.; Byblow, W.D.; Ackerley, S.J.; Barber, P.A.; Smith, M.C. Predicting Recovery Potential for Individual Stroke Patients Increases Rehabilitation Efficiency. *Stroke* **2017**, *48*, 1011–1019. [CrossRef] [PubMed]
32. Belfatto, A.; Scano, A.; Chiavenna, A.; Mastropietro, A.; Mrakic-Sposta, S.; Pittaccio, S.; Tosatti, L.M.; Molteni, F.; Rizzo, G. A multiparameter approach to evaluate post-stroke patients: An application on robotic rehabilitation. *Appl. Sci.* **2018**, *8*, 2248. [CrossRef]
33. Pierella, C.; Pirondini, E.; Kinany, N.; Coscia, M.; Giang, C.; Miehlbradt, J.; Magnin, C.; Nicolo, P.; Dalise, S.; Sgherri, G.; et al. A multimodal approach to capture post-stroke temporal dynamics of recovery. *J. Neural Eng.* **2020**, *17*, 045002. [CrossRef] [PubMed]
34. Bulea, T.C.; Lerner, Z.F.; Gravunder, A.J.; Damiano, D.L. Exergaming with a pediatric exoskeleton: Facilitating rehabilitation and research in children with cerebral palsy. In Proceedings of the International Conference on Rehabilitation Robotics, London, UK, 17–20 July 2017; pp. 1087–1093.
35. Craik, A.; Kilicarslan, A.; Contreras-Vidal, J.L. A translational roadmap for a brain-machine-interface (BMI) system for rehabilitation. In Proceedings of the International Conference on Systems, Man and Cybernetics (SMC), Bari, Italy, 6–9 October 2019; pp. 3613–3618. [CrossRef]
36. Moher, D.; Liberati, A.; Tetzlaff, J.; Altman, D.G. Preferred reporting items for systematic reviews and meta-analyses: The PRISMA statement. *BMJ* **2009**, *339*, 332–336. [CrossRef] [PubMed]
37. Mima, T.; Toma, K.; Koshy, B.; Hallett, M. Coherence between cortical and muscular activities after subcortical stroke. *Stroke* **2001**, *32*, 2597–2601. [CrossRef] [PubMed]
38. Cremoux, S.; Tallet, J.; Dal Maso, F.; Berton, E.; Amarantini, D. Impaired corticomuscular coherence during isometric elbow flexion contractions in humans with cervical spinal cord injury. *Eur. J. Neurosci.* **2017**, *46*, 1991–2000. [CrossRef] [PubMed]
39. Guo, Z.; Qian, Q.; Wong, K.; Zhu, H.; Huang, Y.; Hu, X.; Zheng, Y. Altered Corticomuscular Coherence (CMCoh) Pattern in the Upper Limb During Finger Movements After Stroke. *Front. Neurol.* **2020**, *11*, 410. [CrossRef] [PubMed]
40. She, Q.; Zheng, H.; Tan, T.; Zhang, B.; Fan, Y.; Luo, Z. Time-Frequency-Domain Copula-Based Granger Causality and Application to Corticomuscular Coupling in Stroke. *Int. J. Humanoid Robot.* **2019**, *16*, 1–18. [CrossRef]
41. Sui, J.; Liu, Y.; Ji, L. Effect of unilateral non-rhythmical stimulation on bilateral cerebral cortex and muscle activation in People. *J. Biomim. Biomater. Tissue Eng.* **2014**, *19*, 25–33. [CrossRef]
42. Bao, S.C.; Leung, W.C.; Vincent, V.C.; Zhou, P.; Tong, K.Y. Pathway-specific modulatory effects of neuromuscular electrical stimulation during pedaling in chronic stroke survivors. *J. Neuroeng. Rehabil.* **2019**, *16*, 1–15. [CrossRef]
43. Lai, M.I.; Pan, L.L.; Tsai, M.W.; Shih, Y.F.; Wei, S.H.; Chou, L.W. Investigating the effects of peripheral electrical stimulation on corticomuscular functional connectivity stroke survivors. *Top. Stroke Rehabil.* **2016**, *23*, 154–162. [CrossRef]
44. Xu, R.; Wang, Y.; Wang, K.; Zhang, S.; He, C.; Ming, D. Increased corticomuscular coherence and brain activation immediately after short-term neuromuscular electrical stimulation. *Front. Neurol.* **2018**, *9*, 1–10. [CrossRef]
45. Zhai, T.; Fu, A.; Xu, R.; Qiu, S.; Qi, H.; Zhou, P.; Zhang, L.; Wan, B.; Ming, D. Cortico-muscular coherence analysis under voluntary, stimulated and imaginary neuromuscular activities. *Biomed. Tech.* **2013**, *58*, 3–4. [CrossRef] [PubMed]
46. Zhai, T.; Qi, C.; Xu, R.; Fu, A.; Zhang, L.; Zhao, X.; Zhou, P.; Qi, H.; Wan, B.; Cheng, X.; et al. EEG-EMG Analysis on Corticomuscular Coherence of Middle Finger Flexion Tasks. In *Advances in Cognitive Neurodynamics (IV)*; Springer: Dordrecht, The Netherlands, 2015; pp. 337–344. ISBN 9789401795487.
47. Li, J.; Dimitrakopoulos, G.N.; Thangavel, P.; Chen, G.; Sun, Y.; Guo, Z.; Yu, H.; Thakor, N.; Bezerianos, A. What Are Spectral and Spatial Distributions of EEG-EMG Correlations in Overground Walking? An Exploratory Study. *IEEE Access* **2019**, *7*, 143935–143946. [CrossRef]
48. Li, J.; Thakor, N.; Bezerianos, A. Brain Functional Connectivity in Unconstrained Walking with and without an Exoskeleton. *IEEE Trans. Neural Syst. Rehabil. Eng.* **2020**, *28*, 730–739. [CrossRef] [PubMed]
49. Youssofzadeh, V.; Zanotto, D.; Wong-Lin, K.F.; Agrawal, S.K.; Prasad, G. Directed functional connectivity in Fronto-Centroparietal circuit correlates with motor adaptation in gait training. *IEEE Trans. Neural Syst. Rehabil. Eng.* **2016**, *24*, 1265–1275. [CrossRef] [PubMed]
50. Bartur, G.; Pratt, H.; Frenkel-Toledo, S.; Soroker, N. Neurophysiological effects of mirror visual feedback in stroke patients with unilateral hemispheric damage. *Brain Res.* **2018**, *1700*, 170–180. [CrossRef] [PubMed]
51. Jensen, P.; Jensen, N.J.; Terkildsen, C.U.; Choi, J.T.; Nielsen, J.B.; Geertsen, S.S. Increased central common drive to ankle plantar flexor and dorsiflexor muscles during visually guided gait. *Physiol. Rep.* **2018**, *6*, 1–11. [CrossRef]
52. Park, J.H.; Moon, H.S.; Kim, H.; Chung, S.T. Detection of movement intention for operating methods of serious games. *Appl. Sci.* **2021**, *11*, 883. [CrossRef]
53. Úbeda, A.; Azorín, J.M.; Farina, D.; Sartori, M. Estimation of neuromuscular primitives from EEG slow cortical potentials in incomplete spinal cord injury individuals for a new class of brain-machine interfaces. *Front. Comput. Neurosci.* **2018**, *12*, 3. [CrossRef]

54. Palmer, J.A.; Wheaton, L.A.; Gray, W.A.; Saltão da Silva, M.A.; Wolf, S.L.; Borich, M.R. Role of Interhemispheric Cortical Interactions in Poststroke Motor Function. *Neurorehabil. Neural Repair* **2019**, *33*, 762–774. [CrossRef]
55. Vladimirov, A.A.; Andriyashek, Y.I. Neurophysiological markers of stress within a recovery period after acute complications of cardiovascular diseases. *Neurophysiology* **2014**, *46*, 225–232. [CrossRef]
56. Yilmaz, O.; Birbaumer, N.; Ramos-Murguialday, A. Movement related slow cortical potentials in severely paralyzed chronic stroke patients. *Front. Hum. Neurosci.* **2015**, *8*. [CrossRef] [PubMed]
57. Jacobs, J.V.; Henry, S.M.; Nagle, K.J. Low back pain associates with altered activity of the cerebral cortex prior to arm movements that require postural adjustment. *Clin. Neurophysiol.* **2010**, *121*, 431–440. [CrossRef] [PubMed]
58. Chen, A.; Yao, J.; Dewald, J.P.A. A novel experimental setup combining EEG and robotics to investigate brain activity driving controlled reaching movements in chronic stroke survivors. In Proceedings of the 10th International Conference on Rehabilitation Robotics, Noordwijk, The Netherlands, 13–15 June 2007; pp. 876–882.
59. Major, Z.Z.; Vaida, C.; Major, K.A.; Tucan, P.; Simori, G.; Banica, A.; Brusturean, E.; Burz, A.; Craciunas, R.; Ulinici, I.; et al. The impact of robotic rehabilitation on the motor system in neurological diseases. A multimodal neurophysiological approach. *Int. J. Environ. Res. Public Health* **2020**, *17*, 6557. [CrossRef] [PubMed]
60. Park, T.; Lee, M.; Jeong, T.; Shin, Y.I.; Park, S.M. Quantitative analysis of eeg power spectrum and emg median power frequency changes after continuous passive motion mirror therapy system. *Sensors* **2020**, *20*, 2354. [CrossRef]
61. Chwodhury, A.; Raza, H.; Dutta, A.; Nishad, S.S.; Saxena, A.; Prasad, G. A study on cortico-muscular coupling in finger motions for exoskeleton assisted neuro-rehabilitation. In Proceedings of the Annual International Conference of the Engineering in Medicine and Biology Society, Milano, Italy, 25–29 August 2015; pp. 4610–4614.
62. Gandolla, M.; Ferrante, S.; Coelli, S.; Tacchino, G.; Giglioli, I.A.C.; Cottini, M.C.; Seneci, C.; Scannapieco, A.; Veniero, M.; Pedrocchi, A. Technical validation of an integrated robotic hand rehabilitation device: Finger independent movement, EMG control, and EEG-based biofeedback. In Proceedings of the 2nd International Forum on Research and Technologies for Society and Industry Leveraging a better tomorrow (RTSI), Bologna, Italy, 7–9 September 2016; pp. 5–9.
63. Donati, A.R.C.; Shokur, S.; Morya, E.; Campos, D.S.F.; Moioli, R.C.; Gitti, C.M.; Augusto, P.B.; Tripodi, S.; Pires, C.G.; Pereira, G.A.; et al. Long-Term Training with a Brain-Machine Interface-Based Gait Protocol Induces Partial Neurological Recovery in Paraplegic Patients. *Sci. Rep.* **2016**, *6*, 1–16. [CrossRef]
64. Hashimoto, Y.; Ota, T.; Mukaino, M.; Liu, M.; Ushiba, J. Functional recovery from chronic writer's cramp by brain-computer interface rehabilitation: A case report. *BMC Neurosci.* **2014**, *15*, 1–7. [CrossRef]
65. Leerskov, K.; Rehman, M.; Niazi, I.; Cremoux, S.; Jochumsen, M. Investigating the feasibility of combining EEG and EMG for controlling a hybrid human computer interface in patients with spinal cord injury. In Proceedings of the 20th International Conference on Bioinformatics and Bioengineering, Cincinnati, OH, USA, 26–28 October 2020; pp. 403–410.
66. Kim, B.; Kim, L.; Kim, Y.H.; Yoo, S.K. Cross-association analysis of EEG and EMG signals according to movement intention state. *Cogn. Syst. Res.* **2017**, *44*, 1–9. [CrossRef]
67. Lou, X.; Xiao, S.; Qi, Y.; Hu, X.; Wang, Y.; Zheng, X. Corticomuscular coherence analysis on hand movement distinction for active rehabilitation. *Comput. Math. Methods Med.* **2013**, *2013*, 908591. [CrossRef]
68. Yang, H.; Guan, C.; Chua, K.S.G.; Chok, S.S.; Wang, C.C.; Soon, P.K.; Tang, C.K.Y.; Ang, K.K. Detection of motor imagery of swallow EEG signals based on the dual-tree complex wavelet transform and adaptive model selection. *J. Neural Eng.* **2014**, *11*, 035016. [CrossRef]
69. Krauth, R.; Schwertner, J.; Vogt, S.; Lindquist, S.; Sailer, M.; Sickert, A.; Lamprecht, J.; Perdikis, S.; Corbet, T.; Del Millán, J.R.; et al. Cortico-muscular coherence is reduced acutely post-stroke and increases bilaterally during motor recovery: A pilot study. *Front. Neurol.* **2019**, *10*, 126. [CrossRef]
70. Yao, J.; Dewald, J.P.A. Cortico-muscular communication during the generation of static shoulder abduction torque in upper limb following stroke. In Proceedings of the Annual International Conference of the Engineering in Medicine and Biology, New York, NY, USA, 30 August–3 September 2006; pp. 181–184.
71. Yao, J.; Ellis, M.D.; Dewald, J. Mechanisms and rehabilitation of discoordination following stroke using a cortical imaging method. In Proceedings of the Annual International Conference of the IEEE Engineering in Medicine and Biology, Shanghai, China, 1–4 September 2005; Volume 7, pp. 4138–4141.
72. Dutta, A.; Boulenouar, R.S.; Guiraud, D.; Nitsche, M.A. Delineating the effects of anodal transcranial direct current stimulation on myoelectric control based on slow cortical potentials. In Proceedings of the 36th Annual International Conference of the IEEE Engineering in Medicine and Biology Society, Chicago, IL, USA, 27–31 August 2014; pp. 3094–3097.
73. Li, C.; Jia, T.; Ji, L. *Investigation of the Effects of Game Difficulty on the Engagement Level of Patient with Brain Injury During Rehabilitation Exercise*; International Society for Optics and Photonics: Bellingham, DC, USA, 2019; p. 35.
74. Qidwai, U.; Ajimsha, M.S.; Shakir, M. The role of EEG and EMG combined virtual reality gaming system in facial palsy rehabilitation—A case report. *J. Bodyw. Mov. Ther.* **2019**, *23*, 425–431. [CrossRef]
75. Bao, S.C.; Wong, W.W.; Leung, T.W.H.; Tong, K.Y. Cortico-muscular coherence modulated by high-definition transcranial direct current stimulation in people with chronic stroke. *IEEE Trans. Neural Syst. Rehabil. Eng.* **2019**, *27*, 304–313. [CrossRef] [PubMed]
76. Benninger, D.H.; Iseki, K.; Kranick, S.; Luckenbaugh, D.A.; Houdayer, E.; Hallett, M. Controlled study of 50-Hz repetitive transcranial magnetic stimulation for the treatment of parkinson disease. *Neurorehabil. Neural Repair* **2012**, *26*, 1096–1105. [CrossRef]

77. Olsen, S.; Signal, N.; Niazi, I.K.; Rashid, U.; Alder, G.; Mawston, G.; Nedergaard, R.B.; Jochumsen, M.; Taylor, D. Peripheral electrical stimulation paired with movement-related cortical potentials improves isometric muscle strength and voluntary activation following stroke. *Front. Hum. Neurosci.* **2020**, *14*, 156. [CrossRef]
78. Pan, L.L.H.; Yang, W.W.; Kao, C.L.; Tsai, M.W.; Wei, S.H.; Fregni, F.; Chen, V.C.F.; Chou, L.W. Effects of 8-week sensory electrical stimulation combined with motor training on EEG-EMG coherence and motor function in individuals with stroke. *Sci. Rep.* **2018**, *8*, 9217. [CrossRef]
79. Shendkar, C.V.; Lenka, P.K.; Biswas, A.; Kumar, R.; Mahadevappa, M. Therapeutic effects of functional electrical stimulation on gait, motor recovery, and motor cortex in stroke survivors. *Hong Kong Physiother. J.* **2015**, *33*, 10–20. [CrossRef]
80. Calabrò, R.S.; Naro, A.; Russo, M.; Bramanti, P.; Carioti, L.; Balletta, T.; Buda, A.; Manuli, A.; Filoni, S.; Bramanti, A. Shaping neuroplasticity by using powered exoskeletons in patients with stroke: A randomized clinical trial. *J. Neuroeng. Rehabil.* **2018**, *15*, 35. [CrossRef]
81. Chen, I.H.; Yang, Y.R.; Lu, C.F.; Wang, R.Y. Novel gait training alters functional brain connectivity during walking in chronic stroke patients: A randomized controlled pilot trial 11 Medical and Health Sciences 1109 Neurosciences 11 Medical and Health Sciences 1103 Clinical Sciences. *J. Neuroeng. Rehabil.* **2019**, *16*, 33. [CrossRef] [PubMed]
82. Nikmaram, N.; Scholz, D.S.; Großbach, M.; Schmidt, S.B.; Spogis, J.; Belardinelli, P.; Müller-Dahlhaus, F.; Remy, J.; Ziemann, U.; Rollnik, J.D.; et al. Musical Sonification of Arm Movements in Stroke Rehabilitation Yields Limited Benefits. *Front. Neurosci.* **2019**, *13*, 1378. [CrossRef] [PubMed]
83. Rayegani, S.; Raeissadat, S.; Sedighipour, L.; Rezazadeh, I.; Bahrami, M.; Eliaspour, D.; Khosrawi, S. Effect of neurofeedback and electromyographic-biofeedback therapy on improving hand function in stroke patients. *Top. Stroke Rehabil.* **2014**, *21*, 137–151. [CrossRef] [PubMed]
84. Cisotto, G.; Guglielmi, A.V.; Badia, L.; Zanella, A. Joint Compression of EEG and EMG Signals for Wireless Biometrics. In Proceedings of the Global Communications Conference, GLOBECOM, Abu Dhabi, United Arab Emirates, 9–13 December 2018.
85. Sridhar, S.; Manian, V. Eeg and deep learning based brain cognitive function classification. *Computers* **2020**, *9*, 104. [CrossRef]
86. Suprijanto; Noor, A.S.; Risangtuni, A.G.; Susanti, H. Automatic segmentation scheme for effective synchronization of EMG-EEG quantification. *Int. J. Biol. Biomed. Eng.* **2020**, *14*, 136–143. [CrossRef]
87. Tryon, J.; Friedman, E.; Trejos, A.L. Performance evaluation of EEG/EMG fusion methods for motion classification. In Proceedings of the International Conference on Rehabilitation Robotics, Toronto, ON, Canada, 24–28 June 2019; pp. 971–976.
88. Tryon, J.; Trejos, A.L. Classification of task weight during dynamic motion using EEG-EMG fusion. *IEEE Sens. J.* **2021**, *21*, 5012–5021. [CrossRef]
89. Roy, R.; Sikdar, D.; Mahadevappa, M.; Kumar, C.S. EEG based motor imagery study of time domain features for classification of power and precision hand grasps. In Proceedings of the International IEEE/EMBS Conference on Neural Engineering, Shanghai, China, 25–28 May 2017; pp. 440–443.
90. Bizzi, E.; Cheung, V.C. The neural origin of muscle synergies. *Front. Comput. Neurosci.* **2013**, *7*, 51. [CrossRef]
91. Lee, D.D.; Seung, H.S. Learning the parts of objects by non-negative matrix factorization. *Nature* **1999**, *401*, 788–791. [CrossRef]
92. Horwitz, B. The elusive concept of brain connectivity. *Neuroimage* **2003**, *19*, 466–470. [CrossRef]
93. Delorme, A.; Palmer, J.; Onton, J.; Oostenveld, R.; Makeig, S. Independent EEG Sources Are Dipolar. *PLoS ONE* **2012**, *7*, e30135. [CrossRef]
94. Pascual-Marqui, R.D.; Michel, C.M.; Lehmann, D. Low resolution electromagnetic tomography: A new method for localizing electrical activity in the brain. *Int. J. Psychophysiol.* **1994**, *18*, 49–65. [CrossRef]
95. Matsuya, R.; Ushiyama, J.; Ushiba, J. Inhibitory interneuron circuits at cortical and spinal levels are associated with individual differences in corticomuscular coherence during isometric voluntary contraction. *Sci. Rep.* **2017**, *7*, 44417. [CrossRef]
96. Baccalá, L.A.; Sameshima, K. Partial directed coherence: A new concept in neural structure determination. *Biol. Cybern.* **2001**, *84*, 463–474. [CrossRef]
97. Baccala, L.A.; Sameshima, K.; Takahashi, D.Y. Generalized partial directed coherence. In Proceedings of the 15th International Conference on Digital Signal Processing, Cardiff, UK, 1–4 July 2007; pp. 163–166. [CrossRef]
98. Liu, J.; Sheng, Y.; Liu, H. Corticomuscular coherence and its applications: A review. *Front. Hum. Neurosci.* **2019**, *13*, 100. [CrossRef]
99. Gao, Q.; Dou, L.; Belkacem, A.N.; Chen, C. Noninvasive Electroencephalogram Based Control of a Robotic Arm for Writing Task Using Hybrid BCI System. *Biomed. Res. Int.* **2017**, *2017*, 8316485. [CrossRef] [PubMed]
100. Leeb, R.; Sagha, H.; Chavarriaga, R.; Millán, J.D.R. Multimodal fusion of muscle and brain signals for a Hybrid-BCI. In Proceedings of the Annual International Conference of the Engineering in Medicine and Biology Society, Buenos Aires, Argentina, 31 August–4 September 2010; Volume 2010, pp. 4343–4346.
101. Basteris, A.; Nijenhuis, S.M.; Stienen, A.H.; Buurke, J.H.; Prange, G.B.; Amirabdollahian, F. Training modalities in robot-mediated upper limb rehabilitation in stroke: A framework for classification based on a systematic review. *J. Neuroeng. Rehabil.* **2014**, *11*, 111. [CrossRef]
102. Camona, C.; Wilkins, K.B.; Drogos, J.; Sullivan, J.E.; Dewald, J.P.A.; Yao, J. Improving Hand Function of Severely Impaired Chronic Hemiparetic Stroke Individuals Using Task-Specific Training With the ReIn-Hand System: A Case Series. *Front. Neurol.* **2018**, *9*, 923. [CrossRef]
103. Nascimento, L.R.; Michaelsen, S.M.; Ada, L.; Polese, J.C.; Teixeira-Salmela, L.F. Cyclical electrical stimulation increases strength and improves activity after stroke: A systematic review. *J. Physiother.* **2014**, *60*, 22–30. [CrossRef] [PubMed]

104. Knutson, J.S.; Fu, M.J.; Sheffler, L.R.; Chae, J. Neuromuscular electrical stimulation for motor restoration in hemiplegia. *Phys. Med. Rehabil. Clin. N. Am.* **2015**, *26*, 729–745. [CrossRef] [PubMed]
105. Fitzgibbon, S.P.; Pope, K.J.; MacKenzie, L.; Clark, C.R.; Willoughby, J.O. Cognitive tasks augment gamma EEG power. *Clin. Neurophysiol.* **2004**, *115*, 1802–1809. [CrossRef]

sensors

MDPI

Article

Questioning Domain Adaptation in Myoelectric Hand Prostheses Control: An Inter- and Intra-Subject Study

Giulio Marano [1,2], Cristina Brambilla [3], Robert Mihai Mira [3], Alessandro Scano [3], Henning Müller [1,4,*] and Manfredo Atzori [1,5]

1 Information Systems Institute, University of Applied Sciences Western Switzerland (HES-SO Valais), 3960 Sierre, Switzerland; giulio.marano92@gmail.com (G.M.); manfredo.atzori@hevs.ch (M.A.)
2 Department of Computer, Control, and Management Engineering, La Sapienza University, 00185 Rome, Italy
3 UOS STIIMA Lecco-Human-Centered, Smart & Safe, Living Environment, Italian National Research Council (CNR), 23900 Lecco, Italy; cristina.brambilla@stiima.cnr.it (C.B.); robertmihai.mira@stiima.cnr.it (R.M.M.); alessandro.scano@stiima.cnr.it (A.S.)
4 Department of Radiology, Medical Faculty, University of Geneva, 1211 Geneva, Switzerland
5 Department of Neuroscience, University of Padua, 35122 Padua, Italy
* Correspondence: henning.mueller@hevs.ch

Abstract: One major challenge limiting the use of dexterous robotic hand prostheses controlled via electromyography and pattern recognition relates to the important efforts required to train complex models from scratch. To overcome this problem, several studies in recent years proposed to use transfer learning, combining pre-trained models (obtained from prior subjects) with training sessions performed on a specific user. Although a few promising results were reported in the past, it was recently shown that the use of conventional transfer learning algorithms does not increase performance if proper hyperparameter optimization is performed on the standard approach that does not exploit transfer learning. The objective of this paper is to introduce novel analyses on this topic by using a random forest classifier without hyperparameter optimization and to extend them with experiments performed on data recorded from the same patient, but in different data acquisition sessions. Two domain adaptation techniques were tested on the random forest classifier, allowing us to conduct experiments on healthy subjects and amputees. Differently from several previous papers, our results show that there are no appreciable improvements in terms of accuracy, regardless of the transfer learning techniques tested. The lack of adaptive learning is also demonstrated for the first time in an intra-subject experimental setting when using as a source ten data acquisitions recorded from the same subject but on five different days.

Keywords: machine learning; EMG; biofeedback; transfer learning; random forest classifier

Citation: Marano, G.; Brambilla, C.; Mira, R.M.; Scano, A; Müller, H.; Atzori, M. Questioning Domain Adaptation in Myoelectric Hand Prostheses Control: An Inter- and Intra-Subject Study. *Sensors* **2021**, *21*, 7500. https://doi.org/10.3390/s21227500

Academic Editor: Georg Fischer

Received: 24 September 2021
Accepted: 9 November 2021
Published: 11 November 2021

1. Introduction

Amputation is one of the major reasons of disability [1]: it is estimated that 100.000 people have an upper limb amputation in the United States, and 57% of these are transradial amputees [2]. The principal causes of upper limb loss are traumatic events, followed by vascular diseases, congenic absence and cancer [3]. Upper limb amputation limits the daily life activity of a person heavily [4], although myoelectric prosthesis can restore the functionality of the hand using non-invasive EMG signal of the residual muscles [5]. The use of myoelectic signals has several advantages with respect to body-powered prostheses because the user does not need harnesses, the signal is recorded non-invasively on the skin and the effort required to control it is comparable to the one of an intact limb [6]. However, user acceptance is still low because of a lack of intuitive and dexterous control [7]: the rate of prosthesis abandonment is about 44% [8]. The control should be intuitive for the user, robust to arm and electrode positioning, adaptive to changes such as fatigue or sweating and easy to train [9].

In recent years, thanks to the advancement of robotics [10], control systems [11] and artificial intelligence [12], remarkable improvements have been made in the control of dexterous, robotic hand prostheses [13]. In particular, machine learning techniques allowed for developing surface electromyography (sEMG) prostheses that are capable to learn from each subject the myoelectric patterns corresponding to different hand movements [14]. However, such training procedures can be long (particularly to control a large number of hand movements), they are not robust to electrode re-positioning [1,9] (which can happen, for instance, after removing the prosthesis at night) and they can lead to considerable efforts for the patients. It was also found that inter-subject and inter-session variability are factors that may affect muscle coordination patterns [15]. Young et al. [16] found that a higher inter-electrode distance and a combination of longitudinal and transverse oriented channels can reduce the effects of electrode shift on the classification accuracy. The difficulties related to training myoelectric models increased the interest of scientific researchers for pre-built models [17]. Such models are expected to collect previous experience from several subjects and (through appropriate domain adaptation algorithms) they can be adapted to patients, accelerating model training. However, this approach can lead to divergence errors between different domains [18]. Over the years, several experiments lead to promising results in the domain of transfer learning [17–20].

One of the first studies about myoelectric signal divergence is from Castellini et al. [21]. They observed that myoelectric signals differ considerably between different subjects and that the use of pre-trained models should be bound to subjects who share a sufficient amount of similarities. This observation led to different approaches in order to take advantage of the prior knowledge of different subjects. Sensinger et al. [22] proposed several ways to concatenate source and target data in one model. Then, in order to improve non-adaptive baselines, Hypothesis Transfer Learning algorithms were employed in several studies. The advantage of these algorithms is that they do not need direct access to raw data exploiting models previously achieved from source subjects. Côté-Allard et al. [19] showed that transfer learning can lead to improved performance of hand gesture classifiers in three different datasets of able-bodied subjects using a convolutional network for the target domain, combining networks trained on the source with different activation functions. Kanoga et al. [23,24] acquired the same healthy subject for thirty consecutive days and applied domain adaptation on a linear discriminant analysis classifier, interpolating the mean vector and the covariance matrix of the calibration data of each day with the data recorded on the first day, and concluded that these methods allowed to adjust parameters for changes in positioning of the electrodes between different days. Other studies [25,26] applied transfer learning on convolutional neural networks (CNN) to improve model robustness on electrodes shifts: Ameri et al. [25] used a CNN model trained on data before shifting as a pretrained network and fine-tuned the model using few data of the same user after shifting; Wang et al. [26] transferred the parameters of the recurrent CNN model of the source domain to the EMG feature-extraction module of the target domain. Moreover, Liu et al. [27] applied domain adaptation techniques on a polynomial classifier, using the leave-one-out prediction error as a metric for the optimization algorithm, and on a linear discriminant analysis classifier with the Mahalanobis distance as a metric of consistency between prior models and the current training data. These algorithms were applied on both intact-limbed and transradial-amputee subjects for ten consecutive days, and it was found that the domain adaptation methods outperformed the baseline methods for both classifiers, especially with a small size of training data. Finally, Prahm et al. [28] applied domain adaptation exploiting the relationship between source and target domain: since data were recorded with an electrode grid, the distance between the electrodes was maintained equal even after the electrodes shifting, and therefore, they considered the shift on only one electrode and assumed linear feature changes between neighboring electrodes. They applied these approaches on both able-bodied subjects and transradial amputees and noticed that model performance increased with transfer learning on able-bodied subjects, but no relevant improvements were found in amputees.

Recently, transfer learning algorithms were used to train a model over the source domain to adapt it to a target domain with local adjustments of the tree parameters and its architecture [29].

Although different studies stated the efficacy of the use of domain adaptation on gesture recognition, they were usually applied on able-bodied subjects. Exploiting previously achieved results on intact subjects [30] from the Non-Invasive Adaptive Prosthetics (NinaPro) database [31], Gregori et al. [32] extended the study to amputee subjects and presented a novel framework for a realistic experimental setup. They found that, if the hyperparameters were properly tuned, transfer learning approaches showed the same performance of the standard methods that did not employ prior knowledge.

In this paper, we improve these results by applying two recent domain adaptation algorithms [29] to a random forest classifier (which is normally applied without hyperparameter optimization in the domain). A random forest classifier was already used on healthy subjects in combination with a regressor for discriminating reach to grasp strategies, obtaining good results [33]. These domain adaptation techniques modify the structure of decision trees within the forests generated by the source models in order to refine them on target repetitions. Our aims are: (1) to confirm and extend to a different data analysis workflow the results obtained in previous research; (2) to evaluate the quality of random forests as classifier for domain adaptation problems on sEMG data; (3) to extend the experiments to tests performed on data recorded from the same person but in data acquisition sessions of several days.

2. Materials and Methods

2.1. Domain Adaptation and Transfer Learning Algorithms

This section includes a step-by-step explanation of the models used in the domain adaptation and a description of the transfer learning algorithms used in our experiments.

2.1.1. Source, Target and Test Sets

Given $S = [s_1, s_2, \ldots, s_{n-1}, s_n]$, where n is the number of subjects in our dataset, the first step is the signal feature extraction. We can define all signal features extracted as $D = \{x_i, y_i\}_{i=1}^{N}$, where $x_i \in \mathbb{R}^d$ are the input samples, $y_i \in Y = \{1, \ldots, G\}$ are the paired labels and G is the number of possible classes plus the rest pose. We can split our initial collection of subjects into a target model S_T composed of a single subject and a source model S_S with all remaining subjects.

In the next step, we divide the number of repetitions of S_T, using one part as a test $S_{T_{test}}$ and the remaining as intra-subject training in order to compare the results for domain adaptation. Then, S_S is used as input to train a random forest classifier [34], which is a collection of decision trees. In this way, a forest trained on the source is obtained. In the same way, $S_{T_{train}}$ is passed to another random forest classifier, building another model trained on the target as result.

The domain adaptation step follows, where different algorithms transform the forest trained on the source and refine it on the target.

In conclusion, all previous forests are tested on $S_{T_{test}}$ and compared. This procedure is then repeated k-times, with $k \in S$, so that each subject takes the role of target at most once.

All steps described above are graphically explained in Figure 1.

2.1.2. Domain Adaptation Algorithms

Given a number of sources, a domain adaptation algorithm builds a new classification model refining the source on target. We use three algorithms, all based on random forest classifiers. Structure Expansion/Reduction (SER) and Structure Transfer (STRUT) take forests trained with source model as input and adapt them to a new target domain [29].

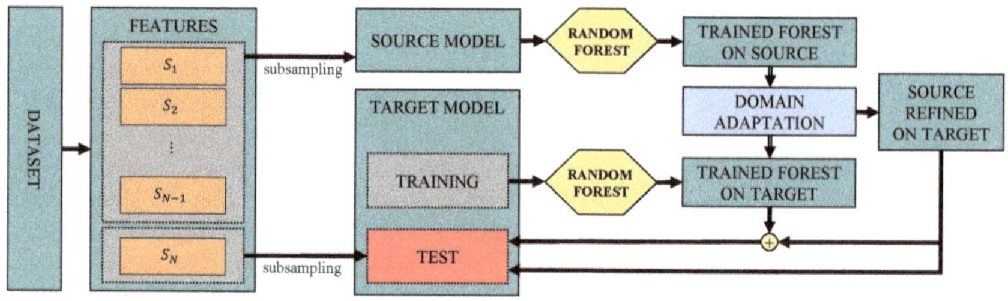

Figure 1. Block scheme for the domain adaptation model.

Then, we define a MIX algorithm that uses ensembles from both SER and STRUT as input and mixes them. More specifically, we can describe domain adaptation algorithms as follows:

(1) *Structure Expansion/Reduction (SER):* Given a random forest RF_S induced using the source data S_S, each decision tree (DT) is processed independently by the SER algorithm. First of all, the set S_v^T of all labeled points in the target data $S_{T_{train}}$ that reach the node v is computed. Then, in the expansion phase, a full tree expands from each leaf v with respect to S_v^T. Lastly, with a bottom-up approach, the algorithm performs a reduction of the structure for each internal node v.
This reduction is determined by two kinds of errors with respect to S_v^T:
- Subtree error E_S;
- Leaf error E_L.

The subtree error is the empirical error of the subtree of which the root is v. The leaf error is defined to be the empirical error on v if it were to be pruned into a leaf. If the following condition holds:

$$E_S > E_L \tag{1}$$

the subtree is pruned into a leaf node. The decision value at each leaf of the DT is obtained using the target (empirical) distribution. The SER algorithm then iterates these operations for each DT contained in the initial forest, building a new forest adapted on source.

(2) *Structure Transfer (STRUT):* While the SER algorithm acts on size of DTs inside S_S, the Structure Transfer algorithm changes the threshold. Since decision trees show similarity for similar problems, the STRUT algorithm exploits a top-down approach, adapting a DT trained on the source samples to the target samples by discarding all numeric threshold values in the tree. The values of the numeric thresholds are substituted by new thresholds $\tau(v)$ for a node v using the subset of target examples S_v^T that reach v.
If S_v^T is empty for a node v, v is pruned because it cannot be reached in the target domain. At each leaf, the final decision value is computed on the target training data. To perform threshold selection for feature ϕ, STRUT uses two parameters:
- Divergence gain *DG*;
- Information gain *IG*.

DG determines the distributional similarity, while *IG* quantifies the informative value of the threshold. The similarity is related only to those thresholds x whose *IG* is larger than the *IG* of any other x' in the ϵ-neighborhood of x for any sufficiently small $\epsilon > 0$. The STRUT algorithm searches for a threshold that gives a high similarity between the induced and the original distributions during the tree induction stage [35].

The selection of the threshold can be considered as an optimization problem:

$$\begin{aligned} &\max_x DG(S_v^T, \phi, x, Q_L, Q_R) \\ &\text{s.t.} \quad x \in \mathbb{R} \\ &\qquad \forall x' \in (x-\epsilon, x+\epsilon) \, : \, IG(S_v^T, \phi, x) \geq IG(S_v^T, \phi, x \pm \epsilon) \end{aligned} \tag{2}$$

where Q_L and Q_R are the left and right distribution, respectively.

(3) *MIX:* Once both SER and STRUT are applied, we obtain two distinct forests as a result. MIX is a combination of the two previous algorithms. This is a simple majority voting ensemble applied to all decision trees of both forests generated by STRUT and SER. As can be seen from the results, MIX does not simply average the results of the previous algorithms but often outperforms both of its constituents and thus is the second best solution. An intuitive explanation described in [29] about this result is given in Results.

2.2. Experimental Setup

The experimental setup is based on the "realistic setting" proposed in a previous work [32]. The setting was considered as "realistic" as it exploits real data coming from the Ninapro Dataset, recorded during the execution of daily life gestures. Section 2.2.1 presents the data used in all experiments. Then, the general settings and details about the experiments are described in Section 2.2.2.

2.2.1. Data

The data used in this work are from the NinaPro database (http://ninapro.hevs.ch/, accessed on 1 November 2021) [31,36], one of the largest publicly available databases that contains sEMG data of a wide range of distal upper limb movements. We use three NinaPro datasets for the experiments (namely NinaPro DB2, DB3 and DB6). As reported in Table 1, in all three cases, the acquisitions were made with Delsys Trigno sEMG sensors. This choice was made at the moment of data acquisition in order to allow the combination of the different datasets in future studies. The following subsections provide a brief presentation of the datasets, which are described in more details in the reference papers [31,36]. An illustration of the available EMG channels in this dataset is presented in Figure 2.

Table 1. NinaPro Data. Data employed in this study come from DB2, DB3, DB6 NinaPro datasets. The 8 electrode-array was used for the analyses.

DB	N° Sensors	N° Gestures	Repetitions
DB2	Delsys Trigno sEMG × 12	40	6
DB3	Delsys Trigno sEMG × 12	40	6
DB6	Delsys Trigno sEMG × 14(8)	7	12 × 2 × 5

(1) *NinaPro DB2 and DB3-Single acquisitions of intact subjects and transradial amputees:* The two datasets include 40 intact subjects and 11 amputees. Three amputees $[s_1, s_6, s_7]$ are excluded because their data acquisitions are not complete. Each subject executed 40 movements 6 times. Each movement repetition lasted 5 s and was followed by 3 s of rest. Twelve Delsys Trigno Wireless electrodes were used to record the sEMG data from the forearm of the subjects. Following a recently employed approach [7], a sliding window of 200 ms and an increment of 10 ms were used for the extraction of signal features. Therefore, the time windows were split into train and test sets for the classifiers, considering repetitions (1,3,4,6) for the training and repetitions 2 and 5 for test. A factor of 10 at regular intervals was used to subsample the training set in order to reduce the computational demands.

(2) *DB6-Intact subjects acquired for 5 consecutive days:* This dataset is composed of 10 intact subjects and targets the analysis of data acquisition repeatability in the same subjects. Each subject executed 7 hand grasps 12 times, two times per day, for 5 consecutive

days. Each grasp is followed with a few seconds of rest. During the acquisition of the movements, fourteen electrodes recorded sEMG data. Eight electrodes are positioned as the first eight electrodes in NinaPro DB2 and DB3 (i.e., equally spaced around the forearm at the height of the radio-humeral joint). The windowing procedure follows the same approach described for the previous datasets. For each session, repetitions (1,3,4,6,7,9,10,12) were dedicated to training, while repetitions (2,5,8,11) were used as test. In this case, the training set was also subsampled by a factor of 10 at regular intervals.

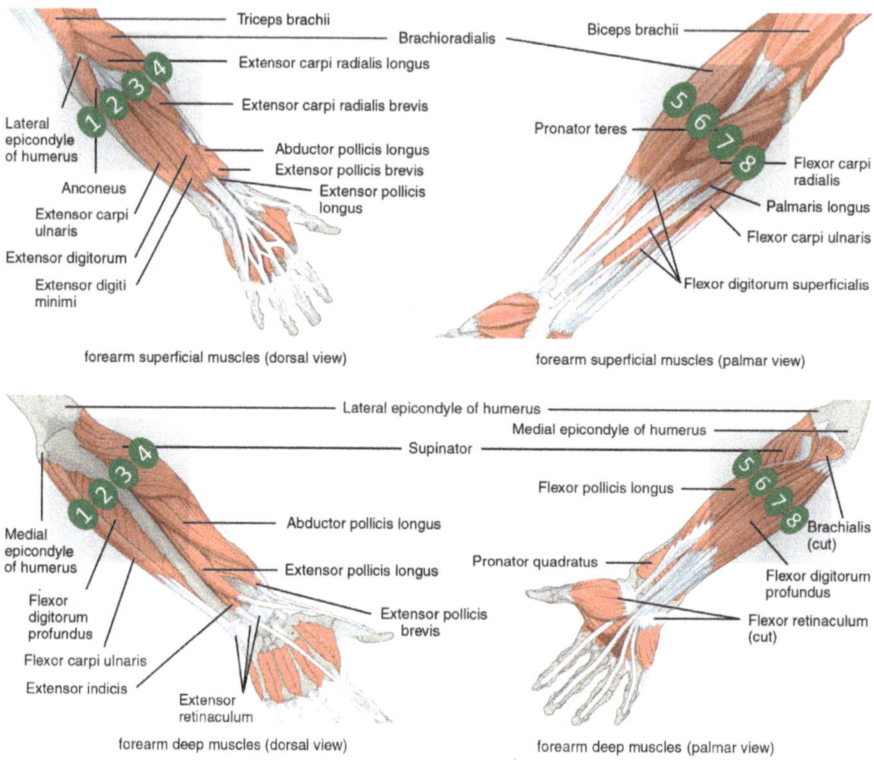

Figure 2. Positioning of the electrodes and underlying specific muscles. The image was adapted using file licensed under the Creative Commons Attribution 4.0 International license (Picture was adapted using https://upload.wikimedia.org/wikipedia/commons/7/73/1120_Muscles_that_Move_the_Forearm.jpg (accessed on 1 November 2021) from the Textbook OpenStax Anatomy and Physiology (source: https://cnx.org/contents/FPtK1zmh@8.25:fEI3C8Ot@10/Preface, accessed on 1 November 2021)).

The standardized data were used according to the protocol already proposed for control by Englehart and Hudgins [7], where features were extracted from a sliding window of 200 ms and an increment of 10 ms. As described in the papers presenting the datasets, sEMG signals were filtered from 50 Hz (and harmonics) power-line interference using a Hampel filter [31,36]. The resulting set of windows was subsequently split in the training set and test set as inputs for the classifier [32]. The sEMG representation used in this setting was the average of the marginal discrete wavelet transform (mDWT), mean absolute value (MAV) and variance (VAR) features [37].

2.2.2. Experiment Settings

One of the novelties of this paper is the use of a random forest classifier for domain adaptation on sEMG data. It was suggested that the incorrect optimization of hyperparameters was the main cause of transfer learning and domain adaptation improvements presented in previous literature [32]. In fact, classifiers such as the SVM need a grid search to find the best hyperparameters. Using random forests makes the optimization phase easier. The number of 100 trees was fixed for each forest and for all the experiments. This setup has shown high level performance (comparable to SVM) in previous results on sEMG data [31]. This approach was used for both the forests generated for the construction of the target model and for the source model.

The same number of trees was also used for the domain adaptation algorithms SER and STRUT.

The following eight classification performances are compared in each experiment:

- Source only;
- Target only;
- SER;
- STRUT;
- MIX;
- Voting ensemble (SER, Target Only);
- Voting ensemble (STRUT, Target Only);
- Voting ensemble (MIX, Target Only).

While the first five values are explained in the previous sections, the last three follow the same observation as for the MIX algorithm, with the aim of exceeding the accuracy obtained by the two individual components separately. Each of them represents a separate voting ensemble of which the underlying model is the union of all decision trees inside $S_{T_{train}}$ and each algorithm presented in Section 2.1.2 independently. The methods are summarized in Figure 3.

The experiments on DB2, DB3 and DB6 are conducted as follows.

(1) *DB2 and DB3:* The first experiments replicate the experiments previously cited:

- Intact–Intact: the classification of each subject from DB2 exploits prior knowledge of remaining subjects of DB2.
- Amputee–Intact: the classification of each subject from DB3 exploits prior knowledge of remaining subjects of DB3 plus all subject of DB2.
- Amputee–Amputee: the classification of each subject from DB3 exploits prior knowledge of remaining subjects of DB3.

In the training set, the subsets from 1 to 4 repetitions were taken into account for training. In each case, the k-fold cross validation was used for the optimization of the target model, with each fold corresponding to samples of one repetition. The source models, instead, were trained using all repetitions.

(2) *DB6:* Due to the very high number of repetitions for each subject (120), two simplified experimental settings were chosen:

- Intra-subject: each subject of DB6 exploits prior knowledge of the remaining repetitions of the same subject.
- Inter-subject: each subject of DB6 exploits prior knowledge of the remaining repetitions of the same subject plus all remaining subjects of DB6.

For both experiments, the target model is composed of 12 repetitions of the fifth afternoon, while the remaining repetitions of each subject are used to build the source model. In the intra-subject setup, we considered almost all possible subsets including 1–8 repetitions. In each case, the target model was optimized using k-fold cross validation, with each fold corresponding to samples of one repetition. In the inter-subject setup, the target model was trained using all 8 repetitions of the same session only. In both cases, the source model was optimized using a k-fold cross validation, where k is the number of repetitions from other sessions of the same subject used as target, plus (for the inter-subject setup) the total number of repetitions of each other subject.

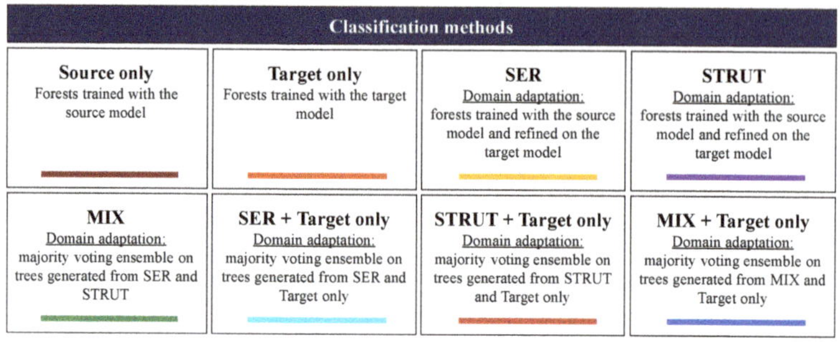

Figure 3. Scheme of the classification methods employed in each experiment.

3. Results

In Figures 4 and 5, the balanced classification accuracy is reported as a function of the number of training repetitions on the target. Domain adaptation does not improve movement classification accuracy in comparison to no-transfer learning, neither when pre-training is performed on different subjects, nor when pre-training is performed on different acquisitions of the same subject.

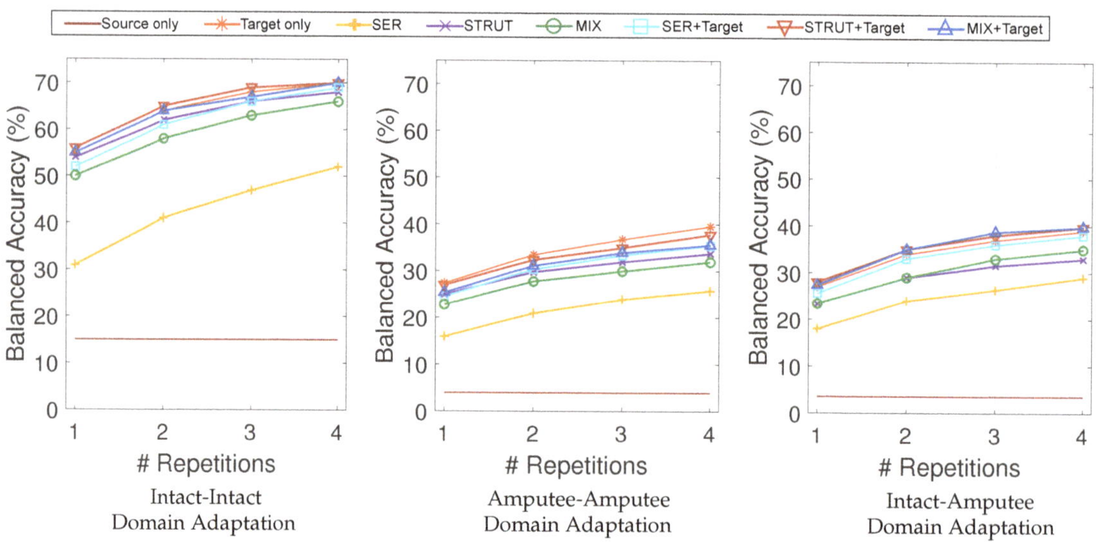

Figure 4. DB2 and DB3 Results: inter-subject balanced classification accuracy as a function of number of training repetitions on target.

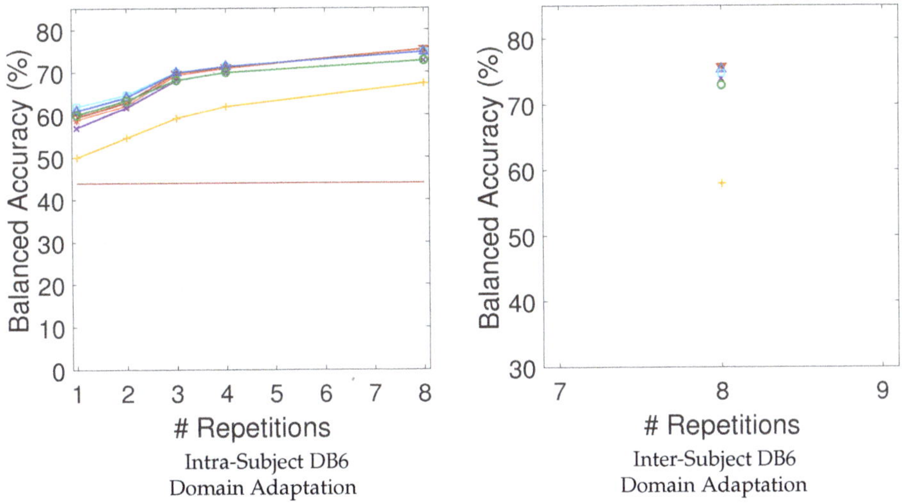

Figure 5. DB6 Results: intra-subject and inter-subject balanced classification accuracy as a function of number of training repetitions on target.

In Figure 4, the first set of experiments, performed on NinaPro DB2 and DB3, extends results obtained previously on SVMs. Results are reported in details in Table 2. Using random forests domain adaptation for these experiments offers a perspective of the problem that is influenced by less variables, since the classifier is normally applied without hyperparameter optimization procedures in the domain. Such procedures had been recently presented as a possible source of errors for domain adaptation works based on SVMs [32].

The SER algorithm has a performance that is lower than the STRUT algorithm, while the latter almost perfectly overlaps with the target-only result. Given the low performances of the SER algorithm, it is not surprising that the MIX algorithm does not give the best results. The plots also include the source model tested directly on the target (the flat series of data with the lowest performance in the plot). This result highlights (especially for amputees) how much the high variability between different subjects affects classification performance. Indeed, the classification accuracy is lower when amputee subjects are included, and it is higher when only intact subjects are considered.

A further novel result is that domain adaptation does not improve movement classification accuracy even when the data come from the same subject. The domain adaptation experiments using several data acquisitions recorded from the same person in different moments show that the "target only" model almost constantly provides results in line with the ones obtained by the domain adaptation models (Figure 5). The voting ensemble between target repetitions combined with the STRUT algorithm obtains a small improvement of the classification accuracy when one or two repetitions are considered in Figure 5 (left).

Finally, domain adaptation from other subjects does not improve classification accuracy even when pre-training on several data acquisitions from different subjects. It is not possible to notice any improvement in Figure 5 (right), showing that the addition of information is ineffective even when relying on such a high number of repetitions of the source. The results are also portrayed in Table 3.

Table 2. Classification methods accuracy on datasets DB2 and DB3 for each combination of subjects and for each repetition.

	DB2 and DB3											
	Intact–Intact				Amputee–Amputee				Intact–Amputee			
Rep / Methods	1	2	3	4	1	2	3	4	1	2	3	4
Source only	0.150	0.150	0.150	0.150	0.040	0.040	0.040	0.040	0.036	0.036	0.036	0.036
Target only	0.550	0.640	0.680	0.700	0.274	0.335	0.360	0.396	0.270	0.340	0.370	0.390
SER	0.310	0.410	0.470	0.520	0.160	0.210	0.240	0.258	0.180	0.240	0.264	0.290
STRUT	0.540	0.620	0.660	0.680	0.250	0.298	0.320	0.338	0.234	0.290	0.316	0.330
MIX	0.500	0.580	0.630	0.660	0.228	0.278	0.300	0.320	0.234	0.290	0.330	0.350
SER+Target	0.520	0.610	0.660	0.690	0.246	0.304	0.334	0.355	0.256	0.330	0.360	0.381
STRUT+Target	0.560	0.650	0.690	0.700	0.270	0.324	0.350	0.378	0.280	0.350	0.380	0.398
MIX+Target	0.550	0.640	0.670	0.700	0.254	0.312	0.340	0.356	0.274	0.350	0.388	0.398

Table 3. Classification methods accuracy on datasets DB6 for each combination of subjects and for each repetition.

	DB6					
	Inter-Subject					Intra-Subject
Rep / Methods	1	2	3	4	8	8
Source only	0.440	0.440	0.440	0.440	0.440	–
Target only	0.588	0.622	0.694	0.710	0.754	0.758
SER	0.500	0.546	0.592	0.620	0.674	0.580
STRUT	0.570	0.618	0.682	0.700	0.728	0.740
MIX	0.600	0.634	0.682	0.700	0.728	0.730
SER+Target	0.620	0.648	0.700	0.714	0.748	0.748
STRUT+Target	0.594	0.630	0.698	0.712	0.754	0.758
MIX+Target	0.610	0.642	0.700	0.714	0.748	0.754

4. Discussion

The results show that inter-subject domain adaptation does not improve classification accuracy, and it extends the result to intra-subject models computed from different acquisitions of the same subject. This result confirms and extends previous results [32] and is in partial disagreement with several previous works on domain adaptation.

While previously this conclusion was explained in relationship to SVM parameter optimization, in our case the result is obtained using a random forest classifier (with a fixed configuration) and several new transfer learning methods. The domain adaptation methods tested in this study performed as well as the target-only baseline, which did not consider the source information. Similarly to previous findings [29], the STRUT algorithm gives better results with respect to the SER when the correspondence between features is maintained, while the SER algorithm outperforms STRUT on inverse problems. From our results, the MIX algorithm obtained a performance that was closer to the best performance, demonstrating that it had not the average performance of the SER and STRUT. SER and STRUT algorithms act differently on the same tree: the SER algorithm changes the size of the original tree, adding depth in the expansion phase and reducing the size of branches in the reduction one; the STRUT algorithm, instead, maintains the original size of the tree, modifying the thresholds. Therefore, the MIX forest turns out to be a more diverse forest, in which the pairwise correlation between two trees, derived from the same original tree, is low.

Our classification accuracy results are lower than some of the results previously proposed [32], but this is probably due to differences in the metrics used. While we preferred to use balanced classification accuracy (due to the unbalanced multi-class nature of the classification problem), unbalanced classification accuracy was most likely used for the realistic setting in the previous work. The difference in accuracy can thus be explained considering the high incidence of the rest in the dataset (which is classified with high accuracy).

Another interesting result is introduced by the experiments performed on the data from the same subject. The intra-subject experiment shows that using the previous experience of the same subject, there is basically no improvement compared to the case of no transfer learning. The most intuitive explanation for this result (also reported by Palermo et al. [36]) is that the re-positioning of the electrodes at each session produces substantially different results, even for the same subject.

The results from this work may impact real-life settings for people with hand prostheses. In fact, a major challenge limiting the use of dexterous robotic hand prostheses controlled via electromyography and pattern recognition relates to the important efforts required to train complex models from scratch. To overcome this problem, several studies in recent years proposed to use transfer learning, combining pre-trained models (obtained from prior subjects) with training sessions performed on a specific user. Differently from several previous papers, our results show that there are no appreciable improvements in terms of accuracy, regardless of the transfer learning techniques employed. The lack of adaptive learning is also demonstrated for the first time in an intra-subject experimental setting, when using as source ten data acquisitions recorded from the same subject but on five different days. This novel result has remarkable repercussions. In fact, in this paper, it was demonstrated for the first time that not only in single-session recordings, but also in an intra-subject experimental setting, adaptive learning is not taking place, and several questions regarding the training of prosthesis with previously acquired data arise. If these results would be confirmed in further studies, the training effort for amputee subjects could not be minimized with the exploiting of the previous knowledge available, at least with algorithms and techniques employed so far. However, authors believe that other strategies (e.g., based on deep neural networks) should be evaluated as well, as they might allow for exploiting prior information thanks to different approaches. These results are in accordance with what was already found when examining the same domain from EMG recordings using other data extraction methods such as muscle synergies [15], in which inter-session analysis was carried out, revealing how data can vary considerably also intra-subject and cannot be used for generalizing intra-subject and inter-subject patterns.

This work has some limitations. First, data from many sessions were used; however, longer time periods could be considered to extend the validity of our results in prolonged recordings. Moreover, while the number of participants is not low for this type of study, it still cannot be considered as representative of all subjects. Future work should expand our results including a higher cohort of volunteers, also divided according to registry data, so that the conclusions could be extended to gender and age differences. Despite our results, we think that classification accuracy of a task may be improved using previous data available from related tasks. Future work needs to consider this problem by conducting experiments with new and different approaches or by using a larger number of acquisitions. Furthermore, it is possible that other classification or pre-processing methods may allow domain adaptation, for instance by taking into account physical constraints (such as physical electrode placement) [38] or by using different transfer learning techniques (e.g., based on deep neural networks).

5. Conclusions

Differently from what has been described in several previous studies on domain adaptation in electromyography, our results show that domain adaptation does not appreciably improve classification accuracy, regardless of the transfer learning techniques tested. The results extend previous studies for a realistic setting by using random forests as classification algorithm and two algorithms for domain adaptation.

The lack of adaptive learning is also demonstrated for the first time in an intra-subject experimental setting, when using as source ten data acquisitions recorded from the same subject but on five different days. The results demonstrate that the use of previous experience does not offer concrete improvement, even when considering data from the same subject and a different classifier, confirming and extending previous achievements and somehow posing alternative interpretations with respect to several previous works on domain adaptation. Future works should consider different approaches or use a higher number of repetitions in order to improve the performance of the classifier by employing prior information from related tasks.

Author Contributions: Conceptualization, M.A., G.M., H.M.; methodology, M.A., G.M., H.M.; software, G.M.; validation, G.M.; formal analysis, G.M., M.A.; investigation, G.M., M.A.; resources, A.S., M.A., H.M.; data curation, G.M., M.A.; writing—original draft preparation, G.M., M.A., H.M.; writing—review and editing, C.B., R.M.M., A.S., G.M., M.A., H.M.; visualization, C.B., R.M.M., A.S., G.M.; supervision, A.S., M.A., H.M.; project administration, H.M., M.A., A.S.; funding acquisition, H.M., M.A., A.S. All authors have read and agreed to the published version of the manuscript.

Funding: This work was partially supported by the Swiss National Science Foundation Sinergia project # 410160837 MeganePro.

Institutional Review Board Statement: This study used the pre-existing publicly available dataset Ninapro.

Informed Consent Statement: Not applicable.

Data Availability Statement: NinaPro data are available at: http://ninaweb.hevs.ch/, accessed on 1 November 2021.

Conflicts of Interest: The authors declare no conflict of interest.

References

1. Young, A.J.; Hargrove, L.J.; Kuiken, T.A. The effects of electrode size and orientation on the sensitivity of myoelectric pattern recognition systems to electrode shift. *IEEE Trans. Biomed. Eng.* **2011**, *58*, 2537–2544. [CrossRef]
2. Waris, A.; Niazi, I.K.; Jamil, M.; Gilani, O.; Englehart, K.; Jensen, W.; Shafique, M.; Kamavuako, E.N. The effect of time on EMG classification of hand motions in able-bodied and transradial amputees. *J. Electromyogr. Kinesiol.* **2018**, *40*, 72–80. [CrossRef] [PubMed]
3. Raichle, K.A.; Hanley, M.A.; Molton, I.; Kadel, N.J.; Campbell, K.; Phelps, E.; Ehde, D.; Smith, D.G. Prosthesis use in persons with lower-and upper-limb amputation. *J. Rehabil. Res. Dev.* **2008**, *45*, 961. [CrossRef]
4. Atzori, M.; Rosanda, E.; Pajardi, G.; Bassetto, F.; Müller, H. Effect of movement type on the classification of electromyography data for the control of dexterous prosthetic hands. In Proceedings of the 2020 8th IEEE RAS/EMBS International Conference for Biomedical Robotics and Biomechatronics (BioRob), New York, NY, USA, 29 November–1 December 2020; pp. 746–751.
5. Yang, D.; Gu, Y.; Thakor, N.V.; Liu, H. Improving the functionality, robustness, and adaptability of myoelectric control for dexterous motion restoration. *Exp. Brain Res.* **2019**, *237*, 291–311. [CrossRef] [PubMed]
6. Parker, P.; Englehart, K.; Hudgins, B. Myoelectric signal processing for control of powered limb prostheses. *J. Electromyogr. Kinesiol.* **2006**, *16*, 541–548. [CrossRef] [PubMed]
7. Englehart, K.; Hudgins, B. A robust, real-time control scheme for multifunction myoelectric control. *IEEE Trans. Biomed. Eng.* **2003**, *50*, 848–854. [CrossRef] [PubMed]
8. Salminger, S.; Stino, H.; Pichler, L.H.; Gstoettner, C.; Sturma, A.; Mayer, J.A.; Szivak, M.; Aszmann, O.C. Current rates of prosthetic usage in upper-limb amputees–have innovations had an impact on device acceptance? *Disabil. Rehabil.* **2020**, 1–12. [CrossRef] [PubMed]
9. Farina, D.; Jiang, N.; Rehbaum, H.; Holobar, A.; Graimann, B.; Dietl, H.; Aszmann, O.C. The extraction of neural information from the surface EMG for the control of upper-limb prostheses: Emerging avenues and challenges. *IEEE Trans. Neural Syst. Rehabil. Eng.* **2014**, *22*, 797–809. [CrossRef]
10. Kyrarini, M.; Lygerakis, F.; Rajavenkatanarayanan, A.; Sevastopoulos, C.; Nambiappan, H.R.; Chaitanya, K.K.; Babu, A.R.; Mathew, J.; Makedon, F. A survey of robots in healthcare. *Technologies* **2021**, *9*, 8. [CrossRef]
11. Roche, A.D.; Rehbaum, H.; Farina, D.; Aszmann, O.C. Prosthetic myoelectric control strategies: A clinical perspective. *Curr. Surg. Rep.* **2014**, *2*, 1–11. [CrossRef]
12. Rong, G.; Mendez, A.; Assi, E.B.; Zhao, B.; Sawan, M. Artificial intelligence in healthcare: Review and prediction case studies. *Engineering* **2020**, *6*, 291–301. [CrossRef]
13. Atzori, M.; Müller, H. Control capabilities of myoelectric robotic prostheses by hand amputees: A scientific research and market overview. *Front. Syst. Neurosci.* **2015**, *9*, 162. [CrossRef] [PubMed]

14. Oskoei, M.A.; Hu, H. Myoelectric control systems—A survey. *Biomed. Signal Process. Control* **2007**, *2*, 275–294. [CrossRef]
15. Pale, U.; Atzori, M.; Müller, H.; Scano, A. Variability of muscle synergies in hand grasps: Analysis of intra-and inter-session data. *Sensors* **2020**, *20*, 4297. [CrossRef]
16. Young, A.J.; Hargrove, L.J.; Kuiken, T.A. Improving myoelectric pattern recognition robustness to electrode shift by changing interelectrode distance and electrode configuration. *IEEE Trans. Biomed. Eng.* **2011**, *59*, 645–652. [CrossRef] [PubMed]
17. Pan, S.J.; Yang, Q. A survey on transfer learning. *IEEE Trans. Knowl. Data Eng.* **2009**, *22*, 1345–1359. [CrossRef]
18. Ben-David, S.; Blitzer, J.; Crammer, K.; Kulesza, A.; Pereira, F.; Vaughan, J.W. A theory of learning from different domains. *Mach. Learn.* **2010**, *79*, 151–175. [CrossRef]
19. Côté-Allard, U.; Fall, C.L.; Drouin, A.; Campeau-Lecours, A.; Gosselin, C.; Glette, K.; Laviolette, F.; Gosselin, B. Deep learning for electromyographic hand gesture signal classification using transfer learning. *IEEE Trans. Neural Syst. Rehabil. Eng.* **2019**, *27*, 760–771. [CrossRef]
20. Zhao, H.; Zhang, S.; Wu, G.; Costeira, J.P.; Moura, J.M.F.; Gordon, G.J. Adversarial multiple source domain adaptation. *Adv. Neural Inf. Process. Syst.* **2018**, *31*, 8559–8570.
21. Castellini, C.; Fiorilla, A.E.; Sandini, G. Multi-subject/daily-life activity EMG-based control of mechanical hands. *J. Neuroeng. Rehabil.* **2009**, *6*, 1–11. [CrossRef]
22. Sensinger, J.W.; Lock, B.A.; Kuiken, T.A. Adaptive pattern recognition of myoelectric signals: Exploration of conceptual framework and practical algorithms. *IEEE Trans. Neural Syst. Rehabil. Eng.* **2009**, *17*, 270–278. [CrossRef]
23. Kanoga, S.; Kanemura, A. Transfer learning over time and position in wearable myoelectric control systems. In Proceedings of the 2018 IEEE International Conference on Industrial Technology (ICIT), Honolulu, HI, USA, 18–21 July 2018; pp. 1478–1483.
24. Kanoga, S.; Matsuoka, M.; Kanemura, A. Assessing the effect of transfer learning on myoelectric control systems with three electrode positions. In Proceedings of the 2018 40th Annual International Conference of the IEEE Engineering in Medicine and Biology Society (EMBC), Lyon, France, 20–22 February 2018; pp. 1–4.
25. Ameri, A.; Akhaee, M.A.; Scheme, E.; Englehart, K. A deep transfer learning approach to reducing the effect of electrode shift in EMG pattern recognition-based control. *IEEE Trans. Neural Syst. Rehabil. Eng.* **2019**, *28*, 370–379. [CrossRef] [PubMed]
26. Wang, W.; Chen, B.; Xia, P.; Hu, J.; Peng, Y. Sensor fusion for myoelectric control based on deep learning with recurrent convolutional neural networks. *Artif. Organs* **2018**, *42*, E272–E282. [CrossRef] [PubMed]
27. Liu, J.; Sheng, X.; Zhang, D.; He, J.; Zhu, X. Reduced daily recalibration of myoelectric prosthesis classifiers based on domain adaptation. *IEEE J. Biomed. Health Inform.* **2014**, *20*, 166–176. [CrossRef] [PubMed]
28. Prahm, C.; Schulz, A.; Paaßen, B.; Schoisswohl, J.; Kaniusas, E.; Dorffner, G.; Hammer, B.; Aszmann, O. Counteracting electrode shifts in upper-limb prosthesis control via transfer learning. *IEEE Trans. Neural Syst. Rehabil. Eng.* **2019**, *27*, 956–962. [CrossRef]
29. Segev, N.; Harel, M.; Mannor, S.; Crammer, K.; El-Yaniv, R. Learn on source, refine on target: A model transfer learning framework with random forests. *IEEE Trans. Pattern Anal. Mach. Intell.* **2016**, *39*, 1811–1824. [CrossRef] [PubMed]
30. Patricia, N.; Tommasit, T.; Caputo, B. Multi-source adaptive learning for fast control of prosthetics hand. In Proceedings of the 2014 22nd International Conference on Pattern Recognition, Stockholm, Sweden, 24–28 August 2014; pp. 2769–2774.
31. Atzori, M.; Gijsberts, A.; Castellini, C.; Caputo, B.; Hager, A.M.; Elsig, S.; Giatsidis, G.; Bassetto, F.; Müller, H. Electromyography data for non-invasive naturally-controlled robotic hand prostheses. *Sci. Data* **2014**, *1*, 1–13. [CrossRef] [PubMed]
32. Gregori, V.; Gijsberts, A.; Caputo, B. Adaptive learning to speed-up control of prosthetic hands: A few things everybody should know. In Proceedings of the 2017 International Conference on Rehabilitation Robotics (ICORR), London, UK, 17–20 July 2017; pp. 1130–1135.
33. Liarokapis, M.V.; Artemiadis, P.K.; Kyriakopoulos, K.J.; Manolakos, E.S. A learning scheme for reach to grasp movements: On EMG-based interfaces using task specific motion decoding models. *IEEE J. Biomed. Health Inform.* **2013**, *17*, 915–921. [CrossRef] [PubMed]
34. Breiman, L. Random forests. *Mach. Learn.* **2001**, *45*, 5–32. [CrossRef]
35. Hunt, E.B.; Marin, J.; Stone, P.J. *Experiments in Induction*; Academic Press: Cambridge, MA, USA, 1966.
36. Palermo, F.; Cognolato, M.; Gijsberts, A.; Müller, H.; Caputo, B.; Atzori, M. Repeatability of grasp recognition for robotic hand prosthesis control based on sEMG data. In Proceedings of the 2017 International Conference on Rehabilitation Robotics (ICORR), London, UK, 17–20 July 2017; pp. 1154–1159.
37. Kuzborskij, I.; Gijsberts, A.; Caputo, B. On the challenge of classifying 52 hand movements from surface electromyography. In Proceedings of the 2012 Annual International Conference of the IEEE Engineering in Medicine and Biology Society, San Diego, CA, USA, 28 August–1 September 2012; pp. 4931–4937.
38. Kim, M.; Kim, K.; Chung, W.K. Simple and fast compensation of sEMG interface rotation for robust hand motion recognition. *IEEE Trans. Neural Syst. Rehabil. Eng.* **2018**,*26*, 2397–2406. [CrossRef]

sensors

MDPI

Article

Experimental Assessment of Cuff Pressures on the Walls of a Trachea-Like Model Using Force Sensing Resistors: Insights for Patient Management in Intensive Care Unit Settings

Antonino Crivello [1,†], Mario Milazzo [2,*,†], Davide La Rosa [1], Giacomo Fiacchini [3], Serena Danti [2,4], Fabio Guarracino [5], Stefano Berrettini [3] and Luca Bruschini [3]

1 ISTI-CNR, Institute of Information Science and Technologies, 56124 Pisa, Italy; antonino.crivello@isti.cnr.it (A.C.); davide.larosa@isti.cnr.it (D.L.R.)
2 The BioRobotics Institute, Scuola Superiore Sant'Anna, 56025 Pontedera, Italy; serena.danti@unipi.it
3 Otolaryngology, Audiology and Phoniatric Operative Unit, Department of Surgical, Medical and Molecular Pathology and Critical Care Medicine, University of Pisa, 56123 Pisa, Italy; giacomo.fiacchini@unipi.it (G.F.); stefano.berrettini@unipi.it (S.B.); luca.bruschini@unipi.it (L.B.)
4 Department of Civil and Industrial Engineering, University of Pisa, 56122 Pisa, Italy
5 Cardiothoracic and Vascular Anaesthesia and Intensive Care, Department of Anaesthesia and Critical Care Medicine, Azienda Ospedaliero Universitaria Pisana, 56123 Pisa, Italy; f.guarracino@ao-pisa.toscana.it
* Correspondence: mario.milazzo@santannapisa.it
† These authors contributed equally to this work.

Abstract: The COVID-19 outbreak has increased the incidence of tracheal lesions in patients who underwent invasive mechanical ventilation. We measured the pressure exerted by the cuff on the walls of a test bench mimicking the laryngotracheal tract. The test bench was designed to acquire the pressure exerted by endotracheal tube cuffs inflated inside an artificial model of a human trachea. The experimental protocol consisted of measuring pressure values before and after applying a maneuver on two types of endotracheal tubes placed in two mock-ups resembling two different sized tracheal tracts. Increasing pressure values were used to inflate the cuff and the pressures were recorded in two different body positions. The recorded pressure increased proportionally to the input pressure. Moreover, the pressure values measured when using the non-armored (NA) tube were usually higher than those recorded when using the armored (A) tube. A periodic check of the cuff pressure upon changing the body position and/or when performing maneuvers on the tube appears to be necessary to prevent a pressure increase on the tracheal wall. In addition, in our model, the cuff of the A tube gave a more stable output pressure on the tracheal wall than that of the NA tube.

Keywords: COVID-19; intubation; tracheoesophageal fistula; tracheal lesions; acute respiratory distress syndrome; modeling; intensive care unit

Citation: Crivello, A.; Milazzo, M.; La Rosa, D.; Fiacchini, G.; Danti, S.; Guarracino, F.; Berrettini, S.; Bruschini, L. Experimental Assessment of Cuff Pressures on the Walls of a Trachea-Like Model Using Force Sensing Resistors: Insights for Patient Management in Intensive Care Unit Settings. *Sensors* **2022**, *22*, 697. https://doi.org/10.3390/s22020697

Academic Editor: James F. Rusling

Received: 23 November 2021
Accepted: 12 January 2022
Published: 17 January 2022

1. Introduction

The coronavirus disease 2019 (COVID-19) outbreak has raised many critical issues in the management of patients affected by acute respiratory distress syndrome (ARDS) in an intensive care unit (ICU) setting [1,2]. Among others, the high incidence of full-thickness tracheal lesions (FTTLs) and tracheoesophageal fistulas (TEFs), and their potential life-threatening complications, such as pneumomediastinum, pneumothorax, and subcutaneous emphysema, have been reported in patients who underwent invasive mechanical ventilation (MV) [3,4]. This procedure consists of ventilating the respiratory apparatus via an endotracheal polymeric tube with an inflatable cuff that seals the tracheal duct. Depending on the targeted application, the endotracheal tube may have an embedded reinforced metal coil to stiffen the structure, making it less likely to be obstructed [5]. However, independent of the tube type, a cuff pressure ranging between 20 and 30 cmH_2O is always recommended to avoid damage or trauma to the host tissue [6,7].

Many etiopathogenetic hypotheses have been proposed to explain the unprecedented increase in complications observed in ARDS patients treated with MV [3,8]; however, to date, a clear explanation has not been found.

One proposed mechanism relies on performing the invasive MV in the prone (P) position with the patient's head laterally rotated. Specifically, by moving the patient from the supine (S) to the P position, the orotracheal tube cuff is supposed to increase its pressure on the tracheal wall, causing tissue lesions. However, there are no studies demonstrating how these lesions are formed at a laryngotracheal level in this body position.

We designed an experimental study to measure the pressure exerted by the cuff on the wall of a test bench mimicking the laryngotracheal tract, placed in different orientations (S and P), with different loading configurations (torsion and bending of the orotracheal tube). The experimental protocol considered both types of endotracheal tubes, namely armored (A) vs. non-armored (NA), under a progressive increase in the internal cuff pressure.

Understanding the causes of FTTL and TEF formation in patients affected by ARDS would greatly improve patient management in the ICU, which can result in faster and less complicated recoveries than those currently experienced.

2. Materials and Methods

The test bench was designed to acquire the pressure exerted by endotracheal tube cuffs inflated inside a cylindrical pipe, simulating an artificial model of a human trachea during the intubation maneuvers.

The endotracheal tubes used in this study were armored (A) (Unomedical UM61214075, 7.5 mm, made by ConvaTec, Deeside, UK) and non-armored (NA) (ETT-P22-75, 7.5 mm, made by Medis Medical, Tianjin, China). The distinctive feature of the A tube is the metal wire coil embedded in its wall, which keeps the lumen of the tube open when it is bent. Moreover, the A tube is more flexible than the NA tube, so it is less prone to kink and/or to being occluded when bent. Because the NA tube is pre-formed and more rigid than the A tube, it does not require the use of a stylet for a successful intubation. In contrast, the A tube, being flexible, always requires the use of a stylet. The artificial laryngotracheal tract used in this study was built using a corrugated plastic tube to replicate the characteristics of the larynx and trachea. To monitor the endotracheal tube cuff pressure exerted on the tracheal segment, we deployed four force sensing resistors (FSRs) along the tracheal tract, positioned on the four sides of the internal tube (i.e., 0°, 90°, 180°, and 270°, clockwise). Figure 1 shows a sketch of the prototype from the front and from a longitudinal cross-section perspective. To assure ease of repeatability and high-precision measurements, the tracheal segment of the prototype where the FSRs were located was made of a transparent material (i.e., plexiglass).

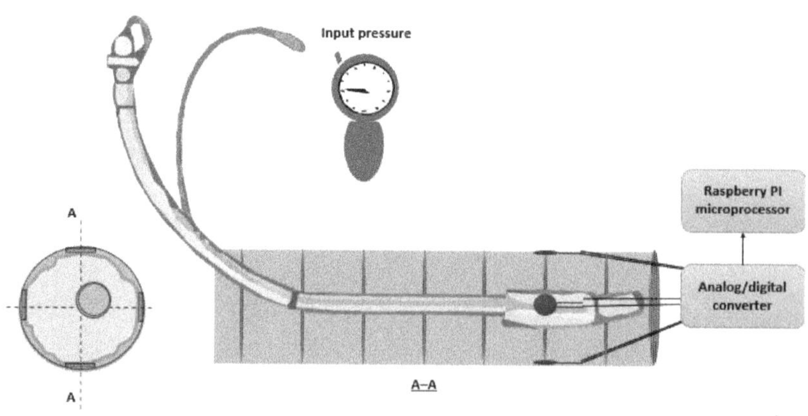

Figure 1. Front view and longitudinal cross-section of the prototype.

The sensors used to measure and record the pressure generated by the inflated cuffs are called FSRs. They are made of a conductive polymer that changes its electrical resistance proportionate to the force applied to its surface. These sensors have been widely applied to acquire different human body functional parameters, such as foot pressure [9], respiration rate [10], finger forces [11], muscle activity [12], and body movements during sleep [13]. The product used in this study was an Interlink FSR® Model 400, which is characterized by its thinness (i.e., 0.3 mm) and circular active area of ~5 mm diameter. These features allow the sensors to be positioned and fastened appropriately onto the inner surface of the tube without creating additional thickness or deformation that may interfere with the endotracheal tube maneuvers, thus influencing the measurements. Figure 2 shows the main dimensions of an FSR and its resistance vs. force characteristics.

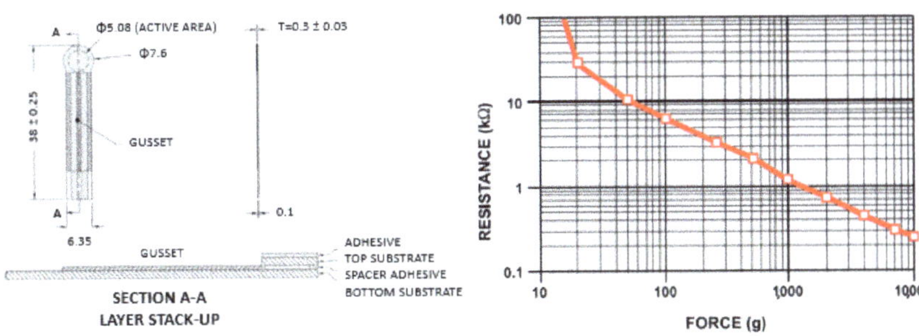

Figure 2. Interlink FSR® Model 400 mechanical data (**left**) and typical force vs. resistance response (**right**) from the datasheet [14].

The FSRs were connected to a Raspberry Pi single-board computer through multiple microchip MCP3424 analog-to-digital converter (ADC) modules, which translate the raw input voltage from the sensors to a digital format data readable by a software application. The MCP3424 is an 18 bit, 4-channel delta-sigma ADC with differential inputs, self-calibration of internal offset, and gain at each conversion. It also provides an internal programmable gain amplifier, an internal voltage reference (i.e., 2.048 V $\pm$ 0.05%, 15 ppm/°C), and a programmable data rate of up to 240 samples per second. The ADC modules communicate with the Raspberry Pi board through the I2C bus in high-speed mode (i.e., 3.4 MHz).

A dedicated application run on the Raspberry Pi oversaw sensor data collection through a sampling frequency of 3 Hz and subsequently stored it in a local file before transmitting it on a remote computer for visualization. The acquired sensor voltage data were analyzed with MATLAB (v2019b), converted into pressure values (in kilopascals), and plotted in real time to provide immediate feedback during the maneuvers.

The FSRs, although versatile, compact, and cheap, are subject to physical effects due to their construction, which necessitates careful calibration before usage to reduce sensing errors. Specifically, it is recommended that the calibration phase of each sensor be conducted in an environment that is close to the final application [15]. Remarkably, FSRs show a power law behavior within the force range from 0 to 20 N. However, the 0–4 N range shows mostly a non-linear relation, thus highlighting the need for accurate calibration of the FSRs in low-pressure range setups [16]. The FSRs were experimentally characterized to obtain the actual response curve, considering the effect of the ADC conditioning circuit. Briefly, 8 sensors were tested by applying a sequence of calibration weights (i.e., 10, 20, 50, 100, and 200 g). After applying each weight, the sensor output was left to stabilize before noting the reading. For each weight, 10 measures were taken to average the outcome. From the experimental readings, we designed the fitting curve using a cubic polynomial regression (Figure 3). Lastly, we computed the inverse function that was used to estimate the force applied on the sensor surface.

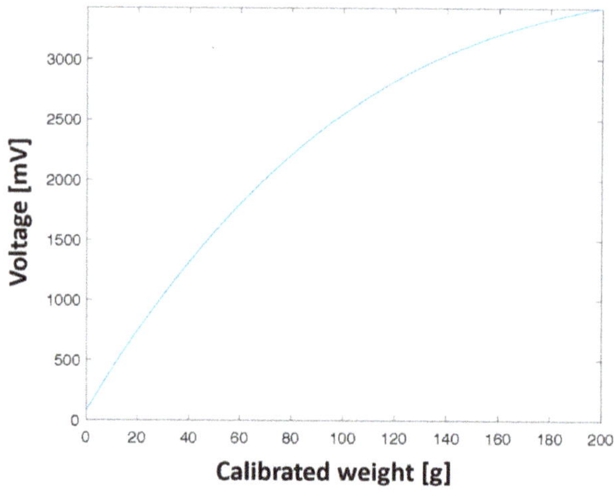

Figure 3. FSRs experimental characterization curve expressed as applied calibrated weight vs. ADC input voltage.

2.1. Measurement Protocol

Pressure values were measured before and after applying a maneuver on two types of endotracheal tubes (NA/A) placed inside two mock-ups resembling the pharyngeal, laryngeal, and tracheal tracts. The mock-ups had two different diameters (i.e., 20 and 25 mm). Increasing pressure values were used to inflate the cuff (30/40/50 cmH_2O). Even though the recommendations suggest limiting cuff pressure to below 30 cmH_2O, we decided to slightly increase this value in order to consider the potential for operator error. For each configuration of the tube-test bench, we inserted the tube with the curvature shown in Figure 1 to replicate the S position of the patient. Then, we verified that the cuff reached the correct position, namely with the middle section of the cuff in contact with the four sensors. We inflated the cuff with the targeted pressure, let the sensor signals stabilize, and then noted the pressure values obtained from each sensor. Before applying a different maneuver, we deflated the cuff, recovered the original geometrical configuration, and reinflated the cuff at the target pressure. Thereafter, we applied two different maneuvers (i.e., torsion (T) and bending (B)) with amplitudes equal to 90°, along two different directions (i.e., clockwise and counterclockwise). To perform the experiments with the test bench replicating the P position, the same set of experiments were performed with the test bench rotated by 180° along the median axis of the mock-up.

2.2. Statistics

We clustered pressure values before and after maneuvers based on the following group types: input pressures (30/40/50 cmH_2O), body position (S/P), maneuver (T/B), and type of endotracheal tube (NA/A). Data from the four sensors were averaged at each measurement. Data processing was carried out using jamovi software (V1.6.16.0).

The mean and standard deviation (SD) for the pressure values before the maneuvers were evaluated. We grouped data from both the torsion and bending maneuvers, giving a comprehensive picture of the initial configuration, before the maneuvers, across input pressures. Later, we performed independent Student's t-test analyses to evaluate the statistical differences, if any, between the groups (NA/A and S/P) at fixed input pressures. The p-value threshold was set at α equal to 0.05.

The differences between the pressure values after and before applying a maneuver were also analyzed. In this case, an analysis separating the outcomes from each type of maneuver was used, gathering the values from the two opposite directions used for each maneuver (T/B). Thus, we performed independent *t*-test analyses to evaluate the statistical differences, if any, between the groups (NA/A and S/P) at fixed input pressures, for each type of maneuver (i.e., T/B). The *p*-value threshold was fixed at α equal to 0.05. In all tests, the null hypothesis concerned the lack of statistical difference between mean values.

3. Results

The dataset was tested to check the normality, which was later confirmed using the Levene's Test. Statistical analyses concerning the pressure values at the initial configuration are reported in Figure 4. Two different representations of the dataset with boxplots are presented, separately comparing the outcomes grouped by the body position (S vs. P) and type of tube (NA vs. A) against the diameter of the trachea (20 mm vs. 25 mm) and input pressure (30–50 cmH_2O). Figure 4A–D shows the proportional increase in the measured pressure against the input pressure using the S vs. P classification. Moreover, significant statistical differences between mean values, independent of input pressures and tracheal diameter, were observed. We noticed that the pressure values measured when using the NA tube were usually higher than those measured with the A tube. This effect is particularly evident in Figure 4B, which shows the results for the prone position with a 20 mm trachea, for which the pressure amplitude reached ~6 kPa. In the 25 mm trachea (Figure 4C,D,G,H), the mock-up was not able to detect any relevant pressure (i.e., contact) in 30–40 cmH_2O. Measurable outcomes were obtained only at 50 cmH_2O, when the cuff was actually compressed. These results are similar to those of the 20 mm tracheal mock-up, although with amplitudes of ~2 kPa for the NA tube and almost negligible for the A tube.

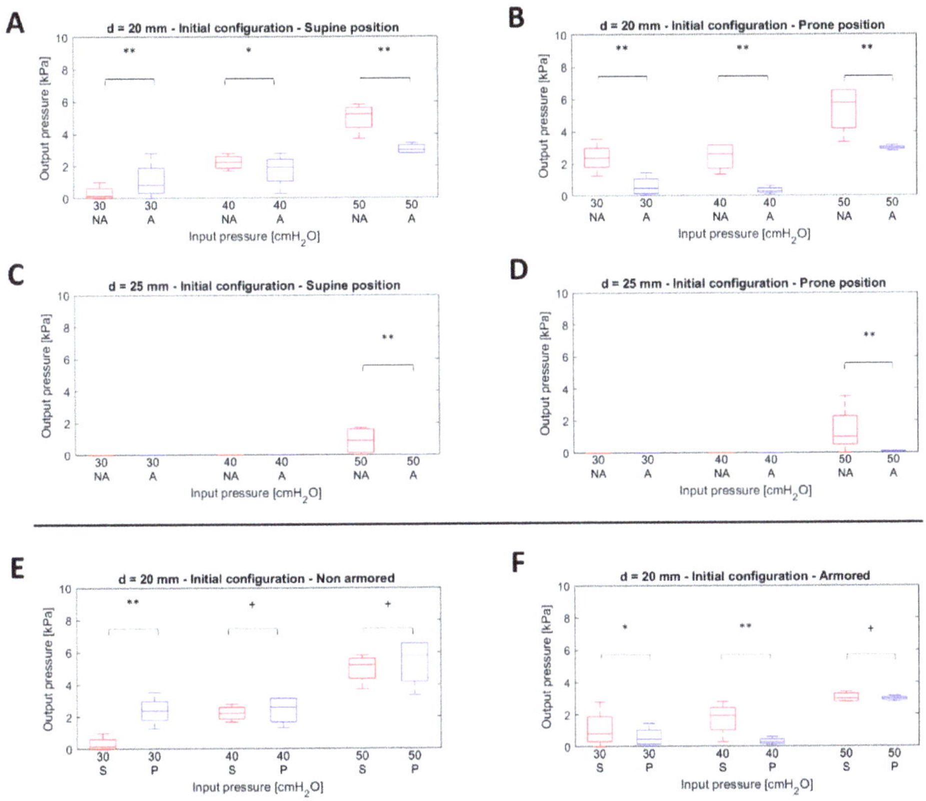

Figure 4. *Cont.*

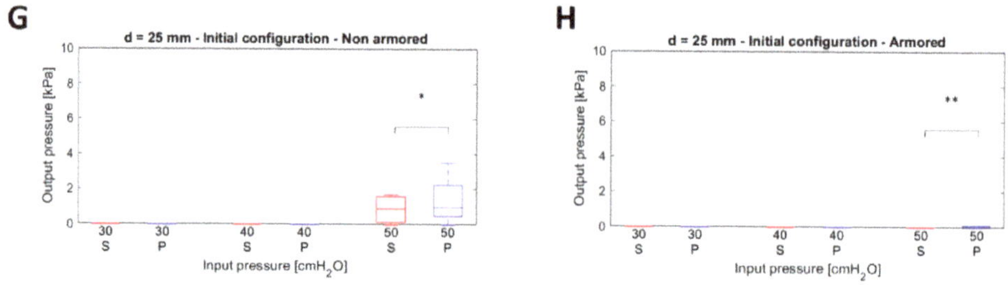

Figure 4. Statistical analysis to determine the differences in pressure, if any, before performing a maneuver. (**A–D**) Box plots showing the comparison between non-armored (NA) and armored (A) tubes across diameters of the trachea and position (S—supine, P—prone). (**E–H**) Box plots showing the comparison between body position (S/P) against diameters and type of endotracheal tube (NA/A). Legend: ** $p < 0.001$, * $0.002 < p \leq 0.05$, and + $p > 0.05$.

Figure 4E–H shows the dataset using a grouping based on the body position (S vs. P). In this case, we had a different scenario for the 20 mm trachea: when increasing the input pressures, the mean values tended to decrease in their significant statistical difference. Focusing on Figure 4E (NA tube—20 mm trachea), at 30 cmH_2O, we estimated a p-value below 0.001 but, already at 40 cmH_2O, the mean values were statistically similar. Concerning the A tube, this effect occurred only at 50 cmH_2O.

By changing the organization of the dataset, we used as a dependent variable the difference in the pressure values before and after performing a maneuver. We compared the effects of two different maneuvers (i.e., T and B) on the tracheal pressures. Figure 5A–D shows the results obtained using a grouping based on the type of tube (NA vs. A), while Figure 5E–H shows the body position (S vs. P).

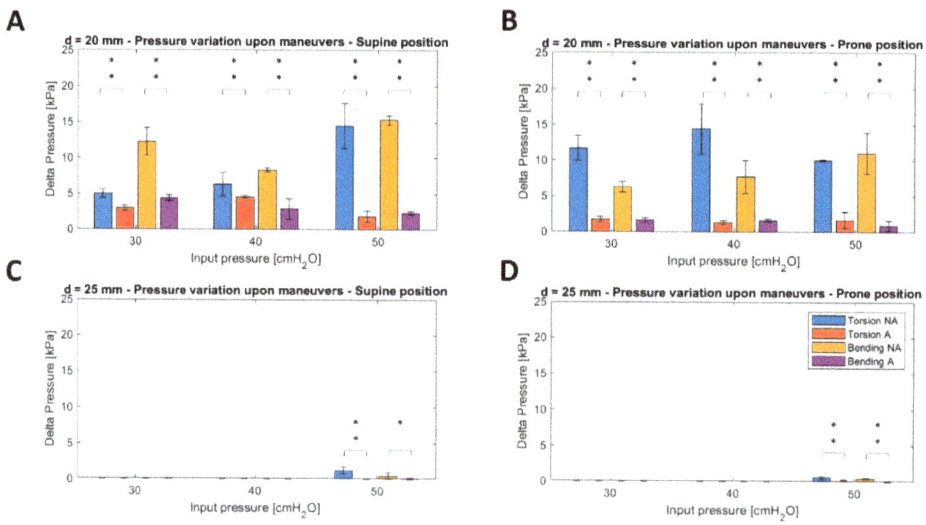

Figure 5. *Cont.*

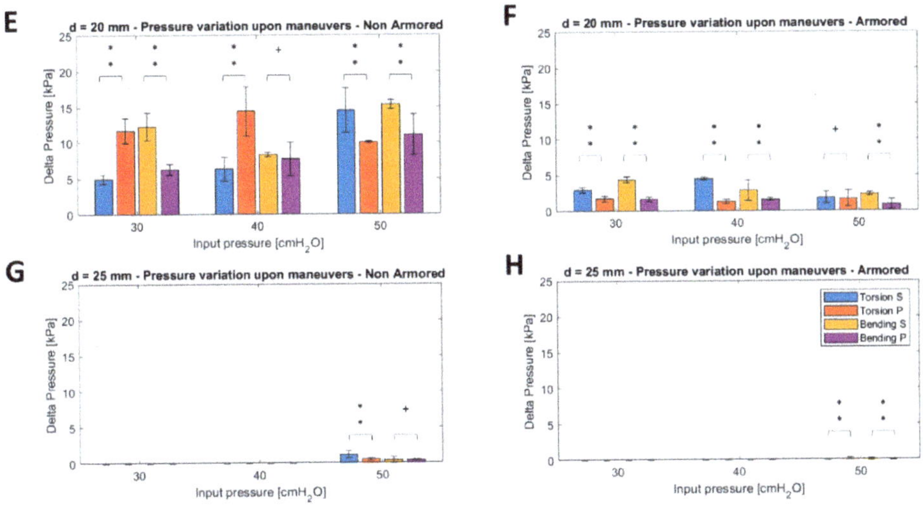

Figure 5. Statistical analysis to determine the differences in pressure variations, if any, after performing a maneuver. (**A–D**) Histograms showing a comparison between non-armored (NA) and armored (A) tubes across diameters of the trachea and position (S—supine, P—prone). (**E–H**) Histograms showing the comparison between body position (S/P) across diameters and type of endotracheal tube (NA/A). Legend: ** $p < 0.001$, * $0.002 < p \leq 0.05$, and + $p > 0.05$.

The comparison between NA and A tubes displays significant differences between the mean values independent of the type of maneuver and input pressure, with amplitudes higher for the NA tube than those associated with the A tube. In contrast, an increasing trend of the values with the input pressure was not detected. As also shown in Figure 4, no relevant pressures were observed using a 25 mm trachea with input pressures below 50 cmH_2O.

A similar scenario was observed by applying the classification based on the body position (Figure 5E–H), even though the statistical differences between mean values were not confirmed in all cases. Specifically, this occurred in relation to the mean values associated with the bending maneuvers at 40 cmH_2O for the 20 mm trachea with the NA tube, torsion maneuvers at 50 cmH_2O for the 20 mm trachea with the A tube, and bending maneuvers at 50 cmH_2O for the 25 mm trachea with the NA tube.

4. Discussion

We investigated the effects of pressure exerted by endotracheal tubes in a mock-up resembling the laryngotracheal tract. Our study aimed to unveil the mechanisms that induce trauma on the tracheal segment of patients suffering from ARDS who were treated with MV. We used two endotracheal tube types (i.e., NA/A) and observed the results by reference to body configuration (i.e., S vs. P) and pressure variations in the cuff. International guidelines recommend keeping the endotracheal tube cuff pressure between 20 and 30 cmH_2O [6,7] to avoid serious complications such as aspiration pneumonia, tracheal ischemia, FTTLs, and TEFs [17,18]. Furthermore, we used values up to 50 cmH_2O to investigate the effect of possible human error. Usually, this pressure is periodically monitored via devices connected to the cuff pilot balloon, which has a mechanical valve that prevents any oxygen leakage [19]. However, the cause–effect relationship of the mechanical pressure on the tracheal tract is still unclear, in particular during the maneuvers exerted on the endotracheal tube or on the patient's body. This issue has become remarkably relevant during the COVID-19 pandemic, as a large number of patients have been treated with prolonged MV in different body configurations. Marti et al. published an in vitro study on the deflation of cuffs placed in a test bench over time [20]; however, to the best of our knowledge, our investigation is the first assessment of the mechanical pressure exerted by a cuff in a simulated environment. The underlying hypothesis of our study is the variation in the pressure exerted by the cuff on the tracheal wall depending on the type of tube (NA/A), the maneuvers performed on the tube (T/B), and on the patient's body position (i.e., S/P), against the same initial insufflation pressure.

The first result concerned the pressure measurements obtained after inserting the tube in the model and before applying the maneuvers. Pressure values measured when using the NA tube were higher than those obtained using the A tube, especially in the P position (Figure 5A–D). This is probably due to the superior stiffness of the NA tube, which exhibited a higher preformed curvature than the A tube. These differences may be compensated in vivo by the viscoelastic behavior of the NA tube polymer at the body temperature, which was not included in our experimental set-up. Therefore, we can consider the measured pressures as instantaneous values that may occur just after the positioning of the tube in the laryngotracheal tube. It is reasonable to hypothesize that the A tube, being more flexible than the NA tube, would require fewer corrections of the cuff. Instead, the NA tube might require careful placement and tuning, especially in the initial period after intubation and before the patient's body temperature induces a shape variation. Another interesting difference between the NA and A tubes was the shape of the cuffs once inflated. The NA cuff took the form of an ellipsoid whereas the A cuff took the form of a regular cylinder. The contact of the surfaces with different curvatures may be another factor affecting the exerted pressures and their distribution on the tracheal wall. This deserves a dedicated investigation.

Another statistically significant result concerned the difference identified in the 20 mm mock-up with the NA tube and the 30 cmH_2O insufflation pressure from the S to P position, which was not observable for higher insufflation pressures and for the A tube (Figure 4E–H). This is probably attributable to the stiffness of the NA tube and to the maneuvers on the external connector of the endotracheal tube during the S–P maneuvers. As mentioned above, this was not evident for the A tube because of its flexibility.

Using the 20 mm mock-up, we also observed significant differences in output pressure when using either the NA or A tube, independent of the type of maneuver (torsion vs. bending) and input pressure, with amplitudes associated with the NA tube being higher than those of the A tube (Figure 4A–D). In our opinion, this is also due to the different flexibility of the endotracheal tubes. Therefore, it is reasonable to recommend that careful tuning and monitoring of cuff pressure should be performed after each maneuver on the endotracheal tube, especially if the NA tube is employed. Body position appeared not to consistently influence the pressure on the tracheal wall when applying torsion and bending maneuvers on the endotracheal tube, regardless of the type of tube and the insufflation pressure used (Figure 4E–H).

In contrast, the results from the 25 mm mock-up were less informative. This is due to the exact nominal dimensions of the inflated cuff and the trachea-like structure producing reduced contact. As a result, contact occurred only at the highest insufflation pressures. This simple deduction opens up an interesting discussion as to the applicability of the most common endotracheal tubes that have a cuff expandable up to 25 mm. The tracheal segment has a highly variable diameter of 10−27 mm [21]. Therefore, the general employment of a 25 mm cuff tube with the recommended input pressure may be either dangerous or inefficient. As a consequence, a preliminary evaluation of the diameter of the tracheal segment (e.g., using parameters such as the weight/height ratio of the patient) should be conducted to fine-tune the input pressure or, if available, to inform the use of an endotracheal tube with a larger cuff diameter. However, it is important to stress that, at this point in time, a scaling factor for input pressures against the diameter of the tracheal segment is not available.

This study has three main limitations. The first is intrinsic to all studies performed on experimental models, as they cannot perfectly replicate the in vivo conditions. In particular, the use of a thermostatic chamber could have obviated the permanent stiffness of the NA endotracheal tube and, with its use, we could have verified the change in cuff pressure exerted on the tracheal wall over time. However, viscous phenomena in polymeric materials are not immediate and we were interested in assessing the pressure values immediately upon insertion in order to evaluate the mechanical effect on the trachea-like wall. A future study will include a thermo-controlled room in which a mechanical assessment will be performed, in order to highlight the contribution of body temperature. The second aspect is the lack of a real tracheal epithelium and endotracheal secretions enabling the sliding of the cuff on the tracheal walls during torsion and bending maneuvers, which could provide insight into the tribological phenomena concurring or preventing damage mechanisms over time. While this limitation cannot be overcome with our current mock-ups, a dedicated study involving ex vivo tissues may help to assess these phenomena. Finally, an improvement in the evaluation of the cuff pressure, over time, is needed to investigate the relaxation phenomena and the effects of maneuvers over time. This would mimic the condition of the current approach for treating COVID-19 patients but, also, in this case, an ex vivo model would better reproduce the tribological mechanisms of the involved tissues.

In addition to the abovementioned limitations, our study provides some relevant take-away messages on the application of endotracheal tubes for MV. In particular, the first important conclusion concerns the need for periodic checks of the cuff pressure upon changing the body position and/or performing maneuvers on the tube. The latter occurs regularly in daily practice even simply by rotating or hyperextending the patient's head. Moreover, the cuff of A tubes appeared to give more stable output pressures on tracheal walls than those of NA tubes. Therefore, instead of the common practice of ICU personnel to use NA tubes, the use of A tubes should be considered. In the specific case of our tertiary referral hospital, patients coming from the operating room intubated with an A tube are promptly reintubated with an NA tube. In the current literature, scientific articles justifying this clinical practice are missing. It seems that the practice has originated from practical experience gained in the field.

5. Conclusions

The outbreak of the COVID-19 pandemic has brought new attention to the well-known practice of ventilating patients affected by respiratory tract pathologies. To the best of our knowledge, our study delivers, for the first time, an investigation on the loads exerted by the contact of endotracheal tube cuffs on the laryngotracheal tract. We used a mock-up of the anatomic system with two different transversal sizes and two different designs of endotracheal tubes (i.e., armored vs. non-armored) to assess the effect of each device when using specific maneuvers and loads, characteristic of common practices. Despite the intrinsic limitations of the model, we unveiled a number of interesting findings. The most important outcome for clinicians concerns the superior wall pressures induced by NA tube cuffs due to their specific design. This effect is more significant for patients placed in a prone position despite the inlet pressure being kept at 30 cmH_2O. Interestingly, A tubes induced more stable wall pressures than those produced by NA tubes. Another interesting point to note relates to the transversal dimension of the laryngotracheal tract. Current procedures and tube designs do not consider anatomical differences among patients, notwithstanding that wall pressures may vary significantly depending on the actual dimensions of the cross-sectional diameter.

In conclusion, although we concede that current clinical practices have not resulted in frequent complications, based on recent scientific evidence, our findings support a reconsideration of the current approach to tracheal intubation aimed at MV in ICU patients. Specifically, A tubes should be preferred to NA tubes, and face-down pillows with a central hole to pass NA tubes should be used when pronation is required.

In view of this, since a full understanding of the damage mechanisms is still missing, we think that future studies in this field should investigate in detail such aspects of MV practices, through both ex vivo and in vivo approaches, to improve patient care.

Author Contributions: Conceptualization, G.F., F.G., S.B. and L.B.; methodology, G.F., M.M., A.C., D.L.R., S.D., S.B. and L.B.; software, M.M., A.C. and D.L.R.; validation, G.F., M.M., A.C., D.L.R., F.G., S.D., S.B. and L.B.; formal analysis, M.M., A.C. and D.L.R.; investigation, M.M., A.C., D.L.R. and S.D.; resources, M.M., A.C. and D.L.R.; data curation, M.M., A.C. and D.L.R.; writing—original draft preparation, G.F., M.M., A.C. and D.L.R.; writing—review and editing, G.F., M.M., A.C., D.L.R., F.G., S.D., S.B. and L.B.; visualization, M.M., A.C. and D.L.R.; supervision, S.D., S.B. and L.B.; project administration, L.B. All authors have read and agreed to the published version of the manuscript.

Funding: This research received no external funding.

Institutional Review Board Statement: Not applicable.

Informed Consent Statement: Not applicable.

Data Availability Statement: Data are available upon request to the corresponding author.

Conflicts of Interest: The authors declare no conflict of interest.

References

1. COVIDSurg Collaborative; GlobalSurg Collaborative. Timing of surgery following SARS-CoV-2 infection: An international prospective cohort study. *Anaesthesia* **2021**, *76*, 748–758. [CrossRef]
2. COVIDSurg Collaborative; GlobalSurg Collaborative. SARS-CoV-2 vaccination modelling for safe surgery to save lives: Data from an international prospective cohort study. *Br. J. Surg.* **2021**, *108*, 1056–1063. [CrossRef] [PubMed]
3. Fiacchini, G.; Tricò, D.; Ribechini, A.; Forfori, F.; Brogi, E.; Lucchi, M.; Berrettini, S.; Bertini, P.; Guarracino, F.; Bruschini, L. Evaluation of the incidence and potential mechanisms of tracheal complications in patients with COVID-19. *JAMA Otolaryngol. Neck Surg.* **2021**, *147*, 70–76. [CrossRef] [PubMed]
4. Belletti, A.; Palumbo, D.; Zangrillo, A.; Fominskiy, E.V.; Franchini, S.; Dell'Acqua, A.; Marinosci, A.; Monti, G.; Vitali, G.; Colombo, S.; et al. Predictors of Pneumothorax/Pneumomediastinum in Mechanically Ventilated COVID-19 Patients. *J. Cardiothorac. Vasc. Anesth.* **2021**, *35*, 3642–3651. [CrossRef] [PubMed]
5. King, H.-K. Pharyngeal obstruction of a reinforced orotracheal tube. *Anesth. Analg.* **1999**, *89*, 261. [CrossRef] [PubMed]
6. Klompas, M.; Branson, R.; Eichenwald, E.C.; Greene, L.R.; Howell, M.D.; Lee, G.; Magill, S.S.; Maragakis, L.L.; Priebe, G.P.; Speck, K.; et al. Strategies to prevent ventilator-associated pneumonia in acute care hospitals: 2014 update. *Infect. Control Hosp. Epidemiol.* **2014**, *35*, S133–S154. [CrossRef] [PubMed]
7. Kalil, A.C.; Metersky, M.L.; Klompas, M.; Muscedere, J.; Sweeney, D.A.; Palmer, L.B.; Napolitano, L.M.; O'Grady, N.P.; Bartlett, J.G.; Carratalà, J.; et al. Management of adults with hospital-acquired and ventilator-associated pneumonia: 2016 clinical practice guidelines by the Infectious Diseases Society of America and the American Thoracic Society. *Clin. Infect. Dis.* **2016**, *63*, e61–e111. [CrossRef] [PubMed]
8. Piazza, C.; Filauro, M.; Dikkers, F.G.; Nouraei, S.A.R.; Sandu, K.; Sittel, C.; Amin, M.R.; Campos, G.; Eckel, H.E.; Peretti, G. Long-term intubation and high rate of tracheostomy in COVID-19 patients might determine an unprecedented increase of airway stenoses: A call to action from the European Laryngological Society. *Eur. Arch. Oto-Rhino-Laryngol.* **2020**, *278*, 1–7. [CrossRef] [PubMed]
9. Malvade, P.S.; Joshi, A.K.; Madhe, S.P. IoT based monitoring of foot pressure using FSR sensor. In Proceedings of the 2017 International Conference on Communication and Signal Processing (ICCSP), Chennai, India, 6–8 April 2017; pp. 635–639.
10. Hesse, M.; Christ, P.; Hormann, T.; Ruckert, U. A respiration sensor for a chest-strap based wireless body sensor. In Proceedings of the IEEE SENSORS 2014 Proceedings, Valencia, Spain, 2–5 November 2014; pp. 490–493.
11. Castellini, C.; Ravindra, V. A wearable low-cost device based upon Force-Sensing Resistors to detect single-finger forces. In Proceedings of the 5th IEEE RAS/EMBS International Conference on Biomedical Robotics and Biomechatronics, Sao Paulo, Brazil, 12–15 August 2014; pp. 199–203.
12. Amft, O.; Troster, G.; Lukowicz, P.; Schuster, C. Sensing Muscle Activities with Body-Worn Sensors. In Proceedings of the International Workshop on Wearable and Implantable Body Sensor Networks (BSN'06), Cambridge, MA, USA, 3–5 April 2006; pp. 138–141.
13. Barsocchi, P.; Bianchini, M.; Crivello, A.; La Rosa, D.; Palumbo, F.; Scarselli, F. An unobtrusive sleep monitoring system for the human sleep behaviour understanding. In Proceedings of the 2016 7th IEEE International Conference on Cognitive Infocommunications (CogInfoCom), Wroclaw, Poland, 16–18 October 2017; pp. 91–96. [CrossRef]
14. Interlink_Electronics FSR 400 Data Sheet. Available online: https://cdn.sparkfun.com/datasheets/Sensors/ForceFlex/2010-10-26-DataSheet-FSR400-Layout2.pdf (accessed on 19 November 2021).

15. Schofield, J.S.; Evans, K.R.; Hebert, J.S.; Marasco, P.D.; Carey, J.P. The effect of biomechanical variables on force sensitive resistor error: Implications for calibration and improved accuracy. *J. Biomech.* **2016**, *49*, 786–792. [CrossRef] [PubMed]
16. Florez, J.A.; Velasquez, A. Calibration of force sensing resistors (fsr) for static and dynamic applications. In Proceedings of the 2010 IEEE ANDESCON, Bogota, Colombia, 15–17 September 2010; pp. 23–28.
17. Nseir, S.; Zerimech, F.; Fournier, C.; Lubret, R.; Ramon, P.; Durocher, A.; Balduyck, M. Continuous control of tracheal cuff pressure and microaspiration of gastric contents in critically ill patients. *Am. J. Respir. Crit. Care Med.* **2011**, *184*, 1041–1047. [CrossRef] [PubMed]
18. Touat, L.; Fournier, C.; Ramon, P.; Salleron, J.; Durocher, A.; Nseir, S. Intubation-related tracheal ischemic lesions: Incidence, risk factors, and outcome. *Intensive Care Med.* **2013**, *39*, 575–582. [CrossRef] [PubMed]
19. Babic, S.A.; Chatburn, R.L. Laboratory evaluation of cuff pressure control methods. *Respir. Care* **2020**, *65*, 62–67. [CrossRef] [PubMed]
20. Marti, J.D.; Isetta, V.; Lazaro, M.R.; Aguilera-Xiol, E.; Comaru, T.; Battaglini, D.; Meli, A.; Ferrer, M.; Navajas, D.; Pelosi, P.; et al. An in-vitro study to evaluate high-volume low-pressure endotracheal tube cuff deflation dynamics. *Minerva Anestesiol.* **2019**, *85*, 846–853. [CrossRef] [PubMed]
21. Breatnach, E.; Abbott, G.C.; Fraser, R.G. Dimensions of the normal human trachea. *Am. J. Roentgenol.* **1984**, *142*, 903–906. [CrossRef] [PubMed]

sensors MDPI

Article

Whole-Body Adaptive Functional Electrical Stimulation Kinesitherapy Can Promote the Restoring of Physiological Muscle Synergies for Neurological Patients

Alessandro Scano [1,*], Robert Mihai Mira [1], Guido Gabbrielli [2], Franco Molteni [3] and Viktor Terekhov [2,*]

1 UOS STIIMA Lecco—Human-Centered, Smart & Safe, Living Environment, Italian National Research Council (CNR), Via Previati 1/E, 23900 Lecco, Italy; robertmihai.mira@stiima.cnr.it
2 VIKTOR S.r.l.—Via Pasubio, 5, 24044 Dalmine (BG), Italy; guido.gabbrielli@gmail.com
3 Villa Beretta Rehabilitation Center, Ospedale Valduce, Via N. Sauro 17, 23845 Costa Masnaga, Italy; franco56.molteni@gmail.com
* Correspondence: alessandro.scano@stiima.cnr.it (A.S.); amministrazione@viktor.physio (V.T.)

Citation: Scano, A.; Mira, R.M.; Gabbrielli, G.; Molteni, F.; Terekhov, V. Whole-Body Adaptive Functional Electrical Stimulation Kinesitherapy Can Promote the Restoring of Physiological Muscle Synergies for Neurological Patients. *Sensors* **2022**, *22*, 1443. https://doi.org/10.3390/s22041443

Academic Editor: Ki H. Chon

Received: 21 December 2021
Accepted: 11 February 2022
Published: 13 February 2022

Abstract: Background: Neurological diseases and traumas are major factors that may reduce motor functionality. Functional electrical stimulation is a technique that helps regain motor function, assisting patients in daily life activities and in rehabilitation practices. In this study, we evaluated the efficacy of a treatment based on whole-body Adaptive Functional Electrical Stimulation Kinesitherapy (AFESK™) with the use of muscle synergies, a well-established method for evaluation of motor coordination. The evaluation is performed on retrospectively gathered data of neurological patients executing whole-body movements before and after AFESK-based treatments. Methods: Twenty-four chronic neurologic patients and 9 healthy subjects were recruited in this study. The patient group was further subdivided in 3 subgroups: hemiplegic, tetraplegic and paraplegic. All patients underwent two acquisition sessions: before treatment and after a FES based rehabilitation treatment at the VIKTOR Physio Lab. Patients followed whole-body exercise protocols tailored to their needs. The control group of healthy subjects performed all movements in a single session and provided reference data for evaluating patients' performance. sEMG was recorded on relevant muscles and muscle synergies were extracted for each patient's EMG data and then compared to the ones extracted from the healthy volunteers. To evaluate the effect of the treatment, the motricity index was measured and patients' extracted synergies were compared to the control group before and after treatment. Results: After the treatment, patients' motricity index increased for many of the screened body segments. Muscle synergies were more similar to those of healthy people. Globally, the normalized synergy similarity in respect to the control group was 0.50 before the treatment and 0.60 after ($p < 0.001$), with improvements for each subgroup of patients. Conclusions: AFESK treatment induced favorable changes in muscle activation patterns in chronic neurologic patients, partially restoring muscular patterns similar to healthy people. The evaluation of the synergic relationships of muscle activity when performing test exercises allows to assess the results of rehabilitation measures in patients with impaired locomotor functions.

Keywords: muscle synergies; whole body FES; neurological patients

1. Introduction

The aging of the population in the Western countries and the increased awareness of the economic and social costs of accidents at work are topical. In fact, it is estimated that in Europe about five million people [1] suffer from pathologies or have suffered trauma of varying severity to the neuro-muscular system. Furthermore, neural aging also leads to the development of various forms and degrees of motor impairment. In 2018, 19.7% of the EU population were 65 or older [2]. A need of advancements in the prevention and cure of neurologic illnesses clearly emerges. In this context, rehabilitation therapies can slow the effects of aging and help improve quality of life [3]. Other than being a physical and psychological burden to the individual, neurological diseases represent also a strain on the community, due to the need to provide aid to impaired individuals either by creating adequate structures for rehabilitation or providing healthcare. According to Eurostat, curative and rehabilitative therapies account for more than 50% of current health expenditure in most EU Member States [4].

In this context, the interest of scientists and practitioners in functional electrical stimulation for the rehabilitation of neurological patients with severe disorders of the musculoskeletal system has grown. Neuromuscular electric stimulation (NMES) has often been used to aid in the recovery of lost motor function [5–8]. The combined action of the patient's neurostimulation and mobilization programs allows the brain to re-educate to recognize muscle stimuli as its own, triggering a series of nervous processes that favor the reactivation of impaired functional capacities (neuroplasticity) [9,10]. Through controlled and synchronized stimulation of specific areas of the body, physicists and therapists can provide functionality to muscle contractions. During the years, this specific branch of NMES has acquired the title of functional electrical stimulation (FES). Many studies have investigated the effects of FES on stroke survivors, in a variety of applications. In gait rehabilitation [11,12], increased stability, improved gait independence and higher gait speed were found after FES treatments. Functional electric stimulation has been also used in rehabilitation of the upper extremity in stroke survivors and allowed to achieve finer hand movements such and finger flexion [13], hand grasping [14] and broader arm movements [15]. In all these studies, the participants regained functionality of the upper extremity confirming the usefulness of FES. Furthermore, in [14] the authors compared the effects of basic electric muscle stimulation with EMG controlled FES and demonstrated that patients who underwent EMG controlled FES treatment performed better than patients who underwent basic electrical stimulation. FES has been also applied in gait rehabilitation for spinal cord injury patients proving its usefulness in aiding the rehabilitation process [16]. Other notable applications of FES have been in aiding full-face transplantation patients regain facial expressions [17]. Ultimately, FES allowed patients to retain functionality even while not using the devices in multiple scenarios [18].

Other studies employed FES for the recovery of upper extremity functionality with the aid of robotic instrumentation [19]. FES was also combined with complex control mechanisms like artificial neural networks trained to mimic natural muscle recruitment patterns, allowing impaired individuals to restore walking patterns [20].

Indeed, the modern view of human movement management is characterized by a multilevel hierarchical system between the brain and the muscular system [21]. These levels are anatomically and functionally connected and communicate through continuous feedback, in order to ensure movement regulation and correct motor performance. The repetition of motor gestures allows the improvement of the execution of the motor task [22–24]. It is known that if the activation of the muscle mass generated by the electrical impulse corresponds to the voluntary physiological activation [25], the brain recognizes stimuli as its own and automatically activates functions that tend to restore the connections that govern the part of the body affected by pathology or trauma and improve its functionality [26].

In rehabilitation scenarios, one of the most promising approaches for improving the prevention of diseases and prescriptions of treatments with novel data for clinicians is the decomposition of the electromyographic signal into muscle synergies [27]. The muscle synergy technique offers the possibility to analyze electromyographic recordings considering the natural couplings between muscles, and thus is a tool useful for the analysis of the modular organization of the human neuro-musculoskeletal system. Muscle synergies propose that the CNS relies on a limited number of modules [28], possibly implemented at the neural level [29], to simplify motion production. Consequently, by appropriately recruiting spatial modules with temporal activation coefficients, the CNS exploits a reduced set of preformed neural pathways, called synergies, to obtain a wide variety of motor outputs. Applications of muscle synergy included, among others, investigations on the muscle synergies of the upper limb in physiological conditions [30,31] and the effect of neurological injuries [32–34]. Synergies have also been applied to investigate locomotion [35–38] and postural control [39,40].

However, currently this evaluation approach has rarely been used to evaluate the efficacy of a rehabilitation program on subjects with CNS lesions based on electrical stimulation, and always on very limited number of subjects.

Given that muscle synergies have proven to be a useful tool to study muscle coordination patterns and that FES is considered a valuable technique to aid motor re-learning [41], it is natural for the two techniques to be used as complementary approaches [42]. In fact, some studies have already used both tools for robot guided rehabilitation [43] where muscle synergies are used to drive a functional electric stimulation system. Researchers have used FES and muscle synergies of healthy people to guide gait rehabilitation for post stroke patients [44] and to study the effects of a FES based rehabilitation technique on post stroke patients during cycling exercises [41]. In their studies, the authors found a significant improvement when comparing synergy similarity to healthy controls before and after the treatment.

In recent years total body electric stimulation (or whole-body electrostimulation) has become a valuable clinical practice [45]. This technique is the natural evolution of FES, it makes use of more electrodes and applies electric stimulation to a wider variety of muscles at once. It was reported that synchronizing the stimuli makes it possible to exercise complete kinetic chains with a synergistic approach guarantying more natural and fluid movements [46].

The overall improvements of whole-body electric stimulation come in the form of the ability to train a vaster array of possible movements and better implement motor control aids to impaired subjects. Another important feature generally observed in whole body electric stimulation is the co-contraction of agonist and antagonist muscles. Antagonist muscles can contribute to the improvement of aerobic strength without presenting damage to the motor patterns [47,48].

The growing interest of neurophysiology in clarifying the physiological mechanisms of the use of electrical stimulation for the treatment of locomotor dysfunctions is known [49]; however, few studies are available on assessing the effects of FES on neurologic patients with the use of muscle synergies when evaluating rehabilitation based on total-body movements. The aim of this study is to propose a pilot study for assessing the effects of a FES-based rehabilitation treatment on neurological patients. We aimed at showing that neuroplasticity can be induced and physiological muscle synergies can be partially restored in chronic neurological patients after a FES-based treatment in patients with various pathologies.

2. Materials and Methods

2.1. Participants

Twenty-four patients were recruited in this study. The included patients were divided into three groups: 8 with hemiplegia/paresis patients; 8 with paraplegia/paresis patients; 8 with tetraplegia/paresis patients. All patients were in the chronic stage of their disease. A control group composed of 9 healthy individuals was also enrolled. Patients with oncological and/or rheumatological and patients which have underwent recent orthopedic surgery and/or recent trauma with respect to the acquisition date were excluded from the study. A further exclusion criterion was established on the homogeneity of the data. In order to be included in the study, a patient had to perform the same exercises and had the same EMG recorded channels in the pre and post treatment assessments.

Two patients were excluded due to inhomogeneous muscle acquisitions between pre and post treatment sessions (at least one different EMG channel, or different executed exercises). The total number of subjects included in the analysis was 22 patients (7 hemiplegia, 7 paraplegia and 8 tetraplegia) and 9 healthy controls. In the CONSORT flow diagram (Figure 1), we illustrate the details of the enrollment procedure.

All patients underwent rehabilitation sessions at the VIKTOR Physio Lab® physiotherapy center. The center independently sought the opinion of the competent Ethics Committee. Each patient (or legal representative) has given consent to the processing of data. The procedures were performed in accordance with ethical standards as set out by institutional and national committee and with the Helsinki Declaration of 1975, as revised in 2000 [50].

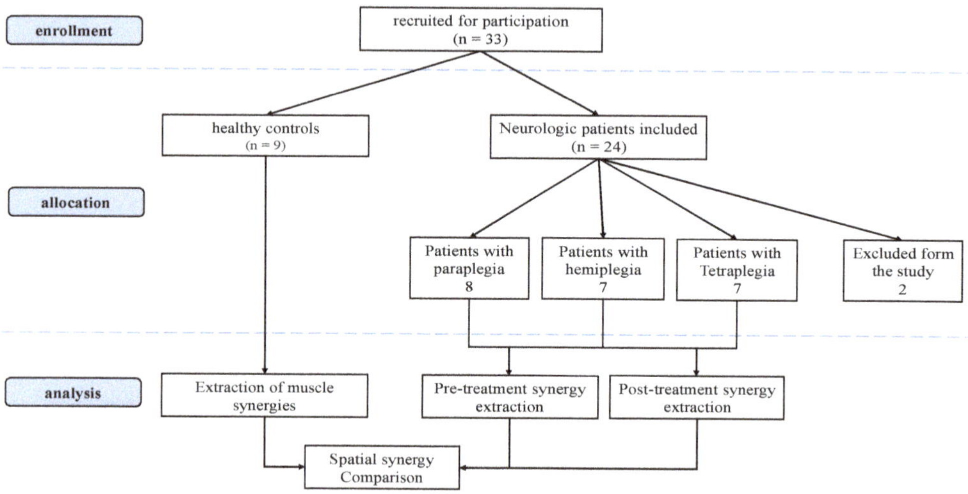

Figure 1. Consolidated Standards of Reporting Trials (CONSORT) flow diagram.

The data used in this retrospective study was collected during the period spanning from November 2018 to December 2020 in the VIKTOR Physio LAB (VIKTOR S.r.l., Milan, Italy). All enrolled patients underwent experimental recordings with a 16-channel surface electromyography (FreeEmg BTS, Milan, Italy) in order to monitor the level of motor functions over the course of the exercises.

The physiotherapy treatment was performed by three qualified physiotherapists. Medical supervision of the treatment was carried out by Dr. Viktor Terekhov and was performed using VIK16 Workstation (VIKTOR S.r.l., Milan, Italy).

Treatment and Device: VIK16 Workstation AFESK™

The rehabilitation treatment was carried out according to the VIKTOR method used with the AFESK™ technology (Adaptive Functional Electrical Stimulation Kinesitherapy).

The VIK16 Workstation technology has been developed exploiting the expertise achieved with more than thirty years of experience in using FES during exercise for the rehabilitation of neurological patients with severe lesions of locomotor functions. The Workstation VIK16 (Figure 2) is a device capable of supporting or partially replacing the CNS in the management of the motor scheme by delivering stimuli of suitable intensity to 16 muscles.

The method is based on percutaneous electrical stimulation of the neuromuscular system during cyclic exercises. For each muscle group involved in cyclic movements, electrical stimuli are given coinciding with the time activation in accordance with the physiological model of the exercise respecting the synergistic, reciprocal and antagonistic relationships between the muscles in each exercise.

In order to synchronize the patient's movement with the supply of an electric stimulus to the muscle, during the exercise, a synchronized sensor or a sound signal were adopted, in order to trigger stimulation with the first muscle group moving during the selected program.

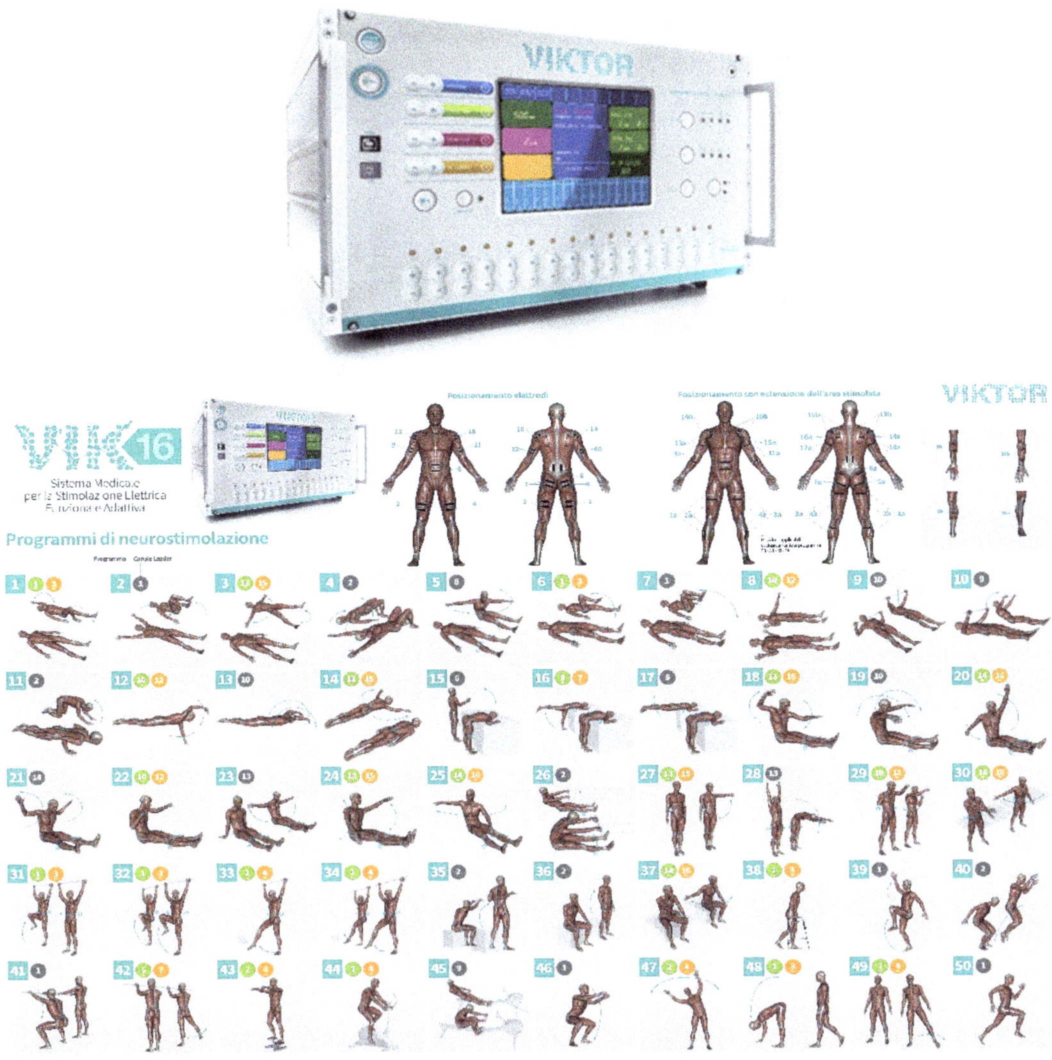

Figure 2. Graphic representation of the VIK16 Workstation and of the set of proposed total-body exercises. Workstation VIK16 has a library of 50 AFESK exercise programs that are used in rehabilitation, athletics and sports training. The programs are created on the basis of polymyographic and biomechanical assessment of the movement of healthy people, considering synergistic, reciprocal and antagonistic relationships of the moments of activation of the main muscle groups of the body. Workstation VIK16 has a wide range of electrical stimulation parameters: including current stabilized in each of 16 channels maximum of 150 mA, duration of a pulse from 100 to 1000 µs, pulse frequency from 50 to 200 Hz, motion cycle time from 200 ms to 10 s, impedance parameters and current level for each muscle group for all exercises performed by the patient, customizable number of cycles (movements) for each program and time for each exercise. In this study, only a subset of the exercises was performed by the enrolled patients.

Thus, with the help of feedback control over the timely and correct performance, the motor function is implemented in the centers for motion control in the cerebral cortex. It is also documented that there is a rationalization of the efferent control of segmental mechanisms at the spinal level with the activation of vegetative support and sewerage of the afferent flow of information through the use of collateral interneuronal connections with adequate electrical excitation of the sensory receptor apparatus of the executive link (muscles, ligaments, joints, skin, etc.) [49,51–57].

When performing a cyclic movement, the electrical stimulation of the neuromuscular system uses movement as a system-forming function that combines the anatomical and physiological connections of the control system from segmental executive to cortical motion control centers [58]. At the same time, the muscle fibers of the muscle performing the cyclic movement are activated and the entire sensitive neuromuscular control complex of the segmental level, transmits afferent signals to the cortical centers of evaluation and movement control [59,60]. In pathology, the coordinated operation of some links in this chain can be disrupted by interrupting or changing the afferent flows of confidential afferent information; the electrical stimulation of peripheral afferents can alter the state of circuits not only within somatosensory cortex, but also within the motor network: It follows that whole body FES is of a multi-stage hierarchical process in which various elements of the cortical motor network are consistently engaged [58,61]. When receiving adequate sensitive information, the cortical centers of motion control begin to restore control of the lost functions by including in the process of reorganizing the compensatory pathological stereotype of movement into a normal one [59,60].

Since each movement is the result of coordinated descending central commands that control the underlying segmental reflex-tuned executive neuromuscular apparatus, the EMG activity of the muscles that implement the movement reflects the frequency-time and amplitude parameters of the activity of these muscles and the objective evaluation of their functional capabilities [62,63]. In the case of adequate electrical stimulation of these muscles, the entire sensory apparatus available in the muscle pool forms an afferent flow of information to the cortical centers, using reflex ascending functionally organized paths. [64–68]. At the end of each session, the workstation VIK16 automatically records in the download the results of work of each patient and treatment.

2.2. Data Acquisition

2.2.1. Patients' Protocol

The data was acquired in VIKTOR Physio Lab® physiotherapy center (Figure 3). All patients underwent two instrumented acquisition sessions: before and after the treatment. In these two sessions, each patient was evaluated with the Arm, Trunk and Leg sections of the Motricity Index (MI). Each patient had his/her own customized FES treatment protocol and thus not all patients performed exactly the same exercises and had EMG recorded on the same muscles. However, the set-up and protocols were kept as homogeneous as possible across groups, compatibly with clinical needs. The muscles acquired were distributed on the whole body of the patient concentrating more on the impaired side of the body. Right hemiplegic patients had a denser EMG mapping on the right side of the body; left hemiplegic patients had more EMG sensors on the left emi-body; paraplegic and tetraplegic subjects were uniformly mapped on both body sides. All EMG probes were placed according to the SENIAM guidelines [69]. The acquired muscles changed between patient groups but were kept as homogeneous as possible in accordance with clinical needs and within patients of the same groups. The average age of participants was: Hemiplegia/paresis group: 52 years (not counting 1 child 6 years old); Paraplegia/paresis group—44 years; Tetraplegia/paresis group—46 years (not counting two children, 6 and 14 years old). The effective time of procedures in each session was on average 45 min.

The following program exercises were performed depending on the rehabilitation cycle:

(1) Introductory, adaptation: 7–10 exercises—on average 3–5 min each
(2) Restorative: 7–10 exercises—on average 3–5 min each

(3) Postural correctional: 5–7 exercises of which 1–2 last for 10–15 min
(4) Speed and endurance: 2–3 exercises of 15–20 min each
(5) Increase the duration of basic exercises: 3–5 exercises according to the program for 10–20 min each

Modality of electrical stimulation parameters were selected in accordance with the functional capability of each patient. Average values for each study group were listed in Table 1.

Table 1. FES parameters for hemiplegic patients.

	Pulse Width (μs)/Pulse Frequency (Hz)		
Hemiplegia	200/100	after adaptation	500/50–100
Hemiparesis	100/50–100	after adaptation	200/100
Paraplegia	200/100	after adaptation	500/50–100
Paraparesis	100/50–100	after adaptation	200/100
Tetraplegia	200/100	after adaptation	500/50–100
Tetraparesis	200/50	after adaptation	200/100

Before the beginning of the rehabilitation course, the sensitivity threshold of each muscle group was measured for each patient. The results obtained were used as reference for determining the level of current in the channels, which was supplied until the appearance of pronounced muscle contraction, without any pain. Usually, the values of the operating current, especially in patients with paresis, did not exceed twice the value of the sensitivity threshold. Stimulation parameters considered that the maximum permissible norms of current density during electrical procedures allow no more than 2 mA/cm^2. Introductory and restorative exercises were performed at the beginning of the course, while postural, speed and endurance and increase of duration exercises were implemented with a proportional increase of time and speed of the exercise in order to increase the summation effects provided with AFESK on both sensory and motor links of neuromuscular regulation of motor functions. The average data for the performed treatments, including number of sessions, average movement per sessions and cycles are shown in Table 2.

Table 2. Rehabilitation treatment data (averages for hemiplegic, paraplegic and tetraplegic groups).

Hemiplegic	N° Sessions	Sessions/Week	Cycles Movements/Session	Total Movement Cycles
Hemiplegic	71	2.25	813	57,720
Paraplegic	80	2.25	705	56,461
Tetraplegic	40	1.4	667	26,709

During the period from November 2018 to December 2020, during which the rehabilitation of these patients was carried out, due to Covid-Sars 2, quarantine measures were repeatedly introduced with the closure of our center. For this reason, most patients, especially those with tetraplegia, reduced the number of visits, which reduced the average number of sessions for tetraplegic patients. In addition, all enrolled patients were in a stable chronic phase, after 2–10 years from the onset of the disease, and had already tried various methods of rehabilitation before admission to the center of the VIKTOR Physio LAB (VIKTOR S.r.l., Milan, Italy). They did not follow other rehabilitative treatments during the period of the FES training.

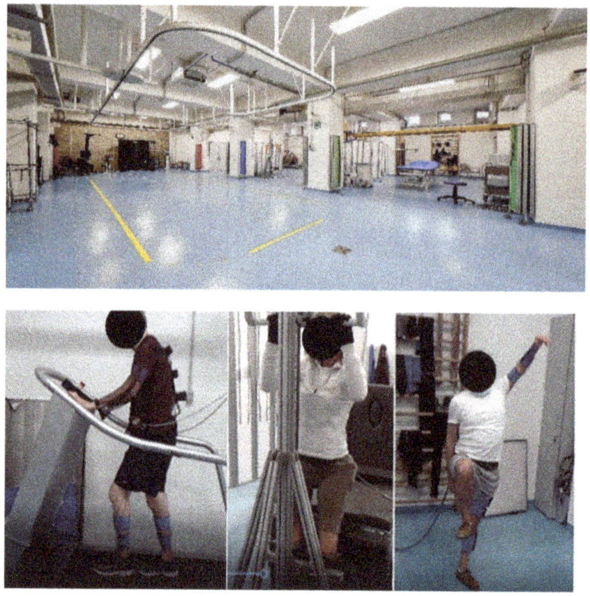

Figure 3. Employed set-ups for training at the VIKTOR Physio LAB.

2.2.2. Control Group Protocol

Healthy control subjects followed an acquisition protocol which encompassed all the set-ups employed with patients' groups. The EMG recording protocol adopted for controls allowed to match muscles and exercises with all patients' recordings. First, healthy controls performed the same exercises performed by patients. Given that previous studies confirmed that there is no major difference in muscle synergies for a wide variety of movements between the left and right limbs on healthy people [70], the muscles recorded on healthy controls were on the right hemi-body to match data for the hemiplegic, tetraplegic and paraplegic groups. Table 3 shows the muscles registered on patients and on healthy subjects to match the data of each patient group.

The exercises were a set of cyclical full body exercises expressively designed to perform active cyclical movements such as walking and specific movements to emphasize either upper-limbs, such as shoulder abduction, or lower-limb exercises, like knee adduction, or both in many cases. The set of the considered exercises could elicit many of the whole-body synergies available to subjects. All the exercises performed in the rehabilitation protocol are presented in Table 4.

2.3. *Data Elaboration*

The acquired EMG data was imported in MATLAB software (MathWorks, Natick, MA, USA) for the pre-processing. The EMG signals were filtered with a band-pass 6th order Butterworth filter covering a bandwidth from 30 Hz to 400 Hz, then they were full wave rectified, filtered with a low-pass 6th order Butterworth filter with cut-off frequency at 10 Hz, according to already employed processing pipelines for muscle synergies applications [71]. Lastly, the electromyographic data amplitude was normalized between zero and one to enable intra and inter subject comparisons, by dividing each channel EMG envelope by the maximum value found for that channel considering all movements performed by that subject in that session [72]. Time normalization was achieved by resampling each acquisition (EMG envelope) at 100 Hz. The elaborated data was organized in 2D arrays containing a concatenation of elaborated EMG data. Each column of the 2D array contained an EMG channel while each row contained the sequence of time samples. All exercises performed by the same subject were concatenated in the an array for the purpose of extracting synergies. A visual summary of the processing stage pipeline is provided in Figure 4.

Table 3. List of muscles acquired for the patients' groups. The green coloured squares indicate which muscles were registered for each group. On healthy controls, EMG was placed on all muscles to match patients' data.

	Hemiplegic Patients	Tetraplegic Patients	Paraplegic Patients
Biceps brachii caput longus	■	■	■
Biceps femoris caput longus	■	■	■
Deltoideus anterior	■	■	■
Deltoideus posterior	■	■	
Erector spinae longissimus	■	■	■
Extensor carpi radialis longus	■	■	
Extensor digitorum	■	■	
Flexor carpi radialis	■	■	
Gastrocnemius lateralis	■	■	■
Gluteus maximus	■	■	■
Latissimus Dorsi			■
Obliquus externus abdominis			■
Obliquus internus abdominis			■
Rectus abdominis			■
Rectus femoris	■	■	■
Tibialis anterior	■	■	■
Trapezius descendens	■	■	
Triceps brachii caput lateralis	■	■	
Triceps brachii caput longus	■	■	■
Vastus lateralis	■	■	■

Table 4. List of exercises performed by the patients' groups. The green coloured squares indicate which exercises were registered for each group. Healthy controls performed all the exercises to match with patients' data.

	Hemiplegic Patients	Tetraplegic Patients	Paraplegic Patients
Walking	■	■	■
Crutch assisted walking			■
Lying down arm and contralateral knee adduction	■	■	■
Lying down Knees and arms abduction	■	■	■
Lying down Jumping jacks		■	
Hip thrust	■	■	■
Torso Torsion		■	■
Knees adduction			■
Prone to cat pose			■
Sitting Punching	■	■	■
Shoulder abduction		■	
Standing Punching	■		
Standing Knee adduction	■		
Sit to stand	■	■	■
March	■		
Jump		■	■
Squat	■	■	
Exoskeleton assisted movement			■
Push		■	

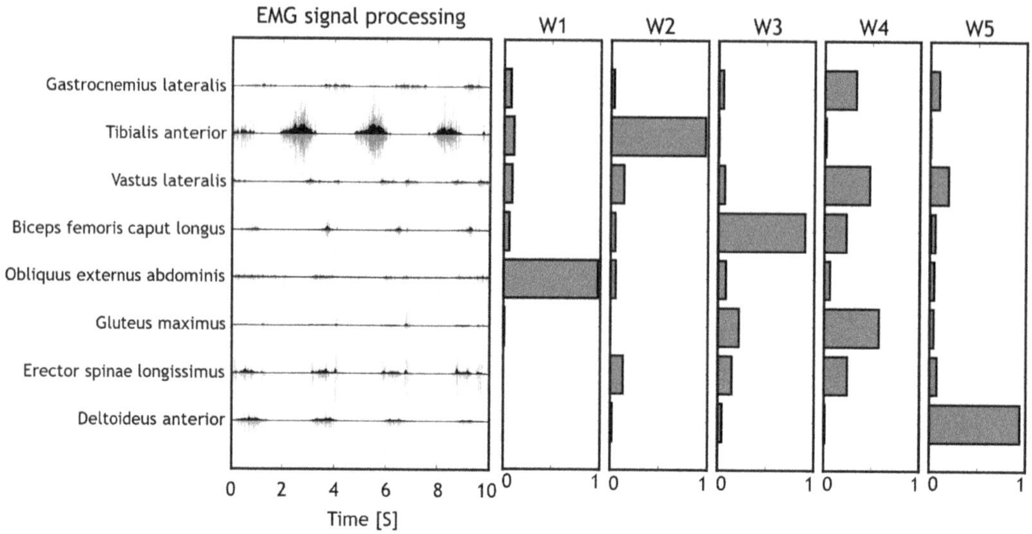

Figure 4. Pipeline for Signal processing. The raw signals (light grey) were filtered to remove movement artefacts and to compute the EMG envelope (dark grey). Muscle synergies were then extracted from the EMG envelope with the NMF algorithm.

2.4. *Synergy Extraction*

Muscle synergies were extracted from the elaborated EMG data using the non-negative matrix factorization algorithm (NMF) which is currently the most used algorithm for muscle synergy extraction. For our study, we used the spatial muscle synergy model, which extracts a set of spatial synergies containing muscle loads and a series of temporal coefficients indicating the time recruitment of each synergy. Synergies were extracted from each patient's dataset, separating the pre-treatment and the post-treatment sessions, for a total of 2 sets of synergies per patient. The EMG electrodes and considered movements were the same for each patient in the two sessions. The number of extracted synergies was chosen by using the first order that reconstructed at least the 0.85 of the reconstruction R^2 of the original signal [73].

Synergy Extraction: Control Group

Since patients from different groups had different EMG acquisition maps and different exercises routines, synergy extraction performed on the control group was repeated individually to match the data for each patient exercise routine and EMG mapping, by concatenating EMG from various repetitions and movements. The corresponding synergies from healthy controls were extracted only on the subset of the muscles and exercises specific for each patient. All healthy subject synergy sets were then averaged across controls and linked to the patient they refer to. Finally, each patient synergy set was ordered and compared to the corresponding healthy synergy set.

2.5. *Outcome Measures*

To compare synergies between healthy controls and patients, a synergy similarity metric was computed. The muscle synergy similarity (*SS*) is the dot product between two-unit norm synergies as shown in Equation (1).

$$SS = W_1 \cdot W_2 \qquad (1)$$

The synergy similarity metric was computed between matched couples of synergies between two sets of synergies (e.g., Hemiplegic patients before treatment and healthy subjects). Patients synergies were compared to healthy subjects' synergies using *SS* both before and after treatment. Then, the mean *SS* (*mSS*) was computed and used as an indicator of the synergy performance of each patient with respect to healthy subjects.

2.6. Statistics

A statistical analysis was implemented in order to verify if after the treatment, the induced synergy modifications were significant. First, all distributions were tested for normality with the Kolmogorov-Smirnov test. Similarity distributions for each patient in both pre and post treatment followed a normal distribution. Pre and post treatment distributions for the Motricity Index was tested with a *t*-test. The significance level was set = 0.05. For muscle synergies, *mSS* were compared using a 1-way ANOVA test to assess if the treatment induced a modification in spatial muscle synergies. The ANOVA test was coupled with a post hoc Tukey-Kramer test. When submitting the retrospective study to the Ethical Committee, assuming a significance level of 0.05 and using a 1-way ANOVA test applied to the outcome variable for comparison, it was verified that with the available dataset, it was possible to obtain a level of statistical power above 0.8. This calculation was performed using GPower software [74].

3. Results

In this section, we first show the results of the treatment found with the Motricity Index (MI) in Table 5. Pre-post improvements were found for motor functions in many items of the motricity index. In Hemiplegic patients, arm MI ($p < 0.0021$) and leg MI ($p < 0.0024$) increased; no differences were found instead for trunk MI ($p = 0.1723$). Paraplegic patients' arm and trunk had already full function at the beginning of the treatment and no change was found; leg MI improved ($p < 0.0183$). Tetraplegic patients' arm MI and trunk MI did not improve ($p = 0.0702$, $p = 0.0523$, respectively); leg MI improved ($p = 0.0446$). For tetraplegic patients, all *p*-values are slightly lower or higher to the threshold for significance.

A typical example of the extracted synergies before and after treatment from a patient with hemiplegia is shown in Figure 5.

Table 5. Motricity Index for Arm, Leg and Trunk in Hemiplegic patients, paraplegic patients and Tetraplegic patients.

Motricity Index		Hemiplegic Patients		Paraplegic Patients		Tetraplegic Patients	
Subject ID	**Body Segment**	**PRE**	**POST**	**PRE**	**POST**	**PRE**	**POST**
S1	ARM	10	34	100	100	40	73
	LEG	70	76	29	67	38	53
	TRUNK	25	25	100	100	61	74
S2	ARM	50	73	100	100	29	29
	LEG	91	100	76	100	1	1
	TRUNK	25	25	100	100	0	0
S3	ARM	34	41	100	100	29	29
	LEG	76	76	1	10	24	24
	TRUNK	25	25	61	61	0	0
S4	ARM	1	18	100	100	19	40
	LEG	24	29	28	28	19	39
	TRUNK	74	74	74	74	0	36

Table 5. *Cont.*

Motricity Index		Hemiplegic Patients		Paraplegic Patients		Tetraplegic Patients	
Subject ID	Body Segment	PRE	POST	PRE	POST	PRE	POST
S5	ARM	10	15	100	100	29	29
	LEG	29	39	1	10	29	29
	TRUNK	100	100	74	74	24	29
S6	ARM	10	29	100	100	29	34
	LEG	19	29	1	10	29	29
	TRUNK	61	74	61	61	48	61
S7	ARM	19	29	100	100	92	100
	LEG	28	38	58	76	92	100
	TRUNK	61	74	87	87	100	100
S8	ARM					34	39
	LEG					38	53
	TRUNK					87	100

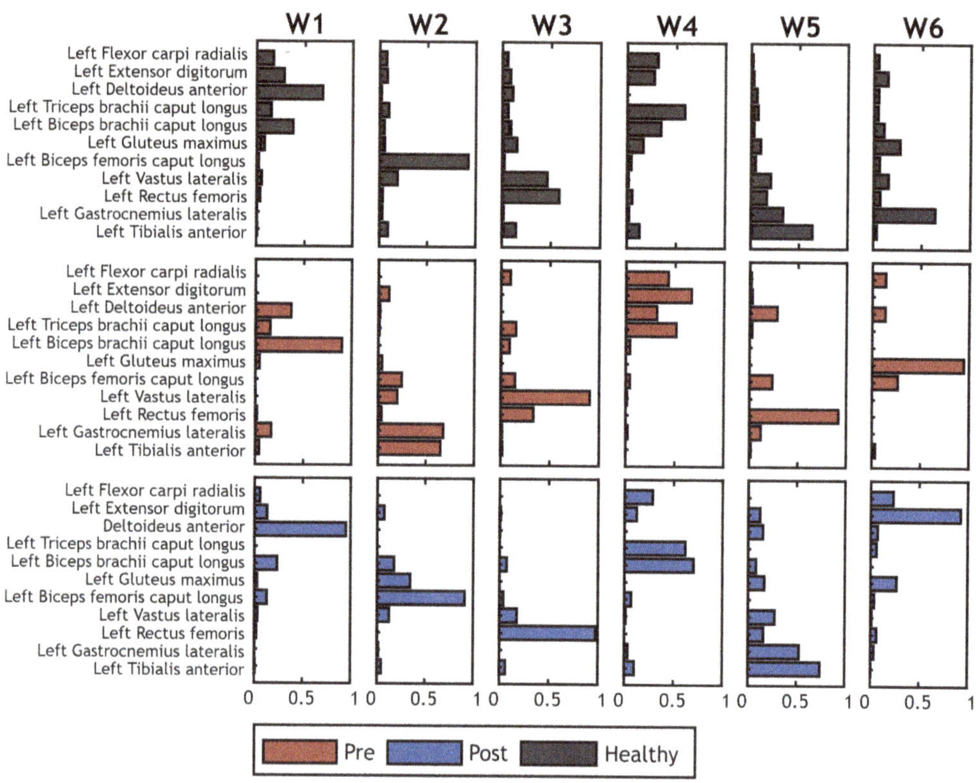

Figure 5. Example of synergies extracted on a hemiplegic patient. Spatial synergies before treatment are represented in red; spatial synergies after treatment are represented in blue. Grey bars show the corresponding reference synergies achived averaging synergies on the control group.

The mSS obtained for all groups of subjects is presented in Figure 6.

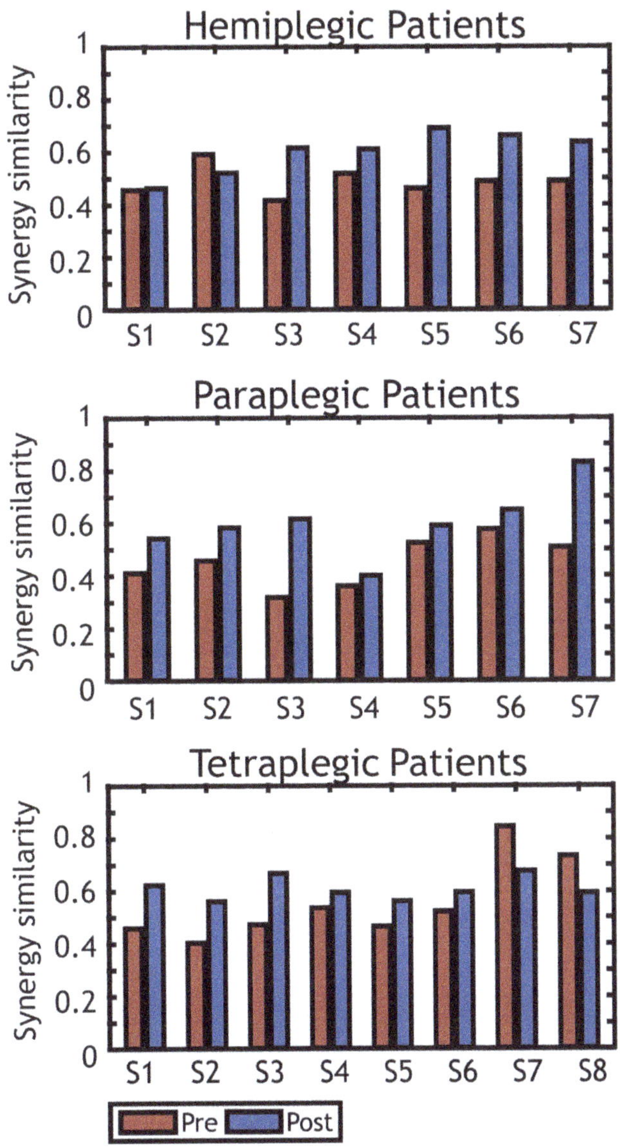

Figure 6. Spatial synergy similarity (healthy vs. tetraplegic) before (Pre) and after (Post) treatment. Graphs represent the similarity of the synergies extracted on each patient with the reference dataset of spatial synergies found on healthy controls. Pre-tratment synergy similarity is represented in red, while post-treatment synergy similarity is represented in blue.

In Figure 7, the results of the statistical analysis are illustrated. The first panel shows the comparison between pre and post treatment for all patients. The other three panels illustrate the comparison for each group of patients separately.

The comparison including all patients showed a difference between pre and post treatment ($p < 0.001$). A median improvement was found increasing mSS from 0.50 in pre-treatment to 0.60 in post treatment. We also show the results achieved when dividing patients according to their disease. The results obtained from comparing pre and post trials for hemiplegic and paraplegic patients ($p = 0.027$ in both cases) showed an improvement in the synergy similarity from 0.45 to 0.60. The comparison between the pre and post trials for the tetraplegic group of patients did not yield a significant result ($p = 0.454$) but there was an improvement in the mSS from 0.57 to 0.61 (even if not significant).

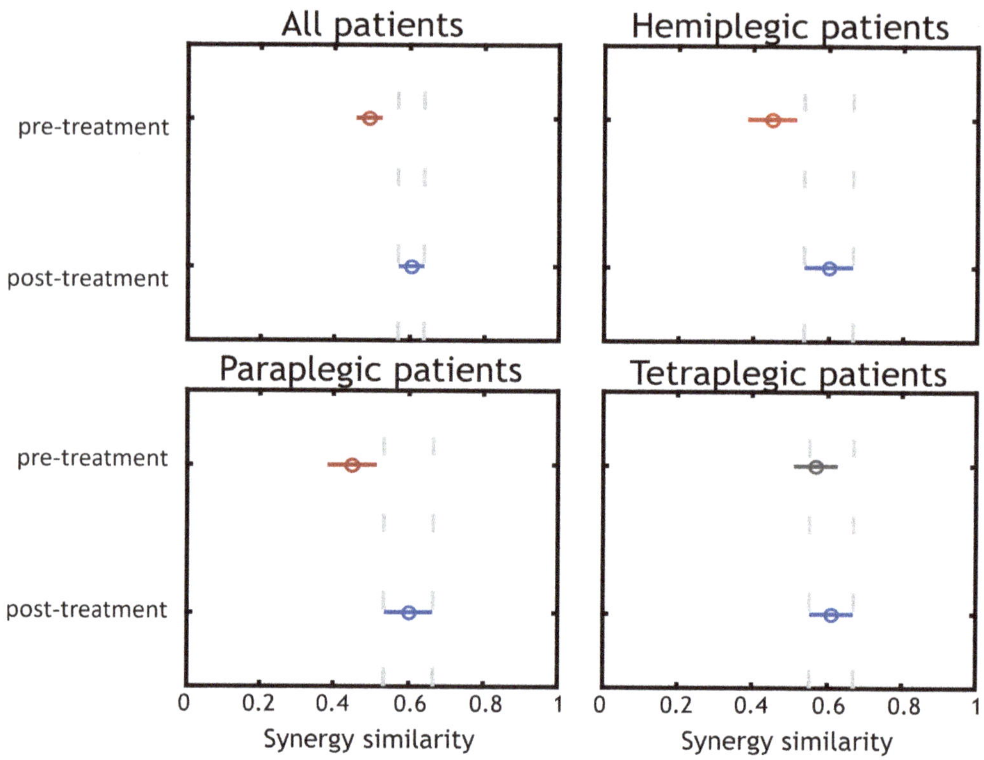

Figure 7. Statistical analysis. Statistical analysis was performed on each group of patients separately and for all subjects in the same group. We found that for the "All patients", "Hemiplegic Patients", "Paraplegic Patients" results were statistically significant (Post treatment synergy similarity in respect to controls increased), while for the "Tetraplegic Patients" group, there was a slight median increase of the MSS which was not statistically significant.

4. Discussion

In this work, we have studied the effects of a total body AFESK treatment method on three groups of neurologic patients, composed of 22 neurologic patients: 7 hemiplegics patients, 7 paraplegic and 8 tetraplegic patients. They all underwent the same rehabilitation intervention protocol, aimed at restoring physiological muscle activation patterns by the means of total-body exercises coupled with multi-channel AFESK. This analysis describes one of the first attempts to combine whole-body FES with the muscle synergy assessment, a relevant biomarker for assessing inter-muscle coordination. Results are confirmed with clinical scales that also show motor improvements.

The results show the for most of the screened body segments, the Motricity Index increased after the treatment, indicating a partial recovery of the motor function.

The results also show a trend towards the restoring of healthy-like synergies was obtained, confirming previous findings achieved with local FES applications [44,75], and extending them to whole body approaches. Previous studies regarding muscle synergy analysis of FES based treatments only analyzed local FES applications, e.g., for walking [44] or for planar upper-limb movements [68].

Both studies have confirmed a tendency of subjects to re-align motor activation patterns to those of healthy subjects. This result is particularly meaningful because it was achieved in different pathologies and with chronic patients, during total-body functional movements strongly related with daily life activities.

Considering each group separately, only the paraplegic and hemiplegic patient groups achieved statistical significance; the tetraplegic group of patients showed also a slight improvement, even if not statistically significant. This result is most likely due to the lower number of sessions and the frequency of visits per week, as well as intervals between treatment sessions due to quarantine measures. We are also aware that this effect is probably related to the limited number of subjects included in the study. Interestingly, a slight improvement was seen both on clinical scales and with muscle synergies, but for both domains, results were mostly close to the threshold for statistical significance. These results should be confirmed on a higher number of subjects. At the same time, despite the fact that the time and frequency of stimulating effects in the tetraplegic group was lower than desirable for the maximum inclusion of reparation processes, positive changes in the level of muscle activity of the muscles were noted in most patients. In fact, while examining the group of patients as a whole, the results indicate a clear improvement in synergy similarity with the control group before and after the treatment.

At the diagnostic level, our results demonstrate the effectiveness of the whole-body FES approach and the appearance of changes at the local level of motor units. With further summation of the positive effects as a result of AFESK, a transition to more refined level of regulation can occur, in which the necessary levels of synergic interaction between the interested muscle groups will be more clearly manifested. The results obtained in this study indicate that whole body FES rehabilitation techniques could in fact be used to realign muscle activation patterns of neurologic patients to those of healthy people and promote neuroplasticity. The groups which benefited the most from the treatment were the group of paraplegic patients and the group of hemiplegic patients.

Despite muscle synergies can capture relevant aspects of muscular coordination patterns, they cannot fully describe the evolution of EMG patterns during the course of the therapy. In fact, for some patients, we did not observe significant changes in muscle synergy recruitment patterns, even though important modifications in clinical outcomes were observed with other methods (such as clinical scales, clinical tests, motor capability, and others).

One can observe that in four out of seven hemiplegic subjects, the treatment brought the synergistic muscular activity to a condition more similar with respect to the activations of the control group. On the contrary, in three patients, the treatment induced a change in the muscular activity, but this did not help the patients to restore muscle activation patterns closer to the control group.

All paraplegic patients underwent improvements in the activation patterns, although to a lesser extent it was expressed also in two patients whose period of injury that caused paraplegia exceeded 10 years, age—39 and 62 years, localization of damage—L2/3 and T 12-L2.

In the tetraplegic group of patients, five out of seven exhibited an improvement in the muscle activation patterns while only two could not. One of these cases had to interrupt the treatment in occasion of the birth of her child, after which the patient's motor capabilities deteriorated, which was confirmed by the results of a repeated myographic examination. The second case is a patient with residual tetraparesis who completed a course of treatment after only 20 sessions.

Comparative myograms before and after the completion of the rehabilitation course of one of the patients with paraplegia level T 12-L2, who did not show modifications in synergic relationships, help clarify the effectiveness of the therapy which was not fully captured with muscle synergies. There was an increase in muscle activity of individual muscle groups, especially the rectus femoris, while walking with an exoskeleton. This result is not highlighted in muscle synergy analysis due to EMG normalization needed to compare synergies across subjects and sessions.

However, an in-depth analysis of changes in muscle activity also revealed a significant increase in the power spectrum of rapid motor units in the absence of significant changes in temporal activation parameters important for analyzing the synergistic relationships between muscle groups underlying movement (not reported here). Probably, the above case is an example of the accumulation of quantitatively functional changes in the neuromuscular apparatus, associated with an increase in the synchronization of the simultaneous inclusion of rapid motor units. The described effect can occur with an insufficient level of reflex regulatory influence on the part of the antagonists of their side, as well as the opposite side, which provides mutual reflex regulation with the participation of specialized interneurons of the segmental level.

With further repetition of AFESK movements according to this program, a further increase in the contractile capabilities of the muscle can occur, which can improve in the synergistic relationship between muscle groups that realize the movement.

Confirmation of the need for prolonged intensive exercises to restore lost functions were found in one of our patients with post-traumatic hemiplegia C1-2 level, who was excluded from the hemiplegia group due to the inconsistency of the protocol of the examined muscle groups that differed when comparing pre and post-therapy;. however, she managed to conduct a long course (190 sessions) with AFESK, including a high frequency of treatment (3–4 times a week), and time of movement execution and speed of movement constantly increasing. Currently, she can perform movement in full capacity.

Limitations and Future Work

While this work provides clear evidence that total body FES helps restore physiological muscle coordination patterns, our results are affected by the low number of subjects involved in the study and non-homogeneous samples. Analyzing cohorts with small sample sizes could lead to non-conclusive results like in the case of the tetraplegic group of patients. Furthermore, non-homogeneity of the studied group should be avoided in future work.

Previous studies have confirmed the heterogeneity between different neurologic patients [44], which reinforces the need to have different protocols for different subjects. However, in order to provide reliable comparisons, a fully consistent protocol needs to be established. Despite this, due to the very low evidences available on total-body FES couples with muscle synergies, our study sets a relevant pilot work for more extensive applications in the future. We in fact noticed that research articles coupling muscle synergies and FES have high innovative approaches but always involve a very low number of subjects (from 2 to 9 patients) [41,42,44,76–83].

In addition to improved homogeneity of the cohorts, analyzing the improved performance of patients with only muscle synergies, one provides a deep, yet partial perspective on the actual quality of motion related to neurologic disorders. One effective way to overcome this limitation is with a conjunct analysis of both EMG and kinematics, for example by detecting the effects on kinematic and muscular patterns; this can be achieved with novel algorithms that allow inter-domain factorization [84]. Multi-domain approaches could be considered to enhance effect of rehabilitation and assessment [85,86].

In addition, given the experience of this study, it should be noted that there is a need for further development of the methodology for assessing synergic relationships of muscle activity in patients with severe neurological disorders of whole body and locomotor functions. The methods currently used in clinical practice do not allow to fully assess the functional nature of pathophysiological disorders of the whole body and locomotor apparatus. At the same time, the methodology used to assess the synergistic relationships of muscle activity during exercise can bring us closer to solving a multi-level assessment of violations in movement control. This was confirmed in our work by the coincidence of the results of the clinical evaluation of the state of patients with the conclusion made on the basis of synergic relationships for each of the interested muscle groups in patients with emi-para-tetraplegia.

If used, such a data collection system will allow to timely receive the necessary information about changes in locomotor functions in the process of rehabilitation timely change the tactics and set of rehabilitation programs, which in turn will certainly enhance the effect of therapy [87–92].

Lastly, it is interesting to evaluate the treatments capability to induce long term changes in patients. Thus, a follow-up session should be included in further studies on the topic.

5. Conclusions

In a few numbers of works, researchers have studied the possibility of either analyzing FES treatments with muscle synergies or using them for control of stimulation patterns. The studies that employed muscle synergies and FES, consistently reported positive outcomes in improvements in muscle synergies patterns for neurologic patients.

This work also adds to this pool of studies by reporting positive changes in patients which underwent whole body FES. It is necessary to be cautious when interpreting our results since more studies need to be performed on the matter and be guided by average indicators on a larger number of cases of the disease for each nosology. In addition, given the prospects of the direction of active whole-body FES in the rehabilitation of patients with severe neurological disorders, it is necessary to develop a comprehensive evaluation system considering clinical practice and objective research methods in the process of implementing locomotor functions.

Author Contributions: Conceptualization, A.S. and V.T.; methodology, A.S., R.M.M., G.G. and V.T.; software, A.S. and R.M.M.; validation, A.S. and R.M.M.; formal analysis, A.S., R.M.M., G.G. and V.T.; investigation, A.S., R.M.M., G.G., F.M. and V.T.; resources, A.S., G.G. and V.T.; data curation, A.S., R.M.M. and V.T.; writing—original draft preparation, A.S., R.M.M., G.G. and V.T.; writing—review and editing, A.S., R.M.M., G.G., F.M. and V.T.; visualization, A.S., R.M.M., G.G. and V.T.; supervision, A.S., G.G., F.M. and V.T.; project administration, A.S., G.G., F.M. and V.T.; funding acquisition, A.S. and V.T. All authors have read and agreed to the published version of the manuscript.

Funding: This research received no external funding.

Institutional Review Board Statement: The study was conducted in accordance with the Declaration of Helsinki, and approved by the Ethical Commettee of the ASST Papa Giovanni XXIII—Ospedale di Bergamo (approved 16 February 2021).

Informed Consent Statement: Written informed consent was obtained from all subjects involved in the study or from their legal representatives.

Data Availability Statement: Data are not available for privacy reasons.

Conflicts of Interest: Viktor Terekhov and Guido Gabbrielli have interests in the exploitation of the AFESK technology: they are shareholders and respectively Chief scientific officer and Chief technology officer of VIKTOR S.r.l.

References

1. Ebels, P. EUObserver. 22 October 2012. Available online: https://euobserver.com/search?query=Disability+in+figures (accessed on 13 September 2021).
2. Eurostat. Population Structure and Ageing. Statistics Explained. 2019. Available online: https://ec.europa.eu/eurostat/statistics-explained/index.php?title=Population_structure_and_ageing (accessed on 13 September 2021).
3. World Health Organization. World Health Organization: Rehabilitation. 16 July 2021. Available online: https://www.who.int/news-room/fact-sheets/detail/rehabilitation (accessed on 14 September 2021).
4. Eurostat. Healthcare Expenditure Statistics. 2021. Available online: https://ec.europa.eu/eurostat/statistics-explained/index.php?title=Healthcare_expenditure_statistics#Healthcare_expenditure (accessed on 10 September 2021).
5. Molteni, F.; Gasperini, G.; Cannaviello, G.; Guanziroli, E. Exoskeleton and End-Effector Robots for Upper and Lower Limbs Rehabilitation: Narrative Review. *PM&R* **2018**, *10*, S174–S188. [CrossRef]

6. Ambrosini, E.; Ferrante, S.; Schauer, T.; Klauer, C.; Gaffuri, M.; Ferrigno, G.; Pedrocchi, A. A myocontrolled neuroprosthesis integrated with a passive exoskeleton to support upper limb activities. *J. Electromyogr. Kinesiol.* **2014**, *24*, 307–317. [CrossRef] [PubMed]
7. Hamzaid, N.A.; Davis, G. Health and fitness benefits of functional electrical stimulation-evoked leg exercise for spinal cord–Injured individuals: A position review. *Top. Spinal Cord Inj. Rehabil.* **2009**, *14*, 88–121. [CrossRef]
8. Jones, S.; Man, W.D.-C.; Gao, W.; Higginson, I.J.; Wilcock, A.; Maddocks, M. Neuromuscular electrical stimulation for muscle weakness in adults with advanced disease. *Cochrane Database Syst. Rev.* **2016**, *2016*, CD009419. [CrossRef]
9. Sheffler, L.R.; Chae, J. Neuromuscular electrical stimulation in neurorehabilitation. *Muscle Nerve Off. J. Am. Assoc. Electrodiagn. Med.* **2007**, *35*, 562–590. [CrossRef] [PubMed]
10. Van Griethuysen, C.M.; Paul, J.P.; Andrews, B.J.; Nicol, A.C. Biomechanics of functional electrical stimulation. *Prosthet. Orthot. Int.* **1982**, *6*, 152–156. [CrossRef]
11. Sharif, F.; Ghulam, S.; Malik, A.N.; Saeed, Q. Effectiveness of Functional Electrical Stimulation (FES) versus Conventional Electrical Stimulation in Gait Rehabilitation of Patients with Stroke. *J. Coll. Physicians Surg. Pak.* **2017**, *27*, 703–706.
12. Dujović, S.D.; Malešević, J.; Malešević, N.; Vidaković, A.S.; Bijelić, G.; Keller, T.; Konstantinović, L. Novel multi-pad functional electrical stimulation in stroke patients: A single-blind randomized study. *NeuroRehabilitation* **2017**, *41*, 791–800. [CrossRef]
13. Knutson, J.S.; Harley, M.Y.; Hisel, T.Z.; Hogan, S.D.; Maloney, M.M.; Chae, J. Contralaterally controlled functional electrical stimulation for upper extremity hemiplegia: An early-phase randomized clinical trial in subacute stroke patients. *Neurorehabilit. Neural Repair* **2012**, *26*, 239–246. [CrossRef]
14. Hara, Y.; Obayashi, S.; Tsujiuchi, K.; Muraoka, Y. The effects of electromyography-controlled functional electrical stimulation on upper extremity function and cortical perfusion in stroke patients. *Clin. Neurophysiol.* **2013**, *124*, 2008–2015. [CrossRef]
15. Alon, G.; Levitt, A.F.; McCarthy, P.A. Functional Electrical Stimulation Enhancement of Upper Extremity Functional Recovery During Stroke Rehabilitation: A Pilot Study. *Neurorehabilit. Neural Repair* **2007**, *21*, 207–215. [CrossRef]
16. Mazzoleni, S.; Battini, E.; Rustici, A.; Stampacchia, G. An integrated gait rehabilitation training based on Functional Electrical Stimulation cycling and overground robotic exoskeleton in complete spinal cord injury patients: Preliminary results. In Proceedings of the 2017 International Conference on Rehabilitation Robotics (ICORR), London, UK, 17–20 July 2017; pp. 289–293.
17. Topçu, Ç.; Uysal, H.; Özkan, Ö.; Polat, Ö.; Bedeloğlu, M.; Akgül, A.; Döğer, E.N.; Sever, R.; Çolak, Ö.H. Recovery of facial expressions using functional electrical stimulation after full-face transplantation. *J. Neuroeng. Rehabil.* **2018**, *15*, 1–7. [CrossRef]
18. Popovic, M.R.; Thrasher, T.A.; Adams, M.E.; Takes, V.; Zivanovic, V.; Tonack, M.I. Functional electrical therapy: Retraining grasping in spinal cord injury. *Spinal Cord* **2006**, *44*, 143–151. [CrossRef]
19. McCabe, J.; Monkiewicz, M.; Holcomb, J.; Pundik, S.; Daly, J.J. Comparison of Robotics, Functional Electrical Stimulation, and Motor Learning Methods for Treatment of Persistent Upper Extremity Dysfunction after Stroke: A Randomized Controlled Trial. *Arch. Phys. Med. Rehabil.* **2015**, *96*, 981–990. [CrossRef] [PubMed]
20. Zhang, D.; Zhu, K. Modeling biological motor control for human locomotion with functional electrical stimulation. *Biol. Cybern.* **2007**, *96*, 79–97. [CrossRef] [PubMed]
21. Kimura, T.; Nakano, W. Repetition of a cognitive task promotes motor learning. *Hum. Mov. Sci.* **2019**, *66*, 109–116. [CrossRef] [PubMed]
22. Dayan, E.; Cohen, L.G. Neuroplasticity Subserving Motor Skill Learning. *Neuron* **2011**, *72*, 443–454. [CrossRef]
23. Doyon, J.; Benali, H. Reorganization and plasticity in the adult brain during learning of motor skills. *Curr. Opin. Neurobiol.* **2005**, *15*, 161–167. [CrossRef]
24. He, K.; Liang, Y.; Abdollahi, F.; Bittmann, M.F.; Kording, K.; Wei, K. The Statistical Determinants of the Speed of Motor Learning. *PLOS Comput. Biol.* **2016**, *12*, e1005023. [CrossRef]
25. Brushart, T. Motor axons preferentially reinnervate motor pathways. *J. Neurosci.* **1993**, *13*, 2730–2738. [CrossRef]
26. Carmel, J.B.; Martin, J.H. Motor cortex electrical stimulation augments sprouting of the corticospinal tract and promotes recovery of motor function. *Front. Integr. Neurosci.* **2014**, *8*, 51. [CrossRef]
27. Lee, W.A. Neuromotor Synergies as a Basis for Coordinated Intentional Action. *J. Mot. Behav.* **1984**, *16*, 135–170. [CrossRef]
28. Bizzi, E.; Cheung, V.C.K.; D'Avella, A.; Saltiel, P.; Tresch, M. Combining modules for movement. *Brain Res. Rev.* **2008**, *57*, 125–133. [CrossRef] [PubMed]
29. Bizzi, E.; Cheung, V.C. The neural origin of muscle synergies. *Front. Comput. Neurosci.* **2013**, *7*, 51. [CrossRef] [PubMed]
30. D'Avella, A.; Portone, A.; Fernandez, L.; Lacquaniti, F. Control of Fast-Reaching Movements by Muscle Synergy Combinations. *J. Neurosci.* **2006**, *26*, 7791–7810. [CrossRef]
31. D'Avella, A.; Fernandez, L.; Portone, A.; Lacquaniti, F. Modulation of Phasic and Tonic Muscle Synergies with Reaching Direction and Speed. *J. Neurophysiol.* **2008**, *100*, 1433–1454. [CrossRef]
32. Cheung, V.C.-K.; Piron, L.; Agostini, M.; Silvoni, S.; Turolla, A.; Bizzi, E. Stability of muscle synergies for voluntary actions after cortical stroke in humans. *Proc. Natl. Acad. Sci. USA* **2009**, *106*, 19563–19568. [CrossRef] [PubMed]
33. Cheung, V.C.-K.; Turolla, A.; Agostini, M.; Silvoni, S.; Bennis, C.; Kasi, P.; Paganoni, S.; Bonato, P.; Bizzi, E. Muscle synergy patterns as physiological markers of motor cortical damage. *Proc. Natl. Acad. Sci. USA* **2012**, *109*, 14652–14656. [CrossRef]
34. Banks, C.L.; Pai, M.M.; McGuirk, T.E.; Fregly, B.J.; Patten, C. Methodological Choices in Muscle Synergy Analysis Impact Differentiation of Physiological Characteristics Following Stroke. *Front. Comput. Neurosci.* **2017**, *11*, 78. [CrossRef]

35. Ivanenko, Y.P.; Poppele, R.E.; Lacquaniti, F. Five basic muscle activation patterns account for muscle activity during human locomotion. *J. Physiol.* **2004**, *556*, 267–282. [CrossRef]
36. Clark, D.J.; Ting, L.H.; Zajac, F.E.; Neptune, R.R.; Kautz, S.A. Merging of Healthy Motor Modules Predicts Reduced Locomotor Performance and Muscle Coordination Complexity Post-Stroke. *J. Neurophysiol.* **2010**, *103*, 844–857. [CrossRef] [PubMed]
37. Dominici, N.; Ivanenko, Y.P.; Cappellini, G.; d'Avella, A.; Mondì, V.; Cicchese, M.; Fabiano, A.; Silei, T.; di Paolo, A.; Giannini, C.; et al. Locomotor primitives in newborn babies and their development. *Science* **2011**, *334*, 997–999. [CrossRef]
38. Lencioni, T.; Jonsdottir, J.; Cattaneo, D.; Crippa, A.; Gervasoni, E.; Rovaris, M.; Bizzi, E.; Ferrarin, M. Are Modular Activations Altered in Lower Limb Muscles of Persons with Multiple Sclerosis during Walking? Evidence from Muscle Synergies and Biomechanical Analysis. *Front. Hum. Neurosci.* **2016**, *10*, 620. [CrossRef] [PubMed]
39. Torres-Oviedo, G.; Ting, L.H. Muscle Synergies Characterizing Human Postural Responses. *J. Neurophysiol.* **2007**, *98*, 2144–2156. [CrossRef]
40. Safavynia, S.A.; Ting, L.H. Task-level feedback can explain temporal recruitment of spatially fixed muscle synergies throughout postural perturbations. *J. Neurophysiol.* **2012**, *107*, 159–177. [CrossRef] [PubMed]
41. Ambrosini, E.; Parati, M.; Peri, E.; De Marchis, C.; Nava, C.; Pedrocchi, A.; Ferriero, G.; Ferrante, S. Changes in leg cycling muscle synergies after training augmented by functional electrical stimulation in subacute stroke survivors: A pilot study. *J. Neuroeng. Rehabil.* **2020**, *17*, 1–14. [CrossRef] [PubMed]
42. Muceli, S.; Boye, A.T.; D'Avella, A.; Farina, D. Identifying Representative Synergy Matrices for Describing Muscular Activation Patterns During Multidirectional Reaching in the Horizontal Plane. *J. Neurophysiol.* **2010**, *103*, 1532–1542. [CrossRef] [PubMed]
43. Denis, W.; Brunetti, F.; Piazza, S.; Torricelli, D.; Pons, J.L. Functional Electrical Stimulation Controller Based on Muscle Synergies. *Converging Clin. Eng. Res. Neurorehabil.* **2013**, 283–287. [CrossRef]
44. Ferrante, S.; Bejarano, N.C.; Ambrosini, E.; Nardone, A.; Turcato, A.M.; Monticone, M.; Ferrigno, G.; Pedrocchi, A. A Personalized Multi-Channel FES Controller Based on Muscle Synergies to Support Gait Rehabilitation after Stroke. *Front. Neurosci.* **2016**, *10*, 425. [CrossRef]
45. Pano-Rodriguez, A.; Beltran-Garrido, J.V.; Hernández-González, V.; Reverter-Masia, J. Effects of whole-body ELECTROMYOSTIMULATION on health and performance: A systematic review. *BMC Complement. Altern. Med.* **2019**, *19*, 1–14. [CrossRef]
46. Filipovic, A.; Kleinöder, H.; Dörmann, U.; Mester, J. Electromyostimulation—A Systematic Review of the Effects of Different Electromyostimulation Methods on Selected Strength Parameters in Trained and Elite Athletes. *J. Strength Cond. Res.* **2012**, *26*, 2600–2614. [CrossRef] [PubMed]
47. Hashida, R.; Matsuse, H.; Takano, Y.; Omoto, M.; Nago, T.; Shiba, N. Walking exercise combined with neuromuscular electrical stimulation of antagonist resistance improved muscle strength and physical function for elderly people: A pilot study. *J. Phys. Fit. Sports Med.* **2016**, *5*, 195–203. [CrossRef]
48. Bo, A.P.L.; da Fonseca, L.O.; de Sousa, A.C.C. FES-induced co-activation of antagonist muscles for upper limb control and disturbance rejection. *Med. Eng. Phys.* **2016**, *38*, 1176–1184. [CrossRef] [PubMed]
49. Carson, R.G.; Buick, A.R. Neuromuscular electrical stimulation-promoted plasticity of the human brain. *J. Physiol.* **2019**, *599*, 2375–2399. [CrossRef] [PubMed]
50. Rickham, P. Human experimentation. Code of ethics of the world medical association. Declaration of Helsinki. *Br. Med. J.* **1964**, *2*, 177.
51. Ikuno, K.; Matsuo, A.; Shomoto, K. Sensory Electrical Stimulation for Recovery of Hand and Arm Function in Stroke Patients: A Review of the Literature. *J. Nov. Physiother.* **2012**, *1*, 7. [CrossRef]
52. Karhu, J.; Tesche, C.D. Simultaneous early processing of sensory input in human primary (SI) and secondary (SII) somatosensory cortices. *J. Neurophysiol.* **1999**, *81*, 2017–2025. [CrossRef]
53. del Vecchio, M.; Caruana, F.; Sartori, I.; Pelliccia, V.; Russo, G.L.; Rizzolatti, G.; Avanzini, P. Ipsilateral somatosensory responses in humans: The tonic activity of SII and posterior insular corte. *Brain Struct. Funct.* **2019**, *224*, 9–18. [CrossRef]
54. Nas, K.; Yazmalar, L.; Şah, V.; Aydın, A.; Öneş, K. Rehabilitation of spinal cord injuries. *World J. Orthop.* **2015**, *6*, 8. [CrossRef]
55. Klingner, C.M.; Huonker, R.; Flemming, S.; Hasler, C.; Brodoehl, S.; Preul, C.; Burmeister, H.; Kastrup, A.; Witte, O.W. Functional deactivations: Multiple ipsilateral brain areas engaged in the processing of somatosensory information. *Hum. Brain Mapp.* **2010**, *32*, 127–140. [CrossRef]
56. Jovanovic, I.L.; Kapadia, N.; Lo, L.; Zivanovic, V.; Popovic, R.M.; Marquez-Chin, C. Restoration of Upper Limb Function After Chronic Severe Hemiplegia: A Case Report on the Feasibility of a Brain-Computer Interface-Triggered Functional Electrical Stimulation Therapy. *Am. J. Phys. Med. Rehabil.* **2020**, *99*, e35–e40. [CrossRef]
57. Halonen, J.-P.; Jones, S.; Shawkat, F. Contribution of cutaneous and muscle afferent fibres to cortical SEPs following median and radial nerve stimulation in man. *Electroencephalogr. Clin. Neurophysiol. Potentials Sect.* **1988**, *71*, 331–335. [CrossRef]
58. Avanzini, P.; Abdollahi, R.O.; Sartori, I.; Caruana, F.; Pelliccia, V.; Casaceli, G.; Mai, R.; Russo, G.L.; Rizzolatti, G.; Orban, G.A. Four-dimensional maps of the human somatosensory system. *Proc. Natl. Acad. Sci. USA* **2016**, *113*, E1936–E1943. [CrossRef]
59. Kaas, J.; Pons, T. The somatosensory system of primates. In *Neurosciences: Comparative Primate Biology*; The Univeristy of Chicago Press Journals: Chicago, IL, USA, 1988; pp. 421–468.
60. Hardy, S.G.P.; Spalding, T.B.; Liu, H.; Nick, T.G.; Pearson, R.H.; Hayes, A.V.; Stokic, D. The Effect of Transcutaneous Electrical Stimulation on Spinal Motor Neuron Excitability in People Without Known Neuromuscular Diseases: The Roles of Stimulus Intensity and Location. *Phys. Ther.* **2002**, *82*, 354–363. [CrossRef] [PubMed]

61. Driver, J.; Noesselt, T. Review Multisensory Interplay Reveals Crossmodal Influences on 'Sensory-Specific' Brain Regions, Neural Responses, and Judgments. *Neuron* **2008**, *57*, 11–23. [CrossRef] [PubMed]
62. Rath, M.; Vette, A.H.; Ramasubramaniam, S.; Li, K.; Burdick, J.; Edgerton, V.R.; Gerasimenko, Y.P.; Sayenko, D.G. Trunk Stability Enabled by Noninvasive Spinal Electrical Stimulation after Spinal Cord Injury. *J. Neurotrauma* **2018**, *35*, 2540–2553. [CrossRef] [PubMed]
63. Tan, Z.; Liu, H.; Yan, T.; Jin, D.; He, X.; Zheng, X.; Xu, S.; Tan, C. The Effectiveness of Functional Electrical Stimulation Based on a Normal Gait Pattern on Subjects with Early Stroke: A Randomized Controlled Trial. *BioMed Res. Int.* **2014**, *2014*, 1–9. [CrossRef] [PubMed]
64. Arendsen, L.J.; Guggenberger, R.; Zimmer, M.; Weigl, T.; Gharabaghi, A. Peripheral Electrical Stimulation Modulates Cortical Beta-Band Activity. *Front. Neurosci.* **2021**, *15*, 339. [CrossRef] [PubMed]
65. Cebolla, A.; Cheron, G. Sensorimotor and cognitive involvement of the beta–gamma oscillation in the frontal N30 component of somatosensory evoked potentials. *Neuropsychologia* **2015**, *79*, 215–222. [CrossRef]
66. Cooke, S.F.; Bliss, T.V. Plasticity in the human central nervous system. *Brain* **2006**, *129*, 1659–1673. [CrossRef]
67. Howlett, O.A.; Lannin, N.; Ada, L.; McKinstry, C. Functional Electrical Stimulation Improves Activity After Stroke: A Systematic Review with Meta-Analysis. *Arch. Phys. Med. Rehabil.* **2015**, *96*, 934–943. [CrossRef] [PubMed]
68. Guertin, P.A. Central Pattern Generator for Locomotion: Anatomical, Physiological, and Pathophysiological Considerations. *Front. Neurol.* **2013**, *3*, 183. [CrossRef]
69. Stegeman, D.; Hermens, H. Standards for surface electromyography: The European project Surface EMG for non-invasive assessment of muscles (SENIAM). *Enschede Roessingh Res. Dev.* **2007**, *10*, 108–112.
70. Duthilleul, N.; Pirondini, E.; Coscia, M.; Micera, S. Effect of handedness on muscle synergies during upper limb planar movements. In Proceedings of the 2015 37th Annual International Conference of the IEEE Engineering in Medicine and Biology Society (EMBC), Milano, Italy, 25–29 August 2015; pp. 3452–3455.
71. Steele, K.M.; Rozumalski, A.; Schwartz, M. Muscle synergies and complexity of neuromuscular control during gait in cerebral palsy. *Dev. Med. Child Neurol.* **2015**, *57*, 1176–1182. [CrossRef] [PubMed]
72. Scano, A.; Dardari, L.; Molteni, F.; Giberti, H.; Tosatti, L.M.; D'Avella, A. A Comprehensive Spatial Mapping of Muscle Synergies in Highly Variable Upper-Limb Movements of Healthy Subjects. *Front. Physiol.* **2019**, *10*, 1231. [CrossRef] [PubMed]
73. Torres-Oviedo, G.; MacPherson, J.M.; Ting, L.H. Muscle Synergy Organization Is Robust Across a Variety of Postural Perturbations. *J. Neurophysiol.* **2006**, *96*, 1530–1546. [CrossRef]
74. Mayr, S.; Erdfelder, E.; Buchner, A.; Faul, F. A short tutorial of GPower. *Tutor. Quant. Methods Psychol.* **2007**, *3*, 51–59. [CrossRef]
75. Niu, C.M.; Bao, Y.; Zhuang, C.; Li, S.; Wang, T.; Cui, L.; Xie, Q.; Lan, N. Synergy-Based FES for Post-Stroke Rehabilitation of Upper-Limb Motor Functions. *IEEE Trans. Neural Syst. Rehabil. Eng.* **2019**, *27*, 256–264. [CrossRef]
76. Galeano, D.; Brunetti, F.; Torricelli, D.; Piazza, S.; Pons, J.L. A Tool for Balance Control Training Using Muscle Synergies and Multimodal Interfaces. *BioMed Res. Int.* **2014**, *2014*, 1–13. [CrossRef]
77. Li, S.; Zhuang, C.; Niu, C.M.; Bao, Y.; Xie, Q.; Lan, N. Evaluation of Functional Correlation of Task-Specific Muscle Synergies with Motor Performance in Patients Poststroke. *Front. Neurol.* **2017**, *8*, 337. [CrossRef]
78. Cole, N.M.; Ajiboye, A.B. Muscle synergies for predicting non-isometric complex hand function for commanding FES neuroprosthetic hand systems. *J. Neural Eng.* **2019**, *16*, 56018. [CrossRef] [PubMed]
79. Chou, C.-H.; Wang, T.; Sun, X.; Niu, C.M.; Hao, M.; Xie, Q.; Lan, N. Automated functional electrical stimulation training system for upper-limb function recovery in poststroke patients. *Med. Eng. Phys.* **2020**, *84*, 174–183. [CrossRef] [PubMed]
80. Lim, J.; Lim, T.; Lee, J.; Sim, J.; Chang, H.; Yoon, B.; Jung, H. Patient-specific functional electrical stimulation strategy based on muscle synergy and walking posture analysis for gait rehabilitation of stroke patients. *J. Int. Med. Res.* **2021**, *49*. [CrossRef] [PubMed]
81. Matsui, K.; Hishii, Y.; Maegaki, K.; Yamashita, Y.; Uemura, M.; Hirai, H.; Miyazaki, F. Equilibrium-point control of human elbow-joint movement under isometric environment by using multichannel functional electrical stimulation. *Front. Neurosci.* **2014**, *8*, 164. [CrossRef] [PubMed]
82. Lan, Y.; Yao, J.; Dewald, J.P. The impact of shoulder abduction loading on EMG-based intention detection of hand opening and closing after stroke. In Proceedings of the 2011 Annual International Conference of the IEEE Engineering in Medicine and Biology Society, Boston, MA, USA, 30 August–3 September 2011; pp. 4136–4139.
83. Williams, M.R. A pilot study into reaching performance after severe to moderate stroke using upper arm support. *PLoS ONE* **2018**, *13*, e0200787. [CrossRef] [PubMed]
84. Scano, A.; Mira, R.M.; D'Avella, A. Mixed matrix factorization: A novel algorithm for the extraction of kinematic-muscular synergies. *bioRxiv* **2022**. [CrossRef]
85. Belfatto, A.; Scano, A.; Chiavenna, A.; Mastropietro, A.; Mrakic-Sposta, S.; Pittaccio, S.; Tosatti, L.M.; Molteni, F.; Rizzo, G. A Multiparameter Approach to Evaluate Post-Stroke Patients: An Application on Robotic Rehabilitation. *Appl. Sci.* **2018**, *8*, 2248. [CrossRef]
86. Brambilla, C.; Pirovano, I.; Mira, R.M.; Rizzo, G.; Scano, A.; Mastropietro, A. Combined Use of EMG and EEG Techniques for Neuromotor Assessment in Rehabilitative Applications: A Systematic Review. *Sensors* **2021**, *21*, 7014. [CrossRef] [PubMed]

87. Hofstoetter, U.S.; Krenn, M.; Danner, S.M.; Hofer, C.; Kern, H.; McKay, W.B.; Mayr, W.; Minassian, K. Augmentation of Voluntary Locomotor Activity by Transcutaneous Spinal Cord Stimulation in Motor-Incomplete Spinal Cord-Injured Individuals. *Artif. Organs* **2015**, *39*, E176–E186. [CrossRef]
88. Kimura, T.; Kaneko, F.; Iwamoto, E.; Saitoh, S.; Yamada, T. Neuromuscular electrical stimulation increases serum brain-derived neurotrophic factor in humans. *Exp. Brain Res.* **2018**, *237*, 47–56. [CrossRef]
89. Korvenoja, A.; Huttunen, J.; Salli, E.; Pohjonen, H.; Martinkauppi, S.; Palva, J.M.; Lauronen, L.; Virtanen, J.; And, R.J.I.; Aronen, H.J. Activation of multiple cortical areas in response to somatosensory stimulation: Combined magnetoencephalographic and functional magnetic resonance imaging. *Hum. Brain Mapp.* **1999**, *8*, 13–27. [CrossRef]
90. McGie, S.C.; Zariffa, J.; Popovic, M.R.; Nagai, M.K. Short-Term Neuroplastic Effects of Brain-Controlled and Muscle-Controlled Electrical Stimulation. *Neuromodul. Technol. Neural Interface* **2015**, *18*, 233–240. [CrossRef] [PubMed]
91. Nascimento, L.; Michaelsen, S.M.; Ada, L.; Polese, J.C.; Teixeira-Salmela, L.F. Cyclical electrical stimulation increases strength and improves activity after stroke: A systematic review. *J. Physiother.* **2014**, *60*, 22–30. [CrossRef] [PubMed]
92. Olsen, S.; Signal, N.; Niazi, I.K.; Christensen, T.; Jochumsen, M.; Taylor, D. Paired Associative Stimulation Delivered by Pairing Movement-Related Cortical Potentials with Peripheral Electrical Stimulation: An Investigation of the Duration of Neuromodulatory Effects. *Neuromodul. Technol. Neural Interface* **2018**, *21*, 362–367. [CrossRef] [PubMed]

sensors

MDPI

Article

Applications of Laser-Induced Fluorescence in Medicine

Mirosław Kwaśny and Aneta Bombalska *

Institute of Optoelectronics, Military University of Technology, 00-908 Warsaw, Poland; miroslaw.kwasny@wat.edu.pl
* Correspondence: aneta.bombalska@wat.edu.pl; Tel.: +48-222-6183-7514

Abstract: Fluorescence is the most sensitive spectroscopic method of analysis and fluorescence methods. However, classical analysis requires sampling. There are new needs for real-time analyses of biological materials, without the need for sampling. This article presents examples of proprietary applications of laser-induced fluorescence (LIF) in medicine with such methods. A classic example is the analysis of photosensitizers using the photodynamic treatment method (PDT). The level and kinetics of accumulation and excretion of sensitizers in the body are examined, as well as the optimal exposure time after the application of compounds. The LIF method is also used to analyze endogenous fluorophores; it has been used to detect neoplasms, e.g., lung cancer or gynecological and dermatological diseases. Furthermore, it is used for the diagnosis of early stages of tooth decay or detection of fungi. The article will present the construction of sensors based on the LIF method—fiber laser spectrometers and investigated fluorescence spectra in individual applications. Examples of fluorescence imaging, e.g., dermatological, and dental diagnostics and measuring systems will be presented. The advantage of the method is it has greater sensitivity and easily detects lesions early compared to the methods used in observing the material in reflected light.

Keywords: photodynamic therapy; fluorescence; laser; fluorophores; enamel

Citation: Kwaśny, M.; Bombalska, A. Applications of Laser-Induced Fluorescence in Medicine. *Sensors* **2022**, *22*, 2956. https://doi.org/10.3390/s22082956

Academic Editors: Alfonso Mastropietro, Alessandro Scano and Massimo W. Rivolta

Received: 18 March 2022
Accepted: 11 April 2022
Published: 12 April 2022

1. Introduction

Fluorescence methods have played an important role in medicine and biochemistry for 50 years. DNA sequence analyses, immunofluorescence methods, flow cytometry, and analyses of vitamins, amino acids, porphyrins, pharmaceuticals and cations are among the classic examples of fluorescence technique applications.

The advantages of the method include its sensitivity, due to the intensity of fluorescence being proportional to the intensity of the excitation light, selectivity and ability to separate the emission spectra and excitation signals from the background. Another feature of modern methods is the possibility of using a variety of laser sources and optical fibers that transmit excitation and fluorescence radiation from anywhere in the human body or from the external environment.

The use of the LIF method for analyzing the state of biological tissues began in the 1990s. This method has been used for the diagnosis of skin diseases, atherosclerosis, kidney and urolithiasis and early stages of cancer [1,2]. "Optical biopsies", as opposed to histopathological examinations, are non-invasive, do not require material sampling by fine-needle biopsy, the amount of analyzed material is unlimited, radiation is supplied and received via optical fibers, signals are measured in real time and the same areas can be analyzed repeatedly.

The mechanism of changes in the "autofluorescence" spectra of endogenous fluorophores is explained by their quantitative and qualitative differences in tissues, a change in their redox balance and depth of location, different content in tissues that absorb but do not fluoresce chromophores, changes in the extracellular matrix structure and the number of epithelial cell layers. Emission spectra of individual fluorophores in tissues are modified by the phenomena of light scattering and absorption of blood, which absorbs light in the visible part of the spectrum, and local changes in environmental parameters (pH, redox potential).

Quantitative laser-induced fluorescence (QLF) is a new diagnostic technique for enamel caries evaluation and the monitoring of mineral changes in initial caries [3]. The level of the fluorescence intensity of enamel changed in vivo is lower compared to healthy enamel. The decrease in fluorescence effect is mainly due to the scattering of excitation and emission light on damaged surfaces of hydroxyapatite. The coefficient of light scattering on decalcified enamel is 5–10 times greater than that on normal enamel. Research on the use of LIF has been ongoing for 40 years and was started by Bjelkhagen and Sundström [4]. They induced the fluorescence of a tooth material with nitrogen (N: 337 nm) and argon (Ar: 488, 514 nm) lasers. The intensity of the fluorescence emission in the blue–green range decreased with the increased degree of carious enamel decalcification for the excitation wavelengths given above. The field of this research is still continuing and is the subject of many current scientific articles [5–11].

In our research, a laser with a wavelength of 405 nm was used to excite the fluorescence of the enamel, and these studies showed a high correlation of the results with the measurements of mineral losses ($R = 0.97$).

The classic application of the LIF method is the quantitative assessment of the concentrations of photosensitizers in the photodiagnostics and photodynamic therapy methods.

The PDT method relies on the selective photooxidation of biological tissues by reactive oxygen species (ROS). A combination of an external photosensitizer, endogenous oxygen and red light produces singlet oxygen and free radicals, leading to necrosis and apoptosis of diseased cells [12,13]. The method of photodiagnostics is based on the localization of selectively absorbed PS in tissues using fluorescence methods.

Photodynamic therapy used with 5-aminolevulinic acid (ALA) or its methyl ester (MAL) is accepted worldwide for the treatment of skin cancers, non-cancerous diseases and photodiagnosis [14,15] This method has been implemented in urology, gynecology, neurosurgery, pulmonology and gastroenterology [16–22].

When applied exogenously, ALA, or its derivatives, is selectively metabolized to protoporphyrin IX (PpIX), a compound that gives a strong fluorescence, which serves as a basis for diagnosis [23]. The accumulation of PpIX in tissues occurs by avoiding the feedback control in the pathway of hem biosynthesis. Topical photodynamic therapy with ALA (ALA–PDT) reached approval status for actinic keratosis (AK) in US and Canada, whereas MAL–PDT is approved worldwide for AK, Bowen's disease and Morbus Bowen in Europe and Australia.

Fluorescence methods are used to detect materials of biological origin. Our research indicates that the fluorescence of fungi also provides clinically useful information.

We presented methods for the quantification of sensitizer levels as well as examples of fluorescence imaging in in vivo studies of patients. The article presents the basics of the LIF method, the construction of apparatus, the characteristics of the organism's endogenous fluorophores, examples of our own research on the use of the LIF method in medicine and further development directions.

2. Materials and Methods

2.1. Materials

Fungi samples (*Candida albicans*—ATTC 18804, *Aspergillus flavus*—ATTC 16883, *Penicillium chrysogenum*—ATTC 9179) were prepared as suspensions in water at the Military Institute of Hygiene and Epidemiology in Warsaw [24].

Photosensitizers—5-aminolevulinic acid hydrochloride (ALA) and amino acid derivatives protoporphyrin IX (PPIX)—were synthesized and purified at the Institute of Optoelectronics of the Military University of Technology (IOE MUT, Warsaw, Poland). Final preparations with a concentration of 10% ALA were prepared in the form of creams with the LIPOBAZA base. PPIX derivatives were used in the form of injection solutions in doses of 2.5 mg/kg body.

In vitro studies in dentistry were carried out using human teeth removed for various dental indications. Teeth with a completely preserved crown, without clinical caries changes and with carious spots were qualified for the study. The level of fluorescence was tested with an LESA 6 laser analyzer. Microradiography method was used to measure the depth of the lesion and loss of minerals. The teeth were sectioned for transverse microradiography. Tooth slices (250-thick) were sawn perpendicular to the enamel surface and then were manually ground to 70–80 thickness. Mineral content depth was measured with a microscope densitometer.

2.2. Measuring Apparatus

The analyzer was LESA5 spectrometer (BioSpec, Moscow, Russia) (Figure 1) [9], which was installed on the computer card, laser, fiber optic sensor (catheter), optical filters. Depending on the application, the following lasers were used: He-Ne (λ = 632 nm, 25 mW, BioSpec), II harmonic Nd: YAG (λ = 532 nm, 10 mW, BioSpec), semiconductor lasers (λ = 375 nm, 15 mW and λ = 405 nm, 25 mW, Power Technology, Little Rock, AR, USA), He-Cd (λ= 442 nm, 100 mW, Omnichrom, Rochester, NY, USA).

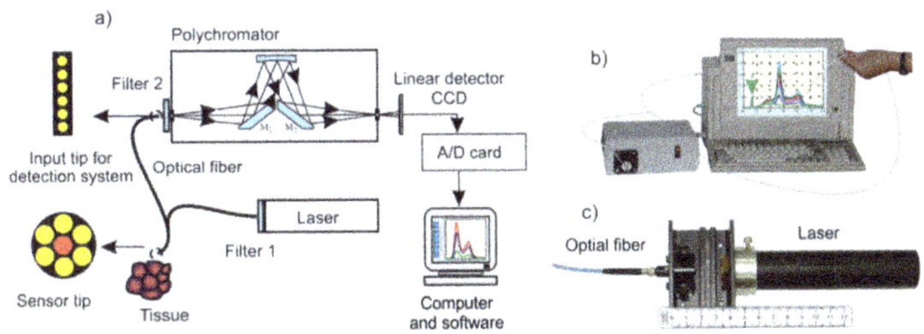

Figure 1. Fiberoptic fluorescence analyzer: (**a**) optical scheme, (**b**) view, (**c**) laser with input optics.

Fluorescence imaging system (Figure 2) consists of the following main components: CCD camera GP-KS162 (Panasonic, Osaka, Japan), xenon lamp with liquid fiber 300 W (Lasar, Warsaw), endoscope (Storc) and optical filters (IOE MUT).

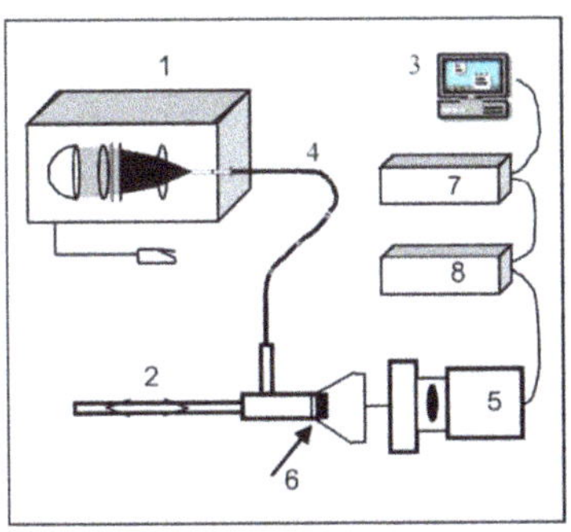

Figure 2. Optical diagram of the fluorescence imaging system: 1—light source with a violet filter (λ = 405 ± 25 nm), 2—endoscope, 3—monitor, 4—liquid optical fiber, 5—CCD camera, 6—optical filter, 7—computer, 8—video.

The LIF method was used for the diagnosis of pathological changes in in vivo conditions on patients in Polish clinics that had approvals of the relevant bioethical committee.

3. Results

3.1. Spectral Characteristics of Fluorophores

Figure 3 shows the collective absorption and emission characteristics of the most important fluorophores (fluorescent chromophores) found in biological systems [25]. The tryptophan bands (components of elastin, collagen), FAD and NADH coenzymes and endogenous porphyrins can be clearly distinguished from the field of fluorescence excitation.

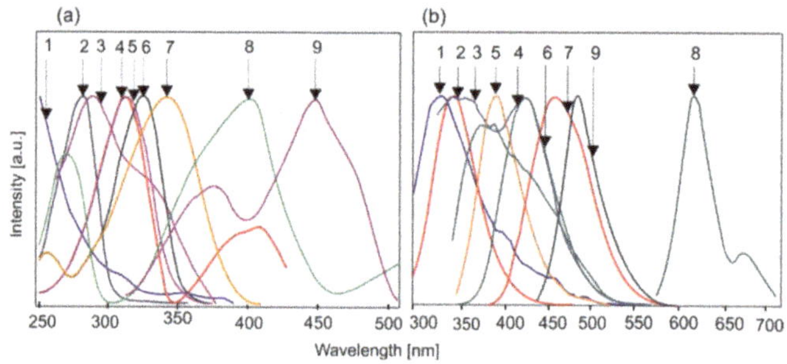

Figure 3. Spectral characteristics of potential endogenous fluorophores: 1—collagen, 2—tryptophan, 3—elastin, 4—pyridoxamine phosphate, 5—pyridoxine, 6—pyridoxal phosphate, 7—NADH, 8—protoporphyrin IX, 9—FAD. (**a**)-absorpton spectra, (**b**) emission spectra.

The sources of endogenous fluorescence in cells and biological tissues are aromatic amino acids, which are used to build proteins and coenzymes. Among the 20 amino acids from which proteins are built, only tryptophan (TRP), tyrosine (TYR) and phenylalanine (PHE) have fluorescence in the UV region 1.

The main component of bone, hydroxyapatite, has strong fluorescence properties in hard tissues. In many disease cases, increased levels of metalloporphyrins are observed [26].

The coenzymes FAD and FMN and vitamin B2 absorb light with a wavelength of around 450 nm and emission at a wavelength of around 530 nm. Unlike NADH, only the oxidized form of FAD shows fluorescence.

The phosphoryl derivative of vitamin B6 is another fluorescent coenzyme [27]. Vitamin B6 occurs in three forms with the same biological activity as pyridoxine, pyridoxal and pyridoxamine. Biologically active forms are phosphate derivatives of pyridoxamine and pyridoxal, which interact with enzymes active mainly in the transformation of amino acids (including racemization of optically active amino acids, transamination, decarboxylation, tryptophan synthesis).

An important group of fluorophores are pteridine derivatives, heterocyclic compounds containing several substituents in the basic pterin structure [28]. The pterin is composed of conjugated pyrazine and pyrimidine rings that contain carbonyl oxygen and an amino group. The pteridine system is widespread in nature because its derivatives are the basis for the coloration of the wings and eyes of insects, as well as the skin of amphibians and fish [25]. Folic acid, necessary to produce red blood cells above the bone marrow, is made up of the pteroyl group, p-aminobenzoic acid and glutamic acid. The pteridine system is found in bacteria and fungi.

3.2. Application of LIF in Dental Diagnostics

Typical single fluorescence characteristics consist of an excitation (equivalent to the absorption spectrum) and emission spectra. By changing the wavelength of the excitation radiation in the entire absorption range, an emission–excitation (EM–EX) matrix is obtained. It is the real spectral imprint of the tested sample. The method is of particular interest for the analysis of substances containing various fluorophores. In addition, it enables the selection of appropriate wavelengths for testing. Figure 4 shows the enamel and dentin EM–EX matrices.

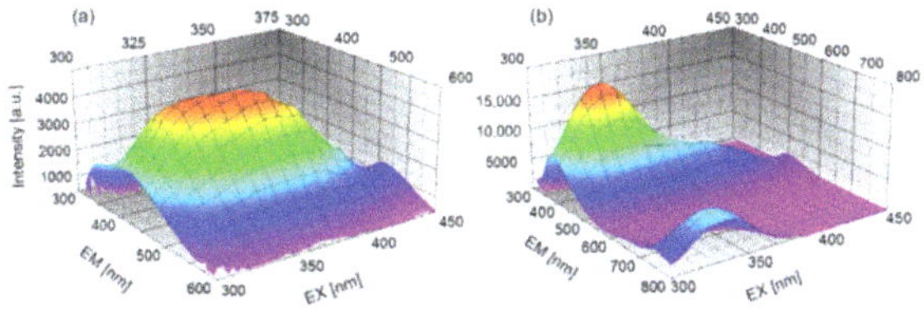

Figure 4. Emission–excitation characteristics (EM–EX) of (**a**) enamel and (**b**) dentin.

The strongest fluorescence of enamel is obtained after excitation with radiation in the range close to UVB and violet.

The LIF spectra with selected excitation wavelengths are shown in Figure 5. The level of fluorescence of the enamel with caries is lower compared to the unchanged enamel. In the case of dental caries, an increased level of fluorescence is observed, which is associated with porphyrin derivatives generated by bacteria.

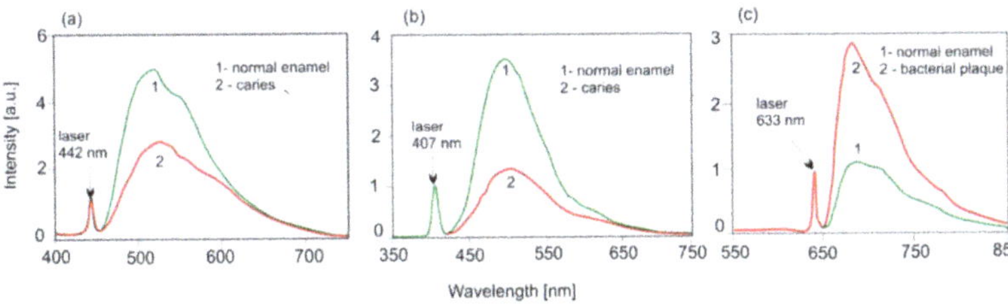

Figure 5. Influence of the excitation wavelength with laser radiation on changes in the level of enamel fluorescence: (**a**) laser 442 nm, (**b**) laser 407 nm, (**c**) spectra of bacterial plaque with excitation 633 nm.

Figure 6 shows the relationship between the decrease in fluorescence intensity and other parameters characterizing the degree of caries: the depth of changes and the degree of mineral loss. Measurements of these parameters were carried out using the microradiography method, and the average decrease in the fluorescence of the demineralized area was determined using a LIF spectrometer with radiation excitation and a wavelength of 407 nm.

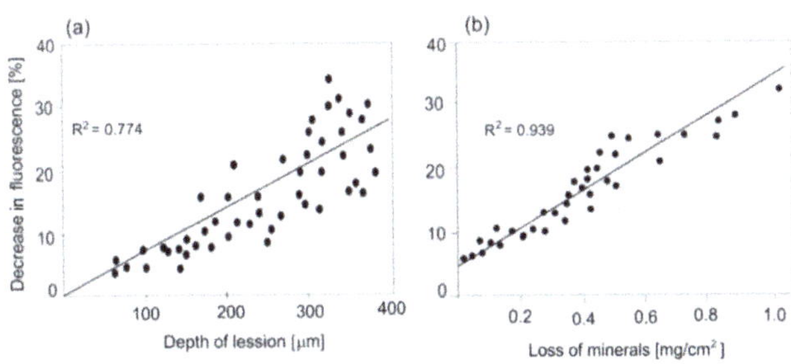

Figure 6. Influence of the (**a**) depth of the lesion and (**b**) loss of minerals on the decrease in fluorescence intensity.

3.3. Clinical LIF Applications Using Endogenous Fluorophores

The LIF method based on the study of endogenous fluorophores is of greatest importance in pulmonology and dermatology. Neoplasms are characterized by lower fluorescence in the green range (about 530 nm) and a higher ratio of fluorescence in the red and green bands compared to healthy tissues. Lowering the level of autofluorescence in neoplastic tissues in the area of FAD emission is related, among other factors, to a greater metabolism of these tissues (an increase in the level of NADH and a decrease in the amount of the oxidized form of FAD). Figure 7 shows the decrease in tissue fluorescence in the case of pleural mesothelioma. LIF studies were conducted on 23 cases of lesions. The data were analyzed by performing ANOVA test comparisons between normal and tumor tissues (significance level $\alpha = 0.05$.). There was statistically significant difference ($p < 0.01$) between these groups of tissues.

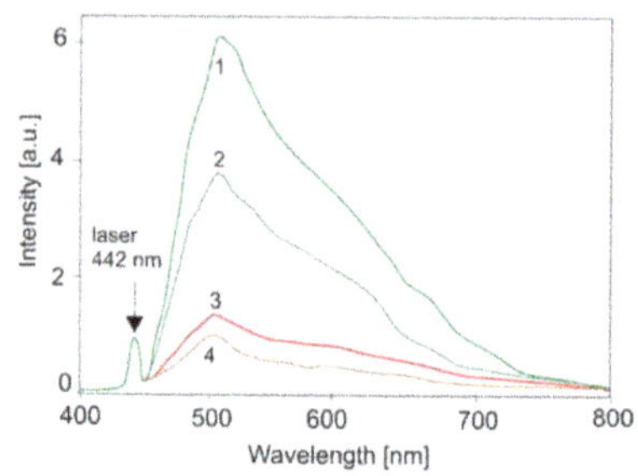

Figure 7. Decrease in autofluorescence in the mesothelium (1,2—normal tissue, 3,4—tumor).

An interesting problem is the presence of increased levels of porphyrins in many diseases. This is evident in the case of porphyria. Increased accumulation of porphyrins in neoplastic tissues has been observed many times by the authors of this work in many skin diseases (senilis keratosis) or in advanced cervical neoplasms. The causes of the fluorescence tissue of the squamous cell carcinoma of the oral cavity are metalloporphyrins contained in bacteria (*Pseudomonas bacteria*).

Examples of the presence of elevated porphyrin levels are shown in Figure 8.

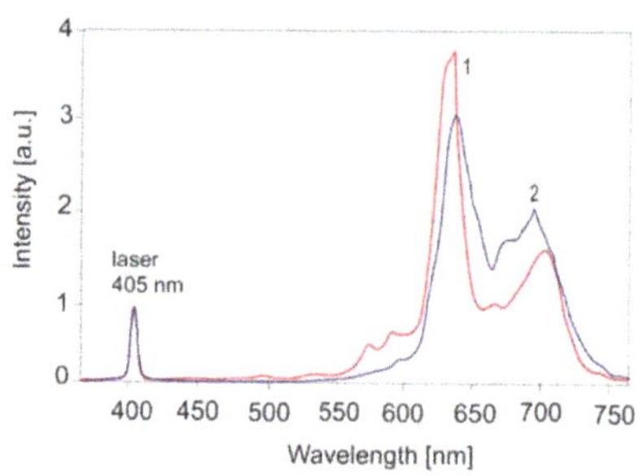

Figure 8. Fluorescence of metalloporphyrins: enamel plaque (1) and skin in senilis keratosis (2).

The conventional diagnosis of oral candidiasis is generally based on biopsy tissue; however, this technique is time-consuming. *Candida* is a pathogenic organism that may cause oral candidiasis upon disruption of the balance of flora. The disease is most commonly caused by an overgrowth of *Candida albicans* in the mouth [29]. Figure 9 shows the fluorescence characteristics of selected fungi.

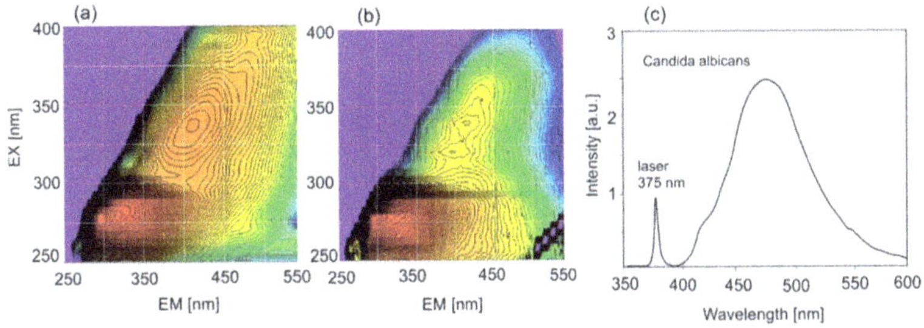

Figure 9. Spectral characteristics of selected fungi: (**a**) EM–EX map of *Penicyllium chrysogenium*, (**b**) EM–EX map of *Aspergillus flavus*, (**c**) LIF spectrum of *Candida albicans*.

3.4. Measurements of Photosensitizers in the PDD/PDT Method

The classic and most important application of the real-time fluorescence method is the analysis of photosensitizers in the photodiagnostics and photodynamic therapy methods. These studies include (i) localization and determination of the level of PS concentrations, (ii) kinetics of their accumulation and excretion over time, (iii) determination of the optimal time of therapeutic irradiation from the moment of introducing compounds into the body, (iv) photochemical distribution of sensitizers and (v) selection of therapeutic irradiation parameters.

An example of the kinetics of PPIX accumulation after ALA application in the case of a change in actinic keratosis is shown in Figure 10. These studies are necessary in respect of introducing new methods to the market to form ALA [30].

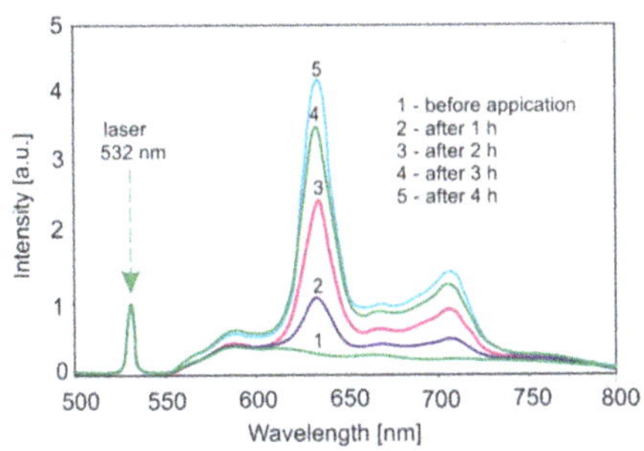

Figure 10. Kinetics of PPIX accumulation in alteration of skin actinic keratosis.

A sufficient level of PPIX for further therapeutic irradiation is obtained at least 2 h after the application of ALA. The topical introduction of an allergic to superficial dermatological changes is the easiest way. In the case of tumors of internal organs or lesions of greater thickness, it is necessary to inject photosensitizers. In cases where it is necessary to analyze changes in tissues of greater thickness, the use of a red laser is a better choice for fluorescence excitation due to greater light penetration.

Figure 11 shows an example of the use of the He-Ne laser for the photodiagnostics of cancer (Merkel carcinoma) 48 h after injecting the amino acid PPIX derivatives (2 mg/kg body mass) into the blood.

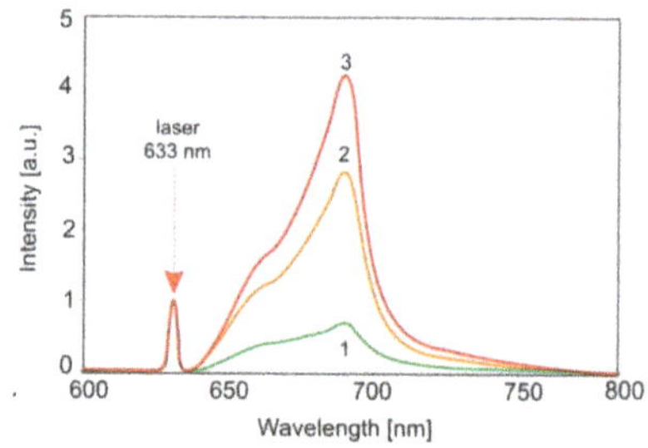

Figure 11. Fluorescence spectra of Merkel tumor with introduced $PP(Ala)_2(Arg)_2$: 1—healthy tissue, 2—tumor on the periphery, 3—tumor in the center of lesions.

The PPD/PDT method has found application in gynecology. Figure 12 shows examples of the use of the LIF method in the treatment of vaginal and cervical lesions.

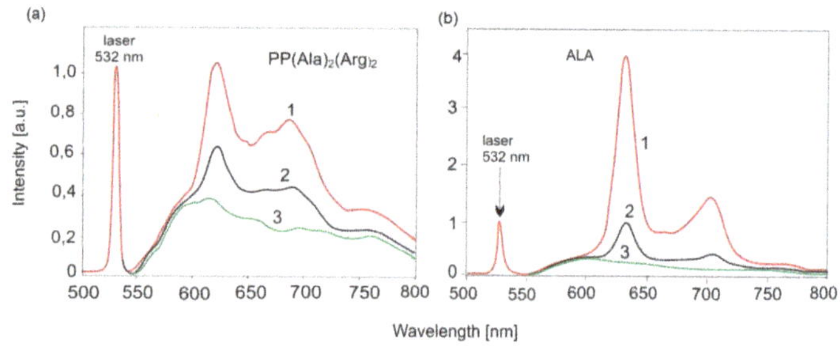

Figure 12. Comparison of accumulated porphyrin concentrations in (**a**) cervical cancer PP $(Ala)_2(Arg)_2$ and (**b**) vaginal (ALA): 1—tumor in the center of lesions, 2—tumor on the border, 3—normal tissue.

The fluorescence images of these changes are shown in Figure 13.

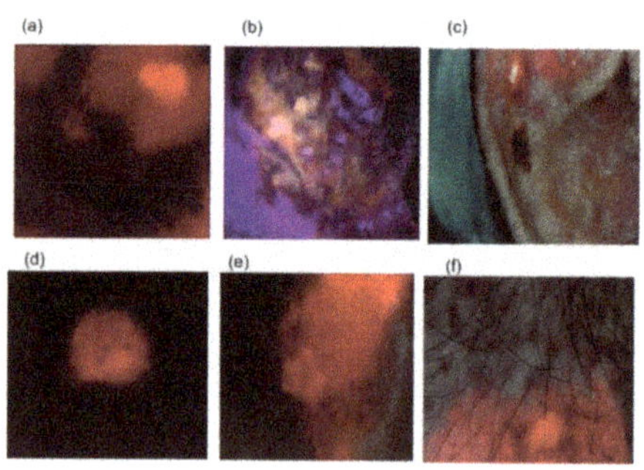

Figure 13. Fluorescent images of (**a**) vaginal, (**b**) cervical, (**c**) basal cell carcinoma of head, (**d**) squamous cell carcinoma of nose, (**e**) actinic keratosis of skin, (**f**) after ALA applications.

Apart from the research on the kinetics of photosensitizers' accumulation in tissues, the LIF method is helpful in determining the light power density in the PDT method. During irradiation, the photochemical decomposition of porphyrins takes place, and this process depends on the intensity of the light. The photobleaching effect of the sensitizer as a function of irradiation is shown in Figure 14. When using a power density of 100 mW/cm^2, the degradation of the sensitizer occurs much faster than at 40 mW/cm^2, the therapeutic effect is insufficient and the treatment procedure must be repeated.

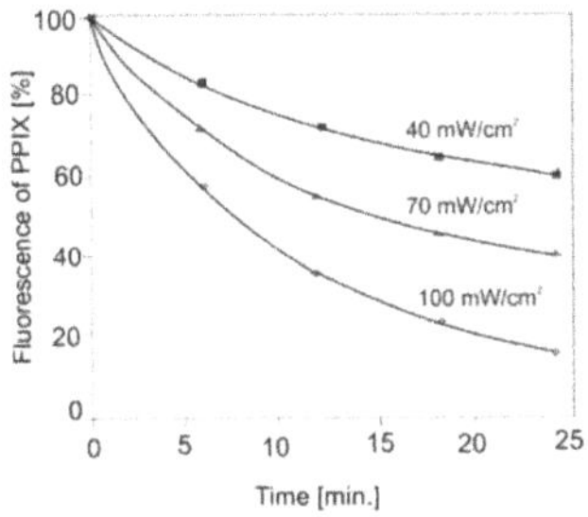

Figure 14. Photobleaching effect during the irradiation of skin actinic keratosis with PPIX.

4. Discussion

Endogenous fluorophores that occur in the body and are the basis of autofluorescence can be divided into three groups based on the spectral range. Absorption in the UVB range (280–325 nm) is demonstrated by amino acids and proteins. Fluorometric analyses of these substances play an important role in biochemistry. In the LIF method, lasers such as He-Cd (325 nm), Nd YAG (266 nm) or tunable OPO or titanium [31] can be used to excite the fluorescence. The systems built are used to detect biological agents in the air, which are mainly used in military technology. The UVA (325–380 nm) and blue light ranges include fluorophores that contribute to the metabolism of the organism (NADH, FAD), pterins and porphyrins. For example, the ratio of NADH to FAD fluorescence is an indicator of the metabolic rate. For medical applications, the visible range is the most important. Violet or blue light excitation on tumor tissues in comparison to healthy ones shows a decreased level of fluorescence (Figure 7).

Hydroxyapatite, the tooth component, fluoresces within a wide spectral range, from 350 to 450 nm. Quantitative QLF methods have already found practical application, and imaging systems are already being built (e.g., Inspektor Research system, Bussum, The Netherlands) [32]. LIF spectrometers allow for more accurate analysis, are many times cheaper and allow every part of the mouth to be reached with optical fibers.

Caries is a complex pathological process, which entail the gradual loss of minerals from the hard tissues of the tooth. Under the conditions of ionic equilibrium, normal enamel undergoes continuous de- and remineralization processes, which do not cause changes in the enamel structure. If the pH drop in the oral cavity is not balanced, it causes disturbances in the biochemical balance and initiation of the destructive process under the influence of acid metabolites. The clinical symptom of early carious lesions is the appearance of whitish, opalescent spots on the enamel surface. However, the diagnostic effectiveness of most of the clinical methods used so far is unsatisfactory. Modern methods of radiological diagnostics currently available in clinical practice do not allow for the detection of changes related to the very early phase of enamel demineralization. The methods with high expectations include fluorescence induced by lasers.

The level of autofluorescence of healthy enamel in comparison to the carious enamel in vivo is higher when excited with lasers with wavelengths of 405 and 442 nm (Figure 5) In imaging systems, the decrease in the fluorescence of the carious area is visible as a dark contrast against the bright background of healthy enamel, which greatly facilitates the diagnosis. One of the important achievements is showing the influence of the depth of the lesion and loss of minerals in enamel on the decrease in fluorescence intensity (Figure 6). A measurable decrease in the fluorescence intensity is already visible at a 5% loss of enamel mineral. In such cases, it is possible to effectively remineralize the enamel with appropriate dental pastes, without the need for drilling. Very early caries diagnosis is the main advantage of the LIF method.

LESA is a PC-based spectroscopy system consisting of a laser source for fluorescence excitation, a miniature monochromator, multichannel CCD detector, an optic fiber sensor and a computer for data acquisition and processing (Figure 1). The entire fluorescence spectra range is recorded simultaneously. The intensity of fluorescence depends on the irradiance, the distance between the sensor and the light, and the position of the sensor. Thus, it is important to normalize the signals. The monochromator receives fluorescence emission and is scattered on the tissue by laser radiation, which is attenuated 10^3 times by an appropriate optical filter. The monochromator receives fluorescence radiation and laser radiation scattered on the tissue, which is attenuated 10^3 times with an appropriate optical filter. The obtained spectra are normalized to the laser signal. It is a convenient reference signal for the fluorescence measurements. The laser emission to fluorescence area ratio depends only on the concentration of the fluorophore (e.g., Figures 8 and 9). Our goal was to develop a technique for quantifying fluorophores in all possible medical cases—for caries testing, PDT sensitizers and cancer by autofluorescence (endogenous sensitizers). The only such commercially available system is the LESA spectrometer, equipped only with an He-Ne laser (633 nm). Therefore, the area of an application was limited only to selected sensitizers. We have modified the system by using many lasers in the UV–VIS range, which allows us to excite the fluorescence of virtually all chemical compounds, including those of biological origin.

The system locally determines in vivo the level of photosensitizer accumulation in any patient's organs and tissues accessible for a fiber optic probe. The system is used during photodynamic therapy of intracavity, interstitial and superficial tumors, and for measurements of biological tissues' autofluorescence.

The LIF method is indispensable for the analysis of photosensitizers that are constantly growing in the market. These include porphyrin derivatives, phthalocyanines, bacteriochlorines. They differ significantly in parameters—dose, time of irradiation commencement and level of accumulation. A properly conducted PDT method requires the control of parameters and conditions.

The LIF method has great potential in pulmonology. Preneoplastic changes (dysplasias) and early neoplastic stages (intraepithelial carcinoma—CIS, microinvasion) are difficult to detect using traditional bronchoscopic methods, as the lesions cover an area up to several millimeters in diameter and several cell layers (0.2–1 mm thick) [33].

In gynecology, photo diagnosis helps to precisely determine the location of precancerous lesions and malignancies of the vulva (e.g., vulvar lichen sclerosis), vagina and cervix. PDD enables the detection of hyperplastic at its early stages (Figure 12).

The autofluorescence method has a good chance of being successful in the diagnosis of various infections, skin pigmentation changes and metabolic disorders. Thus far, the widely used Wood's lamp for observing changes in skin fluorescence is an important tool in dermatology. Currently, changes in fluorescence are determined only visually, which, combined with too low power density of the mercury lamps used, is a big limitation of the method. Some fungal infections caused by pathogenic fungi can be precisely diagnosed by fluorescence methods. The fluorescence spectra depend on the type of disease. The current level of diagnostics allows us to only link the characteristic color of luminescence with the type of infection.

In the case of an infection of the skin with the *Malassezia furfur* fungus, which causes tinea versicolor, the luminescence has a copper-orange color; the light of coral-red emission is characteristic of Erythrasma.

Real-time autofluorescence testing methods cover an increasing range of medical applications. Different fluorescence imaging systems in bronchoscopy (e.g., LIFE, Vancouver, Canada) [34] and dermatology have already been built. A good example is the use of a VELscope (Vancouver, Canada) [35] lamp to evaluate the pathological changes in the mucous membrane. Observation of the changes in the metabolism of the surface layers of the tissues lining the mouth is important because they come into direct contact with many carcinogens and are the starting point of oral cancer. The most common disorders in the oral cavity include leukoplakia, erythroplakia, lichen planus and submucosal fibrosis. The risk of neoplastic metaplasia for these lesions varies, but early detection and prompt treatment can prevent cancer development.

Author Contributions: Conceptualization, M.K.; methodology, M.K. and A.B.; investigation, M.K. and A.B.; resources, M.K.; data curation, M.K.; writing—original draft preparation, M.K.; writing—review and editing, A.B. and M.K.; visualization, M.K.; All authors have read and agreed to the published version of the manuscript.

Funding: This research received no external funding.

Institutional Review Board Statement: The fluorescence studies of the teeth were performed after tooth extraction. No in vitro consent was required. The STORZ fluorescence system is approved for clinical trials and no separate approval for use is required. The ALA-PDT study was conducted according to the guidelines of the Declaration of Helsinki, and approved by by the local ethic committee (KE-0254/286/2019 [30]. This method is approved in UE.

Informed Consent Statement: Informed consent was obtained from all subjects involved in the study.

Conflicts of Interest: The authors declare no conflict of interest.

References

1. Andersson-Engels, S.; Johansson, S.J.; Svanberg, K.; Svanberg, S. Fluorescence imaging and point measurements of tissue: Applications to the demarcation of malignant tumors and atherosclerotic lesion from normal tissue. *Photochem. Photobiol.* **1991**, *53*, 807–814. [CrossRef] [PubMed]
2. Katz, A.; Glassman, S.; Yang, Y.; Tang, G.C.; Budansky, Y.; Celmer, E.; Savage, H.E.; Schantz, S.P.; Alfano, R.R. Optical biopsy fibre-based fluorescence spectroscopy instrumentation. *Proc. SPIE* **1996**, *2679*, 118–123.
3. De Jong, E.D.J.; Sundström, F.; Westerling, H.; Tranaeus, S.; Ten Bosch, J.J.; Angmar-Månsson, B. A new method for in vivo quantification of changes in initial enamel caries with laser fluorescence. *Caries Res.* **1995**, *29*, 2–7. [CrossRef] [PubMed]
4. Bjelkhagen, H.; Sundström, F.; Angmar-Mansson, B.; Ryden, H. Early detection of enamel caries by the luminescence excited by visible laser light. *Swed. Dent. J.* **1982**, *6*, 1–7. [PubMed]
5. Shakibaie, F.; George, R.; Walsh, L.J. Applications of Laser induced Fluorescence in Dentistry. *Int. J. Dent. Clin.* **2011**, *3*, 38–44.
6. El-Sharkawy, Y.H.; Elbasuney, S. Tunable laser-induced fluorescence with signal correlation algorithm for dental caries detection with controlled ablation. *Opt. Laser Technol.* **2020**, *129*, 106299. [CrossRef]
7. Goloshchapov, D.L.; Kashkarov, V.M.; Ippolitov, Y.A. Early screening of dentin caries using the methods of Micro-Raman and laser-induced fluorescence spectroscopy. *Results Phys.* **2018**, *10*, 346–347. [CrossRef]
8. El-Sharkawy, Y.H.; Elbasuney, S. Design and implementation of novel hyperspectral imaging for dental carious early detection using laser induced fluorescence. *Photodiagnosis Photodyn. Ther.* **2018**, *24*, 166–178. [CrossRef]
9. Ko, C.C.; Yi, D.H.; Lee, D.J.; Kwon, J.; Garcia-Godoy, F.; Kwon, Y.H. Diagnosis and staging of caries using spectral factors derived from the blue laser-induced autofluorescence spectrum. *J. Dent.* **2017**, *67*, 77–83. [CrossRef]
10. Jeng, M.J.; Sharma, M.; Chao, T.Y.; Li, Y.C.; Huang, S.F.; Chang, L.B.; Chow, L. Multiclass classification of autofluorescence images of oral cavity lesions based on quantitative analysis. *PLoS ONE* **2020**, *15*, e0228132. [CrossRef]
11. Lalla, Y.; Phil, M.; Matias, M.A.; Farah, C.S. Assessment of oral mucosal lesions with autofluorescence imaging and reflectance spectroscopy. *J. Am. Dent. Assoc.* **2016**, *147*, 650–660. [CrossRef] [PubMed]
12. Correia, J.H.; Rodrigues, J.A.; Pimenta, S.; Dong, T.; Yang, Z. Photodynamic Therapy review: Principles, Photosensitizers, Applications, and Future Directions. *Pharmaceutics* **2021**, *13*, 1332. [CrossRef] [PubMed]
13. Abdel-Kader, M.H. *Photodynamic Therapy. From Theory to Application*; Springer: Berlin/Heidelberg, Germany, 2014.
14. Morton, C.A.; Szeimies, R.M.; Basset-Seguin, N.; Calzavara-Pinton, P.; Gilaberte, Y.; Hædersdal, M.; Hofbauer, G.F.L.; Hunger, R.E.; Karrer, S.; Piaserico, S.; et al. European Dermatology Forum guidelines on topical photodynamic therapy 2019 Part 1: Treatment delivery and established indications-Actinic keratoses, Bowen's disease and basal cell carcinomas. *J. Eur. Acad. Dermatol. Venereol.* **2019**, *33*, 2225–2238. [CrossRef]
15. Morton, C.A.; Braathen, L.R. Daylight photodynamic therapy for actinic keratoses. *Am. J. Clin. Dermatol.* **2018**, *19*, 647–656. [CrossRef] [PubMed]
16. Yan, Y.-J.; Huang, T.-W.; Cheng, N.-L.; Hsieh, Y.-F.; Tsai, M.-H.; Chiou, J.-C. Portable LED-ind uced autofluorescence spectroscopy for oral cancer diagnosis. *J. Biomed. Opt.* **2017**, *22*, 045007. [CrossRef]
17. Francisco, A.L.; Correr, W.R.; Pinto, C.A.; Gonçalves Filho, J.; Chulam, T.C.; Kurachi, C.; Kowalski, L.P. Analysis of surgical margins in oral cancer using in situ fluorescence spectroscopy. *Oral Oncol.* **2014**, *50*, 593–599. [CrossRef]
18. Balasubramaniam, A.M.; Sriraman, R.; Sindhuja, P.; Mohideen, K.; Haris, K.T.M. Autofluorescence based diagnostic techniques for oral cancer. *J. Pharm. Bioallied Sci.* **2015**, *7*, 374–377. [CrossRef]
19. Huang, T.T.; Huang, J.S.; Wang, Y.Y.; Chen, K.C.; Wong, T.Y.; Chen, Y.C.; Wu, C.W.; Chan, L.P.; Lin, Y.C.; Kao, Y.H.; et al. Novel quantitative analysis of autofluorescence images for oral cancer screening. *Oral Oncol.* **2017**, *68*, 20–26. [CrossRef]
20. Ghasemi, F.; Parvin, P.; Lotfi, M. Laser-induced fluorescence spectroscopy for diagnosis of cancerous tissue based on the fluorescence properties of formaldehyde. *Laser Phys. Lett.* **2019**, *16*, 035601. [CrossRef]
21. Andreeva, V.; Aksamentova, E.; Muhachev, A.; Solovey, A.; Litvinov, I.; Gusarov, A.; Shevtsova, N.N.; Kushkin, D.; Litvinova, K. Preoperative AI-Driven Fluorescence Diagnosis of Non-Melanoma Skin Cancer. *Diagnostics* **2022**, *12*, 72. [CrossRef]

22. Dos Santos, A.F.; De Almeida, D.R.Q.; Terra, L.F.; Baptista, M.S.; Labrio, L. Photodynamic therapy in cancer treatment—An update review. *J. Cancer Metastasis Treat.* **2019**, *5*, 25–29. [CrossRef]
23. Loschenov, V.B.; Konov, V.I.; Prokhorov, A.M. Photodynamic Therapy and Fluorescence Diagnostics. *Laser Phys.* **2000**, *10*, 1188–1207.
24. Bombalska, A.; Mularczyk-Oliwa, M.; Kwaśny, M.; Włodarski, M.; Kaliszewski, M.; Kopczyński, K.; Szpakowska, M.; Trafny, E.A. Classification of the biological material with use of FT-IR spectroscopy and statistical analysis. *Spectrochim. Acta A* **2011**, *78*, 1221–1226. [CrossRef] [PubMed]
25. Lakowicz, J. *Principles of Fluorescent Spectroscopy*; Springer: Boston, MA, USA, 2006.
26. Konig, K.; Teschke, M.; Eick, S.; Pfister, W.; Meyer, H.; Halbhuber, K. Photodynamically inactivation of Propionibacterium acnes. *Proc. SPIE* **1994**, *3274*, 106–110.
27. Kwon, O.S.; Blazques, M.; Churchich, J.E. Luminescencje spectroscopy of pyridoxic acid and pyridoxic amid bund to protein. *Eur. J. Biochem.* **1994**, *219*, 807–812. [CrossRef]
28. Blakely, R.L. *The Chemistry of Folic Acid and Related Pteridines*; Wiley Intersciencies: New York, NY, USA, 1969.
29. Yao, Y.; Shi, L.; Zhang, Z.; Sun, H.; Wu, L. Application of fungal fluorescent staining oral candidiasis: Diagnostic analysis of 228 specimens. *BMC Microbiol.* **2019**, *19*, 96. [CrossRef]
30. Bartosińska, J.; Szczepanik-Kułak, P.; Raczkiewicz, D.; Niwiedzioł, M.; Gerkowicz, A.; Kowalczuk, D.; Kwaśny, M.; Krasowska, D. Topical photodynamic therapy with different forms of 5-aminolevulinic acid in the treatment of actinic keratosis. *Pharmaceutics* **2022**, *14*, 346. [CrossRef]
31. Kwaśny, M.; Mierczyk, Z.; Fiedor, P.; Rowiński, W.; Socha, K.; Graczyk, A.; Domaniecki, J.; Mazurek, A.P. Protoporphyrin Derivatives—The Use in Localization of Neoplasmas by Titanium Laser-induced Fluorescence Technique. *Acta Pol. Pharm.* **1997**, *54*, 123–128.
32. Walsh, L.J. Caries Diagnosis Aided by Fluorescence. In *Dental Caries-Diagnosis, Prevention and Management*; Arkanslan, Z., Ed.; IntechOpen: Rijeka, Croatia, 2018; pp. 97–115.
33. Hayata, Y.; Kato, H.; Konaka, C.; Okunaka, T. Photodynamic therapy in early stage cancer. *Lung Cancer* **1993**, *9*, 287–294. [CrossRef]
34. Lam, S.; McAulay, C.; Hung, J.; LeRiche, J.; Profio, A.E.; Palcic, B. Detection of dysplasja and carcinoma in situ by a lung imaging fluorescence endoscope (LIFE) device. *Laser Surg. Med.* **1992**, *4*, 40.
35. Hanken, H.; Kraatz, J.; Smeets, R.; Heiland, M.; Blessmann, M.; Eichhorn, W.; Clauditz, T.S.; Gröbe, A.; Kolk, A.; Rana, M. The detection of oral pre malignant lesions with an autofluorescence based imaging system (VELscope TM)—A single blinded clinical evaluation. *Head Face Med.* **2013**, *9*, 23. [CrossRef] [PubMed]

sensors

MDPI

Article

A Novel Approach for Segment-Length Selection Based on Stationarity to Perform Effective Connectivity Analysis Applied to Resting-State EEG Signals

Leonardo Góngora [1], Alessia Paglialonga [2], Alfonso Mastropietro [3], Giovanna Rizzo [3] and Riccardo Barbieri [1,*]

[1] Department of Electronics, Informatics and Bioengineering, Politecnico di Milano, 20133 Milan, Italy; leonardoandres.gongora@mail.polimi.it
[2] Istituto di Elettronica e di Ingegneria dell'Informazione e delle Telecomunicazioni (IEIIT), Consiglio Nazionale delle Ricerche (CNR), 20133 Milan, Italy; alessia.paglialonga@ieiit.cnr.it
[3] Istituto di Tecnologie Biomediche (ITB), Consiglio Nazionale delle Ricerche (CNR), 20054 Segrate, Italy; alfonso.mastropietro@itb.cnr.it (A.M.); giovanna.rizzo@itb.cnr.it (G.R.)
* Correspondence: riccardo.barbieri@polimi.it

Abstract: Connectivity among different areas within the brain is a topic that has been notably studied in the last decade. In particular, EEG-derived measures of effective connectivity examine the directionalities and the exerted influences raised from the interactions among neural sources that are masked out on EEG signals. This is usually performed by fitting multivariate autoregressive models that rely on the stationarity that is assumed to be maintained over shorter bits of the signals. However, despite being a central condition, the selection process of a segment length that guarantees stationary conditions has not been systematically addressed within the effective connectivity framework, and thus, plenty of works consider different window sizes and provide a diversity of connectivity results. In this study, a segment-size-selection procedure based on fourth-order statistics is proposed to make an informed decision on the appropriate window size that guarantees stationarity both in temporal and spatial terms. Specifically, kurtosis is estimated as a function of the window size and used to measure stationarity. A search algorithm is implemented to find the segments with similar stationary properties while maximizing the number of channels that exhibit the same properties and grouping them accordingly. This approach is tested on EEG signals recorded from six healthy subjects during resting-state conditions, and the results obtained from the proposed method are compared to those obtained using the classical approach for mapping effective connectivity. The results show that the proposed method highlights the influence that arises in the Default Mode Network circuit by selecting a window of 4 s, which provides, overall, the most uniform stationary properties across channels.

Keywords: EEG; effective connectivity; kurtosis; resting-state connectivity; stationarity

Citation: Góngora, L.; Paglialonga, A.; Mastropietro, A.; Rizzo, G.; Barbieri, R. A Novel Approach for Segment-Length Selection Based on Stationarity to Perform Effective Connectivity Analysis Applied to Resting-State EEG Signals. *Sensors* **2022**, *22*, 4747. https://doi.org/10.3390/s22134747

Academic Editors: Andrea Facchinetti and Yvonne Tran

Received: 19 April 2022
Accepted: 20 June 2022
Published: 23 June 2022

Publisher's Note: MDPI stays neutral with regard to jurisdictional claims in published maps and institutional affiliations.

1. Introduction

The analysis of the interactions encompassed by different neural sources in the brain, known as connectivity analysis, has become a topic of great relevance in neuroscience. Specifically, the structural, functional, and causal relationships that take place in the brain during neural activity are considered the building blocks to explain how the brain transmits and retrieves neural information [1–3]. This plays a major role in understanding neurological disorders, providing an overview of the differences that characterize a pathological condition in comparison to a healthy state [4–7].

Like many other topics in neuroscience, connectivity analysis has progressed significantly thanks to the advancements in neuroimaging. Brain imaging techniques allow for expressing neural activity in several ways: considering the temporal variation of bioelectric and magnetic potentials and tracking down the flow, or the light absorbance of the blood circulating in the brain [8]. By measuring such quantities, non-invasive data-acquisition methods such as electroencephalography (EEG), magnetoencephalography (MEG) and functional magnetic resonance imaging (fMRI) provide a way to observe the dynamic behavior of the neural activity expressed as multivariate time series from which connectivity among neural sources can be estimated [9].

Setting aside the anatomical connectivity, which looks for structural connections that physically link groups of neurons [10], in some cases requiring invasive techniques to do so [11], we are left with functional (FC) and effective connectivity (EC) as forms of characterizing neural processes from non-invasive measures. The former explores how spatially remote neural populations are being functionally integrated during a brain process, whereas the latter examines the causal relationships and the directed influences exerted among neural sources over the same kind of process. In this context, estimating effective and functional relationships is highly dependent on the temporal reference from which samples are acquired, hence, neuroimaging approaches such as EEG, MEG, and fMRI (to some extent) are appropriate for such a task.

In this way, to estimate the causal influences in multivariate time series (i.e., EC), EEG and MEG provide well-suited data to develop generative models from which inferences of the coupling of different brain regions are made. In summary, functional connectivity shows the distribution of the brain activity assessed by statistically significant values while EC analysis explains the complex elements of the information processing occurring in the brain, which lead to an understanding of how the brain works [3].

Accordingly, the EC studies comprise a wide range of applications such as theoretical constructions for high-resolution EEG recordings [2], the comparison of different effective connectivity measures according to the neural information to be analyzed [12], or the definition of graphical processing approaches for the coupled systems that can be obtained from multivariate time series [13]. Moreover, a range of disorders has been addressed: for example, studies have focused on resting states in long-standing vegetative-state patients [14], on causal relationships among specific areas in the brain associated with the alpha and beta bands during migraine episodes [15], and on patients with treatment-resistant schizophrenia [5], epilepsy episodes in children [16], autism [7] and drug abuse [17]. Other examples of EC analysis include task classification from features extracted from connectivity relationships and applied to the identification of individuals [18], motor imagery prediction [19], object recognition from visual stimulation [20], and the analysis of the brain response to emotional music [21].

EC analysis applied to resting-state conditions has been also analyzed. The work described in [22] provided a detailed investigation of the connectivity exhibited among EEG sources treated in the channel space, where high-density EEG recordings were analyzed in terms of different EC metrics such as Direct Transfer Function (DTF), Transfer Entropy (TE) and Phase Locking Value (PLV). Here, Olejarczyk and colleagues established the significance of the connections by considering weighted adjacency matrices estimated every 20 s to analyze common brain rhythms comprising the alpha, beta, gamma, delta, and theta frequency bands. From the physiological point of view, the authors found that the information flows from the posterior area of the brain towards the frontal area, exhibiting a marked correlation between the central–posterior to the central–frontal region, suggesting the activation of areas involved in the so-called Default Mode Network (DMN) mostly present over the alpha and beta frequency bands.

A similar work that applied graph theoretical analysis to the connectivity of resting-state conditions with open and closed eyes is presented in [23]. Here, the researchers analyzed the alpha, beta, and theta bands by employing the Synchronization Likelihood (SL) to characterize the connectivity using different topological parameters of the network. They analyzed the network according to the SL to understand where the main nodes are located and how they interact by considering the evolution of the resting-state condition every 10 s. They found that by opening the eyes, the connections in the frontal area for the theta band were decreased, similarly to what was observed for the posterior area connecting in a bilateral way to the surrounding zones for the alpha band. This is different from what was found in [22], where a noticeable cluster of sources offered significant connections over these areas in the posterior region for both open-eyes and closed-eyes conditions.

On the other hand, Chen et al. [24] found direct links in the frontoparietal connections characterizing the resting-state conditions with open and closed eyes. While there was observed suppression of the activity over the alpha band with open eyes, they noticed that the connectivity was strengthened in a significant way between the posterior regions in the left hemisphere in comparison to the right one when the signals were analyzed considering segments of 4 s.

Finally, as an example of the analysis of connectivity in age-related brain degeneration, the work described in [25] describes the employment of segments of 2 s and functional connectivity estimators, from which higher connectivity values quantified by the small-world metrics were observed during the open-eyes condition for the alpha band. This also differs from the observations presented previously and highlights the appreciable differences in the results of similar works, which could be linked to the segment duration employed over the methodologies.

All these research works rely on a suitable framework to obtain connectivity measures that explain the causal influences from the neural information. Such a method comprises several steps including preprocessing, where artifacts, noise, and normalization of the signals are performed, and then, the definition of a working domain from the neural sources, which is established either directly from the multivariate time series [15], regions of interest (ROIs) or dipoles [2,26]. From this working domain, the EC calculation is then performed using different metrics such as Granger causality [15], Directed Transfer Function (DTF) [22], Partial Directed Coherence (PDC) [4], and transfer entropy, among others. After that, in some cases, graph-based metrics are employed to characterize the high-degree network generated by the neural sources [23,24], in order to finally perform the statistical analysis used to test the significant connections of the network, providing the final coupled relationships found across the time series as result of the fitting process of a multivariate autoregressive model (MVAR).

Despite this comprehensive methodology, the approaches described in the literature do not provide a framework for the selection of an appropriate segment length to guarantee stationarity to perform effective connectivity analysis. In general, this matter is not usually addressed, and its influence on the MVAR model regardless of its importance has been overlooked. Moreover, the heterogeneity of segment durations employed to estimate connectivity is so diverse that all the works listed so far employ segments that range in the order of milliseconds [12], up to 100 s [7], which could impact the quality of the results obtained, affecting the analysis of connectivity.

For this reason, in this study, we devise a segment-length-selection method that considers the stationary characteristics of EEG signals based on high-order statistical moments and we assess the influence of the segment length on the MVAR model and its corresponding connectivity results, as compared to the conventional approach based on the framework specified above.

In this study, we employed EEG data acquired in the resting-state conditions with open (R1) and closed (R2) eyes to evaluate the implementation of a segment-length-selection algorithm as a preliminary step for effective connectivity analysis. The objective is to select a segment duration that guarantees constant stationary features from the multivariate time series. To do so, an iterative piecewise segmentation of the EEG signals is performed to divide the time series into smaller portions from which kurtosis values are calculated. Then, distributions of the kurtosis variances from the segments are estimated, and a searching strategy is implemented to find the most common segment duration across the EEG signals that maintain the stationary characteristics, not only on each recording but in different neural conditions and subjects.

2. Materials and Methods

2.1. EEG Dataset

The EEG dataset employed in our study was provided by the Istituto di Technologie Biomediche (ITB) of the Consiglio Nazionale delle Ricerche (CNR). The dataset comprises the EEG time series of 10 healthy subjects. These participants were part of a control group of a clinical research project that evaluated quantitative EEG markers from the brain activity of chronic stroke patients with monolateral upper-limb deficits before and after undergoing a robot-assisted rehabilitation program [27]. The experimental sessions took place at the Presidio di Riabilitazione dell'Ospedale Valduce Villa Beretta, Costa Masnaga (LC), Italy. The project protocol included EEG recordings under resting-state conditions (i.e., relaxation states during open- and closed-eyes conditions) and during motor tasks to characterize and analyze the stroke patients' evolution during the rehabilitation therapy [28,29]. Written informed consent was obtained from each subject before inclusion in the study. The study was reviewed and approved by the local Ethics Committee at A. Manzoni Hospital, Lecco, and was conducted in compliance with the Declaration of Helsinki.

In this study, two EEG recordings of approximately 5 min (4.77 $\pm$ 0.86 min) were analyzed from each subject and comprised the time series of the two resting-state conditions: R1—open-eyes resting state and R2—closed-eyes resting state. This totaled 20 EEG recordings that contained the signals coming from 62 channels that were placed over the scalp using the 10–20 standard system. The Synamps 2/RT system from Compumedics ®Neuroscan™ (Charlotte, NC, USA)was employed for the acquisition and it was configured at a sampling frequency of 1000 Hz with an active power line filtering set at 50 Hz, employing a Notch filter configured at each channel. Out of the 62 channels, the ground electrode CZ was employed to eliminate possible spurious components from the signals, canceling out the noise produced by the ground circuit of the EEG acquisition system.

A preprocessing framework was performed using the EEGLAB toolbox running on MATLAB version 2019a (The Math Works, Inc. MATLAB. Natick, MA, USA) [30], to mitigate noise and artifacts. First, by using EEGLAB's Artifact Subspace Reconstruction tool [31], the artifacts of the signals were reduced and, in those cases where the affected portions of the signals could not be repaired, such segments were eliminated. Then, a data-cleaning stage was performed by setting up a threshold scheme that considered the density power, signal amplitudes, probability of occurrence, and trend analysis. Finally, Independent Component Analysis and the Multiple Artifact Rejection Algorithm (MARA) were employed to discard portions of the signals that were not compliant with the common features of EEG signals; this was evaluated with a custom neural network embedded in the MARA tool [32]. The independent components that were discarded from the EEG datasets were automatically removed by this tool, following the inherent pretrained parameters of the MARA neural network. In addition, to eliminate noisy portions of the signals, an initial epoching that considered epoch durations of 1 s was employed, and EEGLAB considering the ASR, the thresholds, and MARA removed the portions considered as heavily affected by noise; as a result, the EEG recordings were shortened as shown in Table 1. According to this semi-automatic artifact-rejection framework, bad channels were also discarded following the EEGLAB pipeline as explained in [33].

Table 1 summarizes the main characteristics of the EEG signals' duration before and after the preprocessing stage described above. As can be noticed, noise and artifacts heavily affected some of the recordings, resulting in a significant reduction in the signals' duration after the data-cleaning process; in some cases, the proportion of the retained signals was as low as 15%. Hence, in this study, only the clean signals that maintained at least 50% of the original durations were selected to continue the processing. Accordingly, the recordings from subjects 1, 3, 4, and 7 were discarded (highlighted in gray on Table 1), leaving 12 out of 20 recordings from 6 out of 10 subjects available for processing. Table 1 also shows that the number of channels maintained after the artifact rejection was heterogeneous among the recordings and ranged from 55 to 61.

Table 1. Recordings' characteristics before and after preprocessing (Discarded recordings are highlighted in gray).

Subject	R1—Opened Eyes					R2—Closed Eyes				
	Raw Signals Duration [s]	Clean Signals Duration [s]	Clean Signals Percentage	Selected Channels	Channel Selection Percentage	Raw Signals Duration [s]	Clean Signals Duration [s]	Clean Signals Percentage	Selected Channels	Channel Selection Percentage
S1	305	46	15%	51	82%	311	172	55%	53	85%
S2	308	235	76%	59	95%	314	256	82%	60	97%
S3	284	46	16%	59	95%	204	43	21%	58	94%
S4	192	101	53%	54	87%	278	123	44%	53	85%
S5	308	205	67%	57	92%	306	155	51%	59	95%
S6	308	230	75%	57	92%	311	226	73%	60	97%
S7	353	135	38%	54	87%	313	212	68%	55	89%
S8	315	197	63%	61	98%	319	231	72%	62	100%
S9	189	145	77%	55	89%	184	116	63%	58	94%
S10	303	172	57%	60	97%	328	240	73%	61	98%
Mean	286	151.2		57		287	177.4		58	

A resampling step was employed after performing the data-cleaning process on the signals to reduce the sampling frequency from 1000 Hz to 250 Hz, which is an acceptable rate considering the frequency information of the alpha band to accomplish the EC analysis [34]. Then, the resampled signals were band-pass filtered using a Finite Impulse Response (FIR) filter that employed a Kaiser window with cutoff frequencies of 0.5 Hz and 50 Hz [35,36]. Finally, the data were common-average referenced. These steps of resampling, filtering, and referencing conclude the digital conditioning stage of the EEG signals. The following sections explain the segment-length analysis and selection based on kurtosis to perform the EC analysis.

2.2. Segmentation and Kurtosis Estimation

The proposed segment-length analysis is based on an iterative piecewise subdivision of the signals into segments. From such segments, it is possible to obtain estimations of the dynamical properties of the EEG signals and the nonlinear processes behind them by evaluating the effective connectivity.

The segmentation approach is summarized in Figure 1. Let W_L be the basis window length defined as an elemental duration of the segments in seconds, and N_w be the total number of windows (equivalent to the number of segmentation operations) considered to perform the iterative segmentation. Then, according to these parameters, h is defined as the longest segment duration following that $h = N_w \cdot W_L$, thus holding that $0 < W_L \leq h$. By considering the total duration of the recording ($\boldsymbol{t}$), as well as the variable w_{li}, used to keep the value of the segment duration for a specific segmenting step ($i = 1, \ldots, N_w$), at each iteration, a matrix of size t/w_{li} by $w_{li} \cdot fs$ is built and contains the segmented signal with non-overlapping segments. The variable w_{li} refers to the segment duration according to the segmenting step iteration, so that $w_{li} = i \cdot W_L, \forall i = \{1, \ldots, N_w\}$, and fs corresponds to the sample frequency (i.e., 250 Hz in this case).

In summary, the iterative process for the segmentation of a signal is explained by its sequential splitting according to the window length (w_{li}) whose duration is increased at each iteration by a factor defined as a multiple of the basis window (W_L) given i. In this way, the signal of duration $\boldsymbol{t}$ is divided into non-overlapping pieces, each one of length w_{li}. The resulted segments are then stored in matrix form and are organized in chronological order. The procedure is repeated for the original signal N_w times, producing a total of N_w matrices of segments for each of the signals that compose the dataset. Since the number of windows is directly related to the window length, N_w must be chosen according to the physiological characteristics of the brain activity under analysis and the frequency information that we want to cover with the selected windows. However, this process can be trivial if N_w is set large enough so that the different windows comprise the needed frequency components to be analyzed.

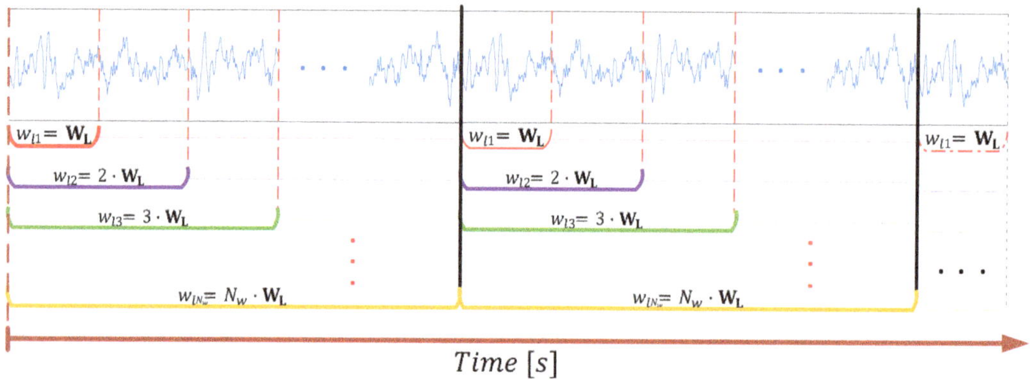

Figure 1. Outline of the segmentation approach performed on EEG signals. (Each color is associated to a specific window length duration: red – $w_{l1,}$ purple – $w_{l2,}$ green – w_{l3}, and so on).

In summary, the iterative process for the segmentation of a signal is explained by its sequential splitting according to the window length (w_{li}) whose duration is increased at each iteration by a factor defined as a multiple of the basis window ($\mathrm{W_L}$) given i. In this way, the signal of duration $\boldsymbol{t}$ is divided into non-overlapping pieces, each one of length w_{li}. The resulted segments are then stored in matrix form and are organized in chronological order. The procedure is repeated for the original signal N_w times, producing a total of N_w matrices of segments for each of the signals that compose the dataset. Since the number of windows is directly related to the window length, N_w must be chosen according to the physiological characteristics of the brain activity under analysis and the frequency information that we want to cover with the selected windows. However, this process can be trivial if N_w is set large enough so that the different windows comprise the needed frequency components to be analyzed.

The sequential process is performed until $i = N_w$, whose value is defined beforehand. From this approach, it can be easily noted that each segment is composed of a sequence of samples generically defined by the vector $w_{li,j} = \left(x_j,\ x_{j+1},\ x_{j+2},\ldots,x_{j+i\cdot \mathrm{SW_L}-1},\ x_{j+i\cdot \mathrm{SW_L}}\right)$, where $\mathrm{x_k}$ (for $k = j,\ \ j+1,\ \ldots\ j+i\ SWL$) generically refers to the components of the EEG segment $w_{li,j}$. Here j corresponds to the index of the sample where the segment starts with respect to its occurrence in time, and $\mathrm{SW_L}$ is the number of samples contained in the basis window $\mathrm{W_L}$ (i.e., $\mathrm{SW_L} = \mathrm{W_L}\cdot fs$). Thus, $w_{li,}$ defines the data vector resulting from a specific segment, a data block formed by a number equal to $i\cdot \mathrm{SW_L}$ samples that initiates at the time instant corresponding to the index j.

The segments that belong to a row of non-overlapping windows (shown in the lower part of Figure 1, represented by the horizontal brackets) form a new matrix containing the windowed signal according to w_{li}. Each row of this matrix encloses a single segment from which different statistical measures such as the mean, variance, skewness, or kurtosis can be estimated. From these statistical moments, it is possible to evaluate the stationary characteristics of a signal as a function of time given the time interval definitions considered in the segmentation approach.

From the previous characterization, each matrix containing the signal's segments associated with a channel (chn) that belongs to the EEG dataset follows the definition shown in Equation (1).

$$W_{matrix} \in \mathbb{R}^{t/w_{li}\ \mathrm{x}\ w_{li}\cdot f_s} \rightarrow s.c \in \mathbb{R}^{t/w_{li}}\ ,\forall chn = \{1,\ldots,M\} \quad (1)$$

where W_{matrix} is a matrix containing the segments of a signal and $s.c$ stands for the statistical characteristic whose values are being mapped into. From Equation (1), it is noted that the statistical characteristic space has a dimension of t/w_{li}, corresponding to a vector that represents the calculated statistical moment and whose components are each related to a segment at a specific time interval, i.e., each vector component is attributed to a time interval represented by a segment. In this way, it is possible to account for the variation over time of these statistical characteristics considering different window ranges and the given matrices.

For our specific case, we rely on the kurtosis (Equation (2)) to account for the stationarity of the segments. This equation calculates the fourth-order central moment by estimating the expected value of the fourth power of the difference between the time series (x) and its mean (μ_x), and applying the normalization by dividing it by the squared variance of x and subtracting the kurtosis value of a pure Gaussian distribution (i.e., 3), so that the offset, known as kurtosis excess, accounts for the difference between the kurtosis of the time-series segment (x) and a strict stationary series that follows Gaussian distribution.

$$K'_x = \frac{E[(x-\mu_x)^4]}{\sigma_x^4} - 3 \tag{2}$$

High-order moments such as kurtosis contribute to the process characterization. Unlike the first- and second-order statistical moments that are limited (in our case) by the zero-mean characteristic of the time series, kurtosis accounts for the existing difference of a normal distribution when it is compared to the Probability Density Function (PDF) formed from the samples that belong to a segment. This fourth-order moment can be employed to determine the non-stationarity behavior exhibited by a segment of fixed duration [37]. Since random processes are assumed to be stationary, and their distributions follow a Gaussian density, then, by evaluating how different segments' PDFs differ from the normal distribution, the non-stationarity proportion of the segment can be estimated.

Under these assumptions, Equation (1) can be rewritten as:

$$W_{i,chn} \in \mathbb{R}^{t/w_{li} \times w_{li} \cdot f_s} \rightarrow K_{i,chn} \in \mathbb{R}^{t/w_{li}} \text{ , } \forall chn = \{1, \ldots, M\} \tag{3}$$

where $K_{i,\ chn}$ is a vector that contains the kurtosis excess estimated at each segment from the $W_{i,chn} = W_{matrix}$ at iteration i and channel chn (i.e., a vector of kurtosis whose components are calculated from each row of the windowed matrix). Then, each component of the vector $K_{i,hn}$ explains the degree of non-stationarity of a segment at a specific time interval bounded by the duration w_{li} on each channel.

2.3. Kurtosis as a Feature

The fourth-order central moment characterizes each segment of the time series in our approach; this is a feature derived from shorter portions of the data and explains the dynamic change of the stationarity from segment to segment. Therefore, different segmentation conditions provide different amounts of information about non-stationary characteristics requiring comparing kurtosis values from a single channel, a complete dataset, and finally, among conditions and subjects.

From Equations (1) and (3), it can be noticed that the resulting kurtosis vectors have dimensions varying according to t/w_{li}; in consequence, an interpolation step is performed to guarantee the same size across the vectors from which the kurtosis PDF is estimated and comprises the kurtosis values of a channel subjected to the N_w segmenting iterations. Figure 2 shows the kurtosis distributions of 4 different channels. As can be observed from Figure 2a, the kurtosis distributions of different channels follow a Gaussian-like density as depicted by the PDFs estimated from the segmentation process of the signals associated with the channels F1, F4, PO5, and PO4. The kurtosis distributions in Figure 2a are the result of the iterative segmentation process of the signals considering a basis window length WL = 1 s, for $1 \leq i \leq 10$, yielding 10 kurtosis vectors as result, from which the PDFs are fitted. The expected values on each distribution correspond to the most likely kurtosis expressed by the signal over different segment lengths and they are used to assess the stationarity of the signal as a function of the window length. The examples in Figure 2a are related to some representative EEG sources, i.e., the frontal (F1, F4) and posterior (PO4, PO5) electrode locations.

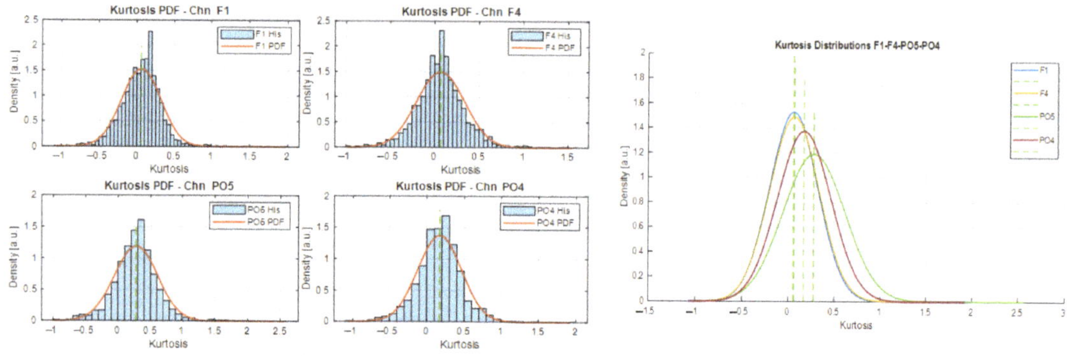

(a) (b)

Figure 2. (**a**) Individual kurtosis distributions for the channels F1, F4, PO5, and PO4. (**b**) Superimposition of the kurtosis distributions. Estimated for subject 10 during closed-eyes resting-state condition.

As can be noted from the densities in Figure 2a, they have similar shapes and their distributions span comparable ranges (i.e., approximately $-1.5 \leq K \leq 2$, with expected values in the range of 0 to 0.5, as shown by the green dashed lines in Figure 2b. Where K corresponds to the kurtosis value). In this specific example, the relationship that exists between the frontal channels F1 and F4 is evident as the PDFs are nearly the same. This is an expected behavior since the electrodes' locations on the F1–F4 channels are close to each other relative to their positions from PO5 and PO4.

Furthermore, considering that the alpha rhythm (comprising frequencies from 8 to 13 Hz) is more noticeable over the occipital area during resting-state conditions, then the EEG data acquired by the channels PO5 and PO4 should evidence an appreciable increment of the power spectrum under this frequency range. Moreover, since the frequency patterns are not uniformly distributed over the scalp, it is expected to find differences of such magnitudes over the same area; in addition, those differences could influence the non-stationarity behavior of the signals, which could be the reason for the slight variations of the kurtosis PDFs in Figure 2b. As explained in [23], the posterior region (covered by the occipital area where PO5 and PO4 are located) has an emergent pattern of connectivity directed to the frontal–parietal regions, suggesting as well non-uniform stationary behavior in this area that may be explained by the kurtosis distributions of these channels.

Moreover, by superimposing the kurtosis distributions from different channels, we can observe how distant their expected values are with respect to each other, from the point of view of each PDF (as shown in Figure 2b). If more distributions are compared, then we can find subsets of channels whose expected kurtosis values are closer than others, hence there exists a probability range that gathers most of these expected values associated with a specific segment duration. With this information, we design a searching strategy to find common segment durations across channels that exhibit similar stationary characteristics.

2.4. Kurtosis Variance and Searching Strategy

The proposed strategy for defining the segment length is based on the search for kurtosis values that are likely to be found across the PDFs estimated from the multivariate time series of the different EEG signals. Specifically, from the time series, different segment lengths are evaluated to find a window duration, common across channels, that guarantees similar stationary characteristics to perform effective connectivity analysis. From the distributions of the kurtosis of different channels (as shown in Figure 2), their corresponding variances are computed as a function of the segment length, i.e., $\sigma^2(K_{i,chn})$. The main advantages of using kurtosis variance are:

1. The kurtosis vector (K_i) generated for each channel is re-expressed as a single value representing the squared deviation from the expected mean magnitude considering a specific segment length. In further processing, this is computationally less expensive than a vector of t/w_{li} components. The data re-expression can be explained as follows:

$$K_{i,chn} \in \mathbb{R}^{t/w_l} \rightarrow \sigma^2(K_{i,chn}) = K_{\sigma^2_{i,chn}} \in \mathbb{R}^1, \ \forall Chn = \{1, \ldots, M\} \tag{4}$$

2. Similar means and variance values from different segment lengths enable a comparison of the dispersion observed on a dataset containing different signals, as exemplified in Figure 2b.

The searching strategy consists of finding the smallest range of kurtosis variance that contains the expected values of the kurtosis PDFs (i.e., the mean) estimated across channels in each recording. As an example, Figure 3b shows the kurtosis variance distribution of the $K_{i,chn}$ vectors from one of the participants of the study. The PDF from the kurtosis variance values is fitted by a Chi-square ($X2$) distribution (black dashed line); therefore, the resulting density is estimated from the $K\sigma^2$ matrix that holds the kurtosis variances with respect to the multiples of the basis segment length (w_{li}, for i = {1, ... ,10}) for each channel of the EEG recording. This means that the matrix $K\sigma^2$ is a result of the concatenation of the $K_{\sigma^2_{i,chn}}$ vectors. This means that $K\sigma^2$ contains the variances estimated from the kurtosis values of each channel's time series segmented at various scales, from 1 s to 10 s, producing a matrix of a maximum size of 62 × 10 components, considering that our dataset comprised a total of 62 EEG signals, but that was reduced in some datasets after performing the signal preprocessing. In this way, the variance searching algorithm is defined as Algorithm 1:

Algorithm 1: Variance searching algorithm

Input: $K\sigma^2$
1. $pdf \leftarrow MLE(K\sigma 2, \chi 2)$
2. $pk \leftarrow max(pdf)$
3. $L_b \leftarrow pk, \ H_b \leftarrow pk$
   Loop
4. $L_b \leftarrow L_b - c$
   $H_b \leftarrow H_b + c$
5. $S_{K\sigma^2} = find(K\sigma^2) \ such \ that: \ L_b \leq K\sigma^2 \leq H_b$
6. $VSelChn \leftarrow S_{K\sigma^2}\left[\min(t_{w,chn})\right]$
   IF $count(VSelChn) > threshold1.$
7. $Avar \leftarrow mean(VselChn)$
   IF $Avar \leq threshold2.$
9. $Append \ Avar \ to \ SELECTION$
   Until *k iterations are reached*

Algorithm 1 shows how to perform the variance range searching strategy. It receives the $K\sigma^2$ matrix as the input, from which the PDF is estimated considering the Maximum Likelihood Estimation (MLE) method by fitting a Gamma distribution. After the peak value of the distribution (pk) is found, it is used to initialize the lower and higher bounds of the variance range searching interval (L_b and H_b). Then, a constant value c, that sets the searching rate, is subtracted and then added to the lower and higher boundaries, respectively, so that a searching interval is initialized. Now, the kurtosis variance values from the $K\sigma^2$ matrix that are under the searching interval $\left(L_b \leq K\sigma^2 \leq H_b\right)$ and that correspond to the shortest segment duration on each of the channels are selected at the actual iteration and stored in the variable ($VSelChn$). Since the rationale is to maximize the number of channels that exhibit the same stationary characteristics, first, the number of channels that have at least one segment within the kurtosis variance searching limits is computed and, if the count is less than 50% (threshold 1) of the total number of channels in the dataset, then the algorithm restarts the searching by increasing the interval limits and setting it up as the new searching range. Then, the average of the variances is calculated, and if its value is lower than a second threshold (set at 40% of the maximum variance per channel registered in the matrix), then it is guaranteed that the selected channels exhibit stationary characteristics bounded by this threshold, assuring only minimum variations of the stationary features across the selected channels, thus allowing to find common features over the channels. The channels that meet these bounds are stored in the SELECTION variable, and their durations are sorted out with respect to their variance values.

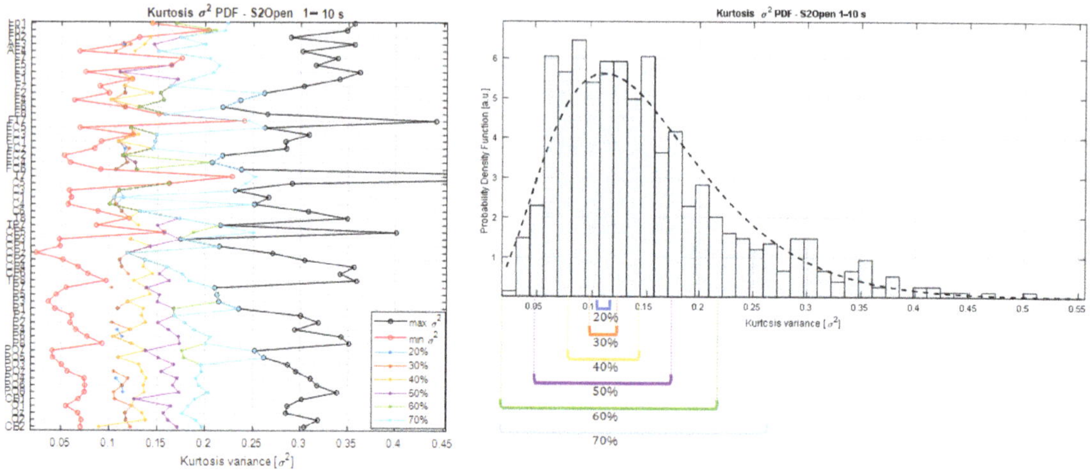

(a) (b)

Figure 3. Searching space boundaries relative to the kurtosis limits, estimated for subject 2 (S2) in open-eyes condition. (**a**) At the channel level. (**b**) From the distribution point of view.

2.5. Searching Domain

Starting with the $K\sigma^2$ matrix, it is possible to rank the variances from the lowest to the highest on each of the channels composing a dataset, and the min and max values per channel are computed to establish the absolute kurtosis variance limits relative to a segment duration. In the same way, the intermediate values are found, and their corresponding durations are associated with a specific proportion of the variance limits, as shown in Figure 3a.

As can be observed from Figure 3a, the lower the percentage threshold, the closer the selected $K\sigma^2$ values are to the min bound and, consequently, the lower the number of channels that will meet the kurtosis variance requirements. A trade-off is necessary as higher percentage threshold values will lead to a higher number of channels but, at the same time, a higher dispersion among the channels, as shown in Figure 3b.

In this sense, sorting the segment durations that are masked out by the percentage thresholds allows us to identify common durations covered by the area of the PDF bounded by the interval limits. As result, we obtain a series of segments that are ordered by both the duration and the relative variance magnitude that they exhibit; this information is used to categorize them accordingly. Then, counting the segment durations that are found at a specific percentage threshold quantifies the number of channels that share similar stationary properties according to their kurtosis variance magnitudes, and, if a cutoff value is established to set a limit on the minimum number of channels expressing the $K\sigma^2$ values within a percentage range, then the searching strategy returns the segment durations meeting these requirements. In this way, the searching strategy shows the segment durations that, at a specific percentage threshold, gather the required number of channels exhibiting similar stationary characteristics. In these terms, the trade-off is solved by performing the selection of the segment length with the lowest percentage threshold reaching the minimum requirement for the number of channels.

To this extent, by considering different percentage thresholds and the minimum number of channels within those limits, it is possible to graphically check the areas covered by the searching interval, and how the selection of the segment duration is derived from the statistical characteristics coming from the dataset, specifically from the kurtosis of the segments.

2.6. Directed Transfer Function

To develop the generative model necessary to explain the directed influences and the relationships that exist between different neural regions, several measures that quantify the existing coupling among sources considering the temporal information from EEG recordings can be employed [1]. Measures based on time series relying on Granger causality and its variants in the frequency domain are the most common choices in effective connectivity analysis for EEG data [12]. In this approach, we employed the Directed Transfer Function (DTF) since it is a measure proven to be more appropriate to be applied to signals registered on the scalp as demonstrated by Ku and colleagues [38]. The quantification of the DTF is defined as shown in Equation (5).

$$DTF_{ij} = \frac{H_{ij}(f)}{\sqrt{\sum_{m=1}^{n} |H_{im}(f)|^2}} \quad (5)$$

where the matrix H contains the spectral and phase information of the sources i, j, from which causality is assessed. The DTF value is a complex measure and provides a metric of the total information that has been flown from the source j to i, being normalized by the total inflow received by i. In this sense, the DTF detects the direct influence of one or several signal sources in the channel of destination [34].

If we consider distinct sources and destinations, defined from the EEG channels, then, Equation (5) is the starting point from which the EC calculation is repeated pairwise on the signals composing the EEG recording. In this way, a relationship among the channels is produced and displayed in matrix form, where each row–column component corresponds to the DTF value estimated on the source and destination signals (j to i), respectively. Since the EEG data are time-dependent, then each of the DTF matrices should refer to a specific time interval to characterize the flow of information at each frequency and comprises components from 1 Hz to $fs/2$. Hence, in this sense, the selected window length derived from our approach is employed to calculate the connectivity matrices along with the signals at every time interval. Figure 4 shows the resulting 3D matrix that comprises the EC values among channels calculated at each frequency on the EEG blocks after segmenting the signals with the selected window duration. The value t_W refers to the duration of the selected window, and the segments indicate the time interval along with the EEG signal product of the segmentation considering a generic selected window.

Finally, to assess the significance of the connectivity among channels, the t-test is applied to the DTF blocks that characterize a desired frequency range (performed by averaging elementwise the EC values of the DTF matrices over a desired frequency range, e.g., alpha band), then, only the connections that have a p-value less than 0.05 are considered significant and their connectivity relationships are maintained for further analysis. Finally, considering the complexity of the DTF matrices, the statistically significant relationships gathered in these matrices are treated as adjacency matrices, from which graph theory indices are calculated to characterize the network of effective connections.

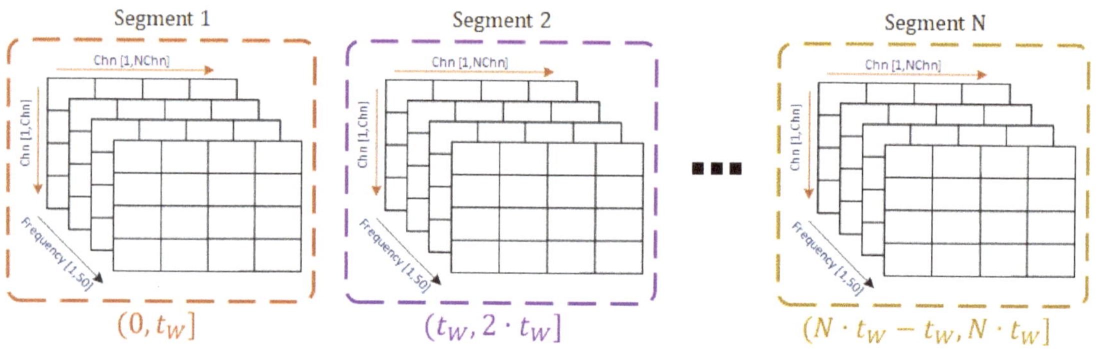

Figure 4. 3D matrix of the segments obtained from a generic selected window ($t_W, \ldots, N{\cdot}t_W$) containing the DTF connectivity values among channels (Chn) at each frequency.

2.7. Network Measures: Graph Theory Indices Applied to EEG Data

The use of the adjacency matrices to encode the significant connections as a result of the DTF estimation across the channels provides the raw structure that shows the effective connections on the channels. This can be difficult to interpret due to the high density of connections that are considered significant from the adjacency matrix. In this way, it is useful to consider graph theory measures to perform a characterization of the connectivity in the network that is able to show hidden structures, and central nodes that participate more in the transmission of information and clusters that could characterize the brain activity that is being investigated, i.e., the resting-state conditions with eyes open (R1) and eyes closed (R2).

As explained in [39], 4 broad classes of graph measures can be distinguished: the basic measures that reflect the importance of a node (channel) in the network by considering the number of connections it has with other nodes (i.e., the degree), the graph density that measures the actual number of connections in the network and that can be expressed as the percentage of links present in the network being 0% when no connections are considered in the graph and 100% when all the significant links are shown, and finally, the strength, which accounts for the amplitude of the connection between two nodes, e.g., the DTF magnitude registered for the pair channel i, j in the matrix.

The second class of measures is the so-called measures of integration, which account for how effortless the communication between the channels is performed. In this category, there are different measures that help to estimate this. The shortest path length between two channels, as its name indicates, calculates the line with the minimum length that connects two nodes on a surface, in this case given the topographic characteristics and the placement of the electrodes over the scalp of a person. Its value is defined for every pair of nodes, and given the high density of nodes that form a network of electrodes, the average shortest path length is used to characterize the typical separation between the nodes.

Conversely, the global efficiency accounts for the inverse of the average shortest path length and indicates the capacity of a network to support the information flow, and in the case where networks are not fully connected, it provides a better representation of the integrative communication characteristics among the nodes since, unlike the average shortest path length, the global efficiency does not diverge to infinity when a connection is not present in the network. This provides a useful way to account for how easy the communication among the present nodes is, since the adjacency matrices in our case are not fully connected, considering that they only contain the statistically significant connections.

The third category of graph parameters is the so-called measures of segregation that characterize the independence of local structures found within the network, given the formation of groups that are interconnected, i.e., clusters of nodes. The clustering coefficient accounts for the channels connected to a node that are interconnected to each other. Another measure of segregation is the local efficiency, which is defined as the efficiency among the neighbors of a node.

Lastly, the importance of a node in the network is estimated by considering the betweenness centrality, a parameter that quantifies how central is a node in the information flow considering the integration and effective connections produced within the structure. This measure calculates the number of the local short paths connected to a node and that represent the importance of a channel in the network.

The density parameter is a basic measure that quantifies the fraction of actual connections that are present on a network. When an adjacency matrix is calculated, it summarizes the effective directed connections among the channels that are significant in statistical terms, as explained above. Then, it is most probable that its density is less than the maximum number of possible connections on the network, defined as $N\ (N-1)$, N being the total number of channels (i.e., 62). Thus, the density of a non-fully connected network, given the statistically significant relationships condensed in the adjacency matrix, will never be equal to $N\ (N-1)$.

In these terms, the density is constrained by the number of significant connections of the adjacency matrix whose elements' magnitudes can be sorted from lowest to highest in order to generate the "cost" variable, which is used as the independent variable from which the remaining graph-based parameters are calculated, and by these considerations, they are defined as a function of the number of actual connections in the network. By sorting out the magnitudes, the cost represented as the proportion of connections encodes a linear scale from the highest to the lowest magnitudes. In this way, 1% of the cost comprises the number of connections in the network that have a magnitude larger or equal to the 99% of the maximum DTF value found in the adjacency matrix. The same applies to the other percentages up to reaching 100%, whose cost comprises all the significant connections regardless of the DTF magnitudes on the matrix.

3. Results

3.1. Selection of the Window Duration

By applying the searching strategy and continuing with the example depicted in Figure 3, the number of channels that meet the kurtosis variance criteria as a function of the corresponding searching interval and the window length can be computed for the two resting-state conditions. The results are shown in Figure 5.

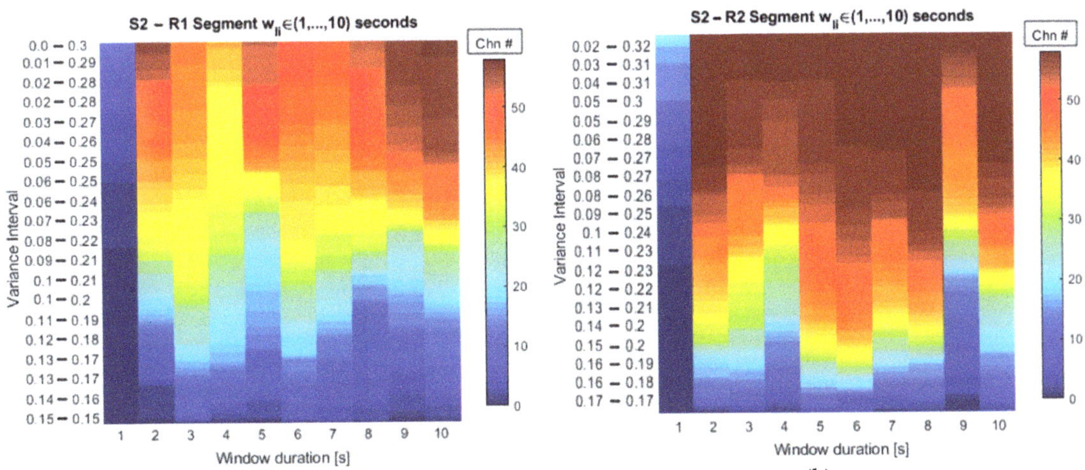

(**a**) (**b**)

Figure 5. The number of channels as a function of the segment length and the searching interval in terms of the kurtosis variance. (**a**) S2 Open-eyes resting state. (**b**) S2 Closed-eyes resting state.

Figure 5 shows that, as discussed previously, the number of selected channels increases if the variance interval is enlarged, reaching the total number of channels at wider variance intervals. Table 2 reports the exact number of channels from a recording that share similar stationary characteristics as a function of the window duration, for open (R1)- and closed (R2)-eyes resting states for the same subject (S2) following what is shown in Figure 5. The numbers in red in Table 2 (51 and 39) correspond to the number of channels with a similar stationary value associated with the selected segment duration for this EEG recording; according to the threshold for the minimum number of channels set at 75% and 65% of the total number of time series of R1 and R2, respectively.

Table 2. Segment duration and the number of channels sharing kurtosis variance features for subject S2 in the resting states. The numbers in red correspond to the channels sharing similar stationary values.

Variance Percentage	10%		20%		30%		40%		50%		60%		70%		80%		90%		100%	
Resting State	**R1**	**R2**	**R1**	**R2**	**R1**	**R2**	**R1**	**R2**	**R1**	**R2**	**R1**	**R2**	**R1**	**R2**	**R1**	**R2**	**R1**	**R2**	**R1**	**R2**
Low lim.	0.15	0.17	0.142	0.168	0.132	0.165	0.113	0.163	0.086	0.131	0.028	0.065	−0.02	0.025	−0.057	−0.01	−0.107	−0.05	−0.16	−0.83
Upp. lim	0.15	0.172	0.158	0.174	0.168	0.177	0.187	0.179	0.214	0.211	0.272	0.277	0.32	0.317	0.357	0.354	0.407	0.389	0.46	1.172
1 s	0	0	0	0	0	0	0	0	0	0	4	6	20	19	37	32	52	46	59	60
2 s	2	0	1	3	5	8	12	12	28	39	47	60	57	60	58	60	59	60	59	60
3 s	3	2	8	4	12	10	24	11	34	44	42	59	45	60	53	60	57	60	59	60
4 s	3	1	5	2	11	5	20	6	29	26	37	57	43	60	47	60	55	60	59	60
5 s	1	1	2	4	8	7	15	9	20	33	51	59	56	60	56	60	56	60	59	60
6 s	1	2	6	4	12	5	22	9	35	48	45	60	50	60	56	60	57	60	59	60
7 s	1	1	5	3	8	4	16	6	30	40	44	60	48	60	50	60	57	60	59	60
8 s	4	2	4	3	6	6	8	9	25	43	50	60	55	60	57	60	59	60	59	60
9 s	1	0	2	1	4	1	12	2	22	13	55	49	58	60	59	60	59	60	59	60
10 s	2	2	4	3	7	7	13	8	28	25	57	60	59	60	59	60	59	60	59	60

As can be observed in Table 2, for the open-eyes state (R1 columns), the number of channels at a percentage level of 40% is significantly inferior compared to wider searching intervals. By considering the limits on the kurtosis density from Figure 3b, the 40% and 50% ranges cover the most probable kurtosis variances of the whole dataset. Hence in Table 2, at 40% with a window length of 3 s, only 24 channels share the kurtosis variance from that range. Following the same logic at 50% of the variance, the number of channels increases to 34, corresponding to ~58% of the total of channels from the dataset. Moreover, by evaluating a segment duration of 5 s, it is observed that 15 and 20 channels are found in the variance intervals of 40% and 50%, respectively. In this way, looking at the $K\sigma 2$ PDF using a segment of 5 s allows the selection of 51 channels (~80% of total channels) that share a variance in the range of $0.028 \leq K\sigma^2 \leq 0.272$. This comprises a probable proportion of the density.

Looking at the data of S2 during the closed-eyes resting-state condition (Table 2 R2 columns), the kurtosis variance proportions for the EEG dataset that was used here are also shown. Following the same analysis, in this case, kurtosis variance intervals comprising relative proportions lower than 50% did not group as many channels as in larger proportions. The windows of 2, 5, and 6 s for the 50% interval ($0.131 \leq K\sigma^2 \leq 0.211$) grouped 65%, 73.3% and 80% of the total of the channels that composed the recording, which suggests that any of these window durations maximize the number of signals with similar stationary characteristics. However, considering the assumptions explained in the methodology, it is required to have a segment length as short as possible that groups many or all the channels. Thus, in this case, a window of 2 s is the one selected for this subject in the closed-eyes resting-state condition. As a reference, the red dashed line in Figure 5 depicts the 50% threshold and graphically shows the number of channels gathered at each window duration for both resting states.

Table 3 summarizes the results obtained for the six subjects. For each of the subjects at each resting-state condition, the table shows the kurtosis variance range limits that provided a sufficiently high number of channels (according to the threshold for the minimum number of channels) with similar stationary characteristics and the resulting window length. The variance percentage indicates what proportion of the maximum kurtosis variance is featured by the selected window duration.

By replicating the same analysis for the remaining 11 recordings composing the overall dataset, the results in Table 3 are obtained. From this table, it is noticed that a window of 4 s is the most common segment length across the EEG data coming from the six subjects. From these results, the kurtosis variance percentage on average corresponds to 40% of the total, providing a significant reduction in the spread of the kurtosis values among the EEG signals from the recording. This, in addition to the 69% of channels that exhibit similar stationary characteristics, reaches the objective of maximizing the number of channels while minimizing the variability of the stationary measures quantified by the kurtosis. This is achieved by analyzing the recordings in relation to the characteristics of each specific dataset.

Table 3. Selected segment lengths for each subject and resting-state condition.

Subject	Condition	Lower Lim.	Upper Lim.	Variance %	Win Length	Remaining channels	Channel %
S2	R1	0.028	0.272	60%	5 s	47	78%
	R2	0.131	0.211	50%	2 s	39	65%
S5	R1	0	0.156	30%	5 s	41	72%
	R2	0	0.197	40%	4 s	29	49%
S6	R1	0	0.138	30%	4 s	35	61%
	R2	0	0.134	40%	4 s	48	80%
S8	R1	0	0.098	30%	4 s	33	54%
	R2	0.032	0.102	30%	4 s	27	44%
S9	R1	0	0.241	50%	3 s	50	91%
	R2	0.135	0.273	67%	2 s	48	83%
S10	R1	0	0.088	30%	4 s	45	75%
	R2	0.006	0.086	30%	4 s	47	77%
			Mean	41%		Mean	69%

By replicating the same analysis for the remaining 11 recordings composing the overall dataset, the results in Table 3 are obtained. From this table, it is noticed that a window of 4 s is the most common segment length across the EEG data coming from the six subjects. From these results, the kurtosis variance percentage on average corresponds to 40% of the total, providing a significant reduction in the spread of the kurtosis values among the EEG signals from the recording. This, in addition to the 69% of channels that exhibit similar stationary characteristics, reaches the objective of maximizing the number of channels while minimizing the variability of the stationary measures quantified by the kurtosis. This is achieved by analyzing the recordings in relation to the characteristics of each specific dataset.

3.2. *Effective Connectivity*

Figure 6 shows the results of the connectivity for both resting-state conditions when a window of 4 s is applied and analyzed for the alpha frequency band (8–13 Hz).

Figure 6 depicts the EC that characterizes each of the resting-state conditions from the 12 recordings considered for our approach. The strength of the connections is color-coded, and the arrows highlight the directionalities and the relationships among the EEG channel sources. Yellow colors transitioning to orange, red, and finally brown/black, show the connectivity strength in the network from low to high according to the magnitudes of the Directed Transfer Function (DTF) [40,41]. Figure 6 shows the results by establishing the statistically significant connections among the channels that monitored the EEG potentials generated over the scalp when subjects were in a resting state (see Section 2.6).

According to the results from the window-selection approach (see Table 3), a window of 4 s was employed to quantify the connectivity and characterize the influence of different neural sources over the scalp since it was the most common segment duration across subjects and conditions. The directional influence of the connectivity was quantified according to different network measures as shown in [39]. From these network measures, basic, integration, segregation, and centrality quantities were calculated from the adjacency matrices considering the DTF estimations from the signals as explained in Section 2.7. The DTF values were estimated considering the window of 4 s using the Source Information Flow Toolbox for EEGLAB in MATLAB [42], and then, the most significant neural sources represented by the EEG channels were found by integrating the network measures and establishing the central nodes according to the neural process under analysis. Such a procedure was carried out considering the methodologies presented in [14,22–24,43].

Conversely, let us now consider the EC patterns by employing a 20 s window according to the methodology explained by Olejarczyk and colleagues [22]. In this regard, Figure 7 shows the connectivity diagrams obtained from such segment length.

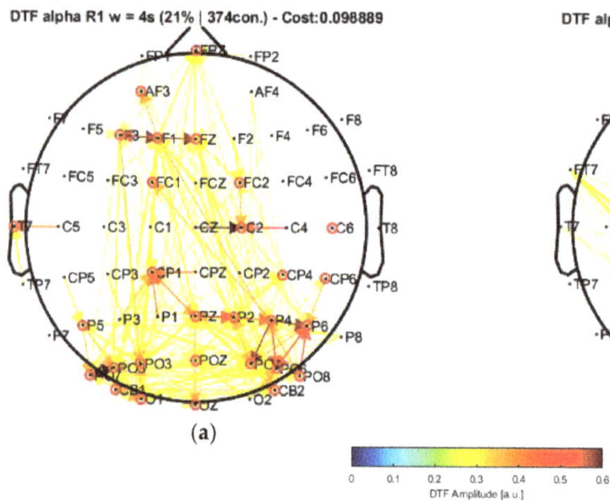

Figure 6. Effective connectivity diagrams considering a window of 4 s (**a**) for the eyes-open state and (**b**) for the eyes-closed state estimated for the 8–13 Hz frequency band (Alpha rhythm).

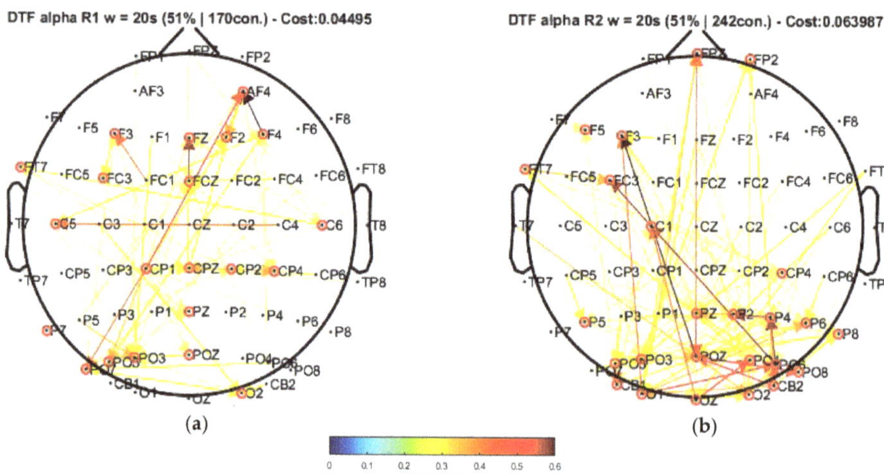

Figure 7. Effective connectivity diagrams considering a window of 20 s. (**a**) for the eyes-open state. (**b**) for the eyes-closed state estimated for the 8–13 Hz frequency band (Alpha rhythm).

The connectivity diagrams in Figures 6 and 7 show the significant effective connections derived from the adjacency matrices formed by the statistically significant relationships at a cost of 21%, 30%, and 51% of the maximum value of the DTF matrix (for R1 and R2 from the windows of 4 s and 20 s, respectively), which means that only the connections that exhibited a DTF magnitude higher or equal these values with respect to the highest DTF element (~0.6 in all conditions) are being plotted in the graph. The intersection of the selected channels from the graph measures according to [14,22–24,43] provides a list of nodes that can be considered as central elements that actively participate in the network. These selected nodes are highlighted by the red circles around their topographic locations in the graphs. Moreover, the directionalities are also depicted in the graph and show how the information is being directed to specific areas from different channels given some identifiable clusters of channels observed in the connectivity diagrams.

From the graphs in Figure 6, it is possible to note that there are clusters formed by some of the electrodes that exhibit a significant increment of their connections. This is the case of the closed-eyes condition in Figure 6b. The electrodes located at the posterior part of the scalp formed by the occipital (Ox), parietal (Px), and central–parietal (CPx) electrodes appear to be more involved in the connectivity process. In this region, the internal connections are very evident in terms of strength and number of connections. In addition, the salient connections with other areas such as the frontal region are shown as well. These connectivity diagrams for the closed-eyes condition provide some insights into the directionality of the connections. In this way, besides the evident internal network occurring in the posterior area of the brain during the eyes-closed condition, an influence is developed from this region towards the frontal area. In the case of the eyes-open condition, the pattern of connections is less consistent, in other words, they are not as structured as in the eyes-closed condition and the strengths from the central channels that participated more in the network were different as well.

For comparison, Figure 7 shows the connectivity diagrams by employing a window of 20 s. As can be observed, the central channels can be grouped to form clusters, which are used to identify the changes in the flow of information not only in the local level given the topographic location of individual electrodes but in a more general view considering complete regions that highlight the active areas in which the connectivity is being produced inside the group of nodes and between these areas. Considering the node grouping, there is observed a set of channels that participate in the EC. Specifically, the channels grouped for the closed-eyes condition (Figure 7b) show a significant influence originating in the posterior region of the brain from which most of the connections are generated. The channels located in this area present local connections and provide an evident influence from the occipital–posterior region towards the channels located in the frontal area of the scalp. Similarly, the significant connections obtained from the window of 4 s (Figure 6b) show similar connectivity patterns exhibiting a greater involvement of the sources located in the posterior part of the scalp, being directed towards the frontal area as well.

In the case of the eyes-open condition for the window of 20 s, the distribution of the channels is different, and by observing its connectivity diagram (Figure 7a), only a few channels show strong connections given the DTF amplitudes, which leads to the idea that the connectivity, in this case, is more uniform among the clusters and tends to have more midrange connectivity amplitudes than the eyes-closed case.

4. Discussion

In this study, we explored the stationary characteristics of EEG signals of resting-state conditions through the iterative segmentation of multivariate time series. It was intended to be an intermediate step in the effective connectivity estimation applied to brain activity. Considering the fourth-order central moment known as kurtosis allowed us to quantify how different the EEG sampling distributions were for a density formed by the samples of a pure stationary time series (i.e., Gaussian distribution). This was performed at different levels of segmentation, and such comparison permitted us to assess the effect of the short-time stationarity characteristics affected by this procedure. In this way, the stationary features did not change drastically over time, ensuring uniform characteristics along with the signal. From this, a searching strategy was designed that was dependent on the number of channels from the EEG recordings and the relative variance of the kurtosis distributions, from which the selection of a common window duration across the signals, subjects, and conditions was performed to then assess the effective connectivity.

As has been highlighted throughout this paper, according to the theoretical considerations from some studies [16,34,44,45], the use of an appropriate segment duration to perform the MVAR model-fitting process for EC is an issue that needs to be considered to guarantee consistent results over different experimental setups and inter-subject analysis. Therefore, this research topic is encouraged since the connectivity results are heavily affected by this parameter. In this context, our approach provides a way to make an informed decision regarding the window duration that could be employed in this regard.

As explained throughout the methods, our approach relied on the piecewise subdivision of the time series. This process was individually performed on each signal composing the EEG recording, which according to the mathematical generalization we presented, also allowed the scalability of multiple signals from which segments′ durations were categorized using a matrix representation. From such a generalization, it was possible to estimate different statistical quantities (see Equation (1)), opening the possibility to describe the data with other metrics or extract features from the short-time sequences that resulted from the segmentation.

As briefly mentioned in this paper, despite the importance of the first- and second-order statistics, these are not useful to characterize the segments. Considering that the signals were high-pass filtered in a previous stage using a cutoff frequency of 0.5 Hz, the first-order moment (i.e., the mean) of a segment was reduced to zero, making its value not useful for the characterization of a process. Moreover, the second-order central moment (i.e., the variance) under the same conditions is equivalent to the mean squared, which does not provide any insights into the dispersion of the samples composing the segment. Similarly, the coefficient variation (σ_x / μ_x) is very sensitive to the changes in the first-order statistics, making it grow abruptly as the mean tends to zero. Thereby, only high-order moments such as skewness and kurtosis contribute to the process characterization. Nevertheless, since the normalized version of the kurtosis accounts for the existing difference with a normal distribution when compared to the PDF that is formed from the samples that belong to a segment, this fourth-order moment can be employed to determine the non-stationarity behavior exhibited by a segment of fixed duration [37].

The representation adopted in our approach based on the kurtosis variance serves as a dimensionality-reduction procedure where a vector composed of t/w_l kurtosis values calculated from a single EEG signal is re-expressed by the data dispersion. By considering this, similar variance values calculated from the multivariate time series at different segment durations could be compared, and a way to do it is by examining the distribution formed by all the kurtosis variances. From such a PDF, a searching algorithm was designed and the segments with similar variance were grouped and organized so that we could examine how many signals shared the dispersion set by the searching interval limits. Since similar variances from different signals and window durations comprise closer expected values in the kurtosis domain, as shown in Figure 2b, it is expected that the deviations in the kurtosis variance domain to be small for the narrower searching intervals, as shown for the 20%, 30%, and 40% trends in Figure 3a. In such conditions, the searching interval limits are considered stiff and as a consequence, the number of signals meeting the searching parameters is not large (e.g., see the 20% where there are scarce dots). In this sense, by setting up a threshold on the minimum number of channels needed per searching interval, we can settle the trade-off and let the algorithm find the variance interval where most of the channels exhibit similar characteristics and, after grouping and ordering according to the window durations, we can find the shorter duration that captures the most common characteristics across channels.

Since our approach relies on the data that are being explored, it could be considered a non-parametric method for estimating an appropriate window duration to perform the MVAR fitting process. In this way, it is possible to consider conditions related to different types of brain activity and not limited to resting-state analysis in "non-epoched" datasets. This is an advantage in the processing scheme; however, it would need to be optimized to be presented as a third-party tool, e.g., a plug-in for EEGLAB.

The representation of the number of channels as a function of the searching interval and the window duration, shown in Figure 5, provides a useful way to perform a visual inspection of the results obtained from a dataset. It provides insights in a graphical way into what segment duration gathers more channels when a searching interval is as narrow as possible. By setting the threshold on the minimum number of channels, the window duration is found and according to the assumptions discussed so far as it ensures similar stationary properties according to the kurtosis values. In this sense, Table 2 complements the information and gives the exact number of channels from the total available in the dataset following the stationary conditions.

Table 2 gathers as a reference the change in the number of channels when fixed variance percentages are evaluated at every 10% increase of the total. These 10% steps allow us to briefly analyze the variation in the number of channels, but this being a nonlinear process, a smoother variation of such a number would be achieved when percentage steps are narrowed and placed between 50% and 60%, for example, so the exact number of channels in between is achieved.

By considering large steps in the variance, the window durations for each of the subjects are shown in Table 3. The statistical mode of the results suggests that a four-second window for the resting-state conditions under analysis is enough to guarantee similar stationary characteristics across subjects. This result is a useful resource for the EC analysis framework since the window size impacts the connectivity analysis as described in [16,44,45]. Moreover, the window size selection is not usually addressed in the state of the art, which is the reason why many research works use different durations to perform EC as evidenced by the references named throughout this paper.

The characterization from the connectivity allowed us to find common observations with some other works that are closely related to the methodologies implemented, as explained in the work performed by Olejarczyk et al. [22]. In this case, by applying their approach to our data and performing the connectivity analysis, the window of 20 s highlighted central nodes located in the posterior, left central–frontal joint, and prefrontal areas, which were not present in their results. Moreover, the statistical analysis allowed us to find a significant increment of the strength parameter over the alpha band by comparing the conditions of eyes closed and eyes open. Such an increase was mainly produced in some of the regions of higher centrality (prefrontal and central areas).

Our results, differently than in other research works [22–24], associated broader scalp areas differentiated by the EEG channels in resting-state conditions with open and closed eyes. In particular, the results point to the participation of more nodes located in the central–parietal and parietal–occipital regions considering the channel network for the alpha frequency band (see Figures 6 and 7). This finding is reflected by the number of connections (i.e., the cost) among the considered channels, suggesting that areas covered by these nodes are possibly involved in facilitating the communication flow between the posterior area and the frontal region. Such speculation could explain the involvement of intermediate structures in the frontal–central joint towards the central–posterior one, thus establishing the importance of the posterior region's influences on the frontal zone.

These observations provide a starting point in the attempt to characterize the awareness state during relaxation, thus without attention or concentration, by considering the EC measurements related to the alpha frequency band. In fact, the alpha rhythms (8–13 Hz) play an important role in resting states (in either open- or closed-eyes conditions), possibly clearing sensory information from distractors, as well as in waiting states before performing attentional or cognitive tasks.

As explained previously, effective connectivity is more noticeable in the posterior areas as evidenced by the increased network cost for the closed-eyes condition compared with the open-eyes case. These are characteristics expected from the alpha frequency components obtained from acquisition settings like ours. Of note, analyzing active thinking or the engagement of cognitive tasks (more noticeable over the beta frequency band (13–40 Hz)), despite being important, is out of the scope of this work, mainly due to the nature of the EEG signals at our disposal, and that our application example was defined to have a glance at the EC results that could be obtained after applying our methodology, from which we obtained consistent results.

In summary, our study was aimed at providing a novel method for the selection of a segment duration that improves the effective connectivity framework by introducing the analysis of stationary characteristics from the EEG signals as an intermediate step of the EC approach. Another advantage of our methodology is that it is not restricted to being used only for signals that measure resting states. Therefore, even though adjustments would be required to optimize the algorithms and provide flexibility in employing different sampling rates and channel sizes to reduce processing times, we have devised a methodological framework that is potentially applicable to any EEG configuration and possibly beyond resting-state conditions. Future studies could be further devised to demonstrate the capabilities of the method in characterizing connectivity in subjects with specific pathologies.

Moreover, following what is described in [22–24], the hypothesis arises of the involvement of the central–posterior and central–frontal regions as the characteristic areas of the resting-state conditions. According to these works, the Default Mode Network (DMN) that comprises those regions is involved in the synchronization over the alpha band. This hypothesis is reinforced by the results obtained in this work as shown in the last section; however, considering only 12 recordings to confirm the hypothesis or not can be misleading, and the EC analysis in our case is limited by the insufficient data available. Despite this limitation and given the consistency of our results, and the resting-state conditions under analysis, our method would provide a feasible approach for the analysis of pathological conditions highlighting statistically significant relationships among brain signal sources that could potentially complement the analysis of clinical conditions.

Under the framework offered by our methodology, when the process is repeated for different EEG recordings acquired from distinct subjects, our approach allows the integration of the data to perform an inter-subject analysis by examining individual results and selecting the window duration that provides common stationary characteristics, firstly for most of the channels on each recording and secondly across subjects selecting the mode, as shown in Table 3. This could be implemented differently, too; similarly to the methods applied in group ICA [46,47], where the brain-activity data from different subjects is concatenated forming a single large block of EEG, MEG, or fMRI time series to find independent components, we could apply this step to analyze the common stationarity of a large multivariate set of signals and obtain a single-window duration from this process. Such a procedure is proposed as a future feature of our approach where we will also analyze the differences in the results with the current method.

Moreover, even though our approach considers a basis window of 1 s for the iterative segmentation, it does not contemplate non-integer values for intermediate segments (e.g., wl = 1.5 s, 2.5 s, 3.5 s, ...) that could improve the results by evaluating stationarity according to the statistical characteristics of the signals. In this way, as future research work, we plan to evaluate the influence of such kinds of segments in our processing scheme, incorporating a larger dataset as well, and improve the characterization of the resting-state conditions through effective connectivity analysis.

Moreover, as explained in [48,49], other metrics such as the Kullback–Leibler (KL) divergence could be employed to estimate the statistical distance between the sample distributions of our segments and a Gaussian PDF. However, pure stationarity exhibited by random time series (where the expected value is zero) in a real case scenario like ours is difficult to achieve, and such statistical distance under these circumstances would constrain our approach, requiring us to find segment durations that exactly follow a normal distribution. In this way, minimizing the distances and setting a threshold that controls the "stationarity level" from the distance difference derived from the window duration would provide a feasible approach to tackle the same problem. This idea is suggested to be implemented as a future work of our research.

Finally, another idea to be implemented is to investigate the signals at source levels. By employing methods such as the low-resolution electromagnetic tomography (LORETA) taking advantage of the multiple EEG recordings, it is proposed to investigate the connectivity relationships directly estimated from underlying brain-activity generators, which could improve our analysis. Being that this work is a starting point of our research, we focused on the methodology to select an appropriate window duration for EC from the EEG signals. This can be enriched by using approaches to solve the inverse problem, providing an electrical source imaging analysis applied to EC.

5. Conclusions

The analysis of stationary characteristics of short-time segments of EEG signals is a topic that, despite its importance in effective connectivity analysis, does not receive enough attention to improve the connectivity results. The use of statistical metrics such as kurtosis to quantify the stationarity of a segment, and by introducing a mathematical description for processing multivariate processes coming from high-density electroencephalography recordings, it contributes to the assessment of the variation of statistical characteristics over time from the different signals. Moreover, including information from different subjects and conditions allows us to make an informed decision of a common segment length that serves to analyze EC in an inter-subject way, guaranteeing uniform conditions among the EEG datasets and conditions.

In addition, the results of our application example showed that uniform characteristics maintained over time with a given segment length provide comparable results to other research works in the literature and other insights that are worth investigating by considering more data, which unfortunately for this case were insufficient to confirm or not the involvement of specific regions in the brain regarding the EC analysis. In this sense, analyzing more data in the analysis is needed to improve the results, in addition to including EEG recordings of other brain activities that could enrich the assessment of the methodology presented here.

Author Contributions: Conceptualization, L.G., A.P. and A.M.; methodology, L.G.; software, L.G.; validation, L.G., A.P. and A.M.; formal analysis, L.G., A.P., A.M., G.R. and R.B; investigation, L.G., A.P. and A.M.; resources, A.M. and G.R.; data curation, L.G.; writing—original draft preparation, L.G.; writing—review and editing, L.G., A.P., A.M., G.R. and R.B.; visualization, L.G.; supervision, A.P., A.M., G.R. and R.B.; project administration, A.P. and R.B. All authors have read and agreed to the published version of the manuscript.

Funding: This research received no external funding.

Institutional Review Board Statement: Not applicable.

Informed Consent Statement: Not applicable.

Data Availability Statement: Not applicable.

Conflicts of Interest: The authors declare no conflict of interest.

References

1. He, B.; Astolfi, L.; Valdes-Sosa, P.A.; Marinazzo, D.; Palva, S.O.; Benar, C.G.; Michel, C.M.; Koenig, T. Electrophysiological Brain Connectivity: Theory and Implementation. *IEEE Trans. Biomed. Eng.* **2019**, *66*, 2115–2137. [CrossRef]
2. Astolfi, L.; Cincotti, F.; Mattia, D.; Marciani, M.G.; Baccala, L.A.; Fallani, F.D.V.; Salinari, S.; Ursino, M.; Zavaglia, M.; Ding, L.; et al. Comparison of Different Cortical Connectivity Estimators for High-Resolution EEG Recordings. *Hum. Brain Mapp.* **2007**, *28*, 143–157. [CrossRef]
3. Friston, K.J. Functional and Effective Connectivity: A Review. *Brain Connect.* **2011**, *1*, 13–36. [CrossRef]

4. Varotto, G.; Visani, E.; Canafoglia, L.; Franceschetti, S.; Avanzini, G.; Panzica, F. Enhanced Frontocentral EEG Connectivity in Photosensitive Generalized Epilepsies: A Partial Directed Coherence Study. *Epilepsia* **2012**, *53*, 359–367. [CrossRef]
5. Wada, M.; Nakajima, S.; Tarumi, R.; Masuda, F.; Miyazaki, T.; Tsugawa, S.; Ogyu, K.; Honda, S.; Matsushita, K.; Kikuchi, Y.; et al. Resting-State Isolated Effective Connectivity of the Cingulate Cortex as a Neurophysiological Biomarker in Patients with Severe Treatment-Resistant Schizophrenia. *J. Pers. Med.* **2020**, *10*, 89. [CrossRef]
6. Maharathi, B.; Loeb, J.A.; Patton, J. Estimation of Resting State Effective Connectivity in Epilepsy Using Direct-Directed Transfer Function. In Proceedings of the Annual International Conference of the IEEE Engineering in Medicine and Biology Society, EMBS, Orlando, FL, USA, 16–20 August 2016; Volume 2016, pp. 716–719.
7. Ghahari, S.; Salehi, F.; Farahani, N.; Coben, R.; Motie Nasrabadi, A. Representing Temporal Network Based on DDTF of EEG Signals in Children with Autism and Healthy Children. *Biomed. Signal Processing Control* **2020**, *62*, 102139. [CrossRef]
8. Tan, D.; Nijholt, A. Brain-Computer Interfaces and Human-Computer Interaction. In *Brain-Computer Interfaces*; Springer: London, UK, 2010; pp. 3–19.
9. Liu, Z.; Zhang, M.; Xu, G.; Huo, C.; Tan, Q.; Li, Z.; Yuan, Q. Effective Connectivity Analysis of the Brain Network in Drivers during Actual Driving Using Near-Infrared Spectroscopy. *Front. Behav. Neurosci.* **2017**, *11*, 211. [CrossRef]
10. Lazar, M. Mapping Brain Anatomical Connectivity Using White Matter Tractography. *NMR Biomed.* **2010**, *23*, 821–835. [CrossRef]
11. Rossini, P.M.; di Iorio, R.; Bentivoglio, M.; Bertini, G.; Ferreri, F.; Gerloff, C.; Ilmoniemi, R.J.; Miraglia, F.; Nitsche, M.A.; Pestilli, F.; et al. Methods for Analysis of Brain Connectivity: An IFCN-Sponsored Review. *Clin. Neurophysiol.* **2019**, *130*, 1833–1858. [CrossRef]
12. Bakhshayesh, H.; Fitzgibbon, S.P.; Janani, A.S.; Grummett, T.S.; Pope, K.J. Detecting Connectivity in EEG: A Comparative Study of Data-Driven Effective Connectivity Measures. *Comput. Biol. Med.* **2019**, *111*, 103329. [CrossRef]
13. Eichler, M. A Graphical Approach for Evaluating Effective Connectivity in Neural Systems. *Philos. Trans. R. Soc. B Biol. Sci.* **2005**, *360*, 953–967. [CrossRef]
14. Varotto, G.; Fazio, P.; Rossi Sebastiano, D.; Duran, D.; D'Incerti, L.; Parati, E.; Sattin, D.; Leonardi, M.; Franceschetti, S.; Panzica, F. Altered Resting State Effective Connectivity in Long-Standing Vegetative State Patients: An EEG Study. *Clin. Neurophysiol.* **2014**, *125*, 63–68. [CrossRef]
15. de Tommaso, M.; Stramaglia, S.; Marinazzo, D.; Trotta, G.; Pellicoro, M. Functional and Effective Connectivity in EEG Alpha and Beta Bands during Intermittent Flash Stimulation in Migraine with and without Aura. *Cephalalgia* **2013**, *33*, 938–947. [CrossRef]
16. Rotondi, F.; Franceschetti, S.; Avanzini, G.; Panzica, F. Altered EEG Resting-State Effective Connectivity in Drug-Naïve Childhood Absence Epilepsy. *Clin. Neurophysiol.* **2016**, *127*, 1130–1137. [CrossRef]
17. Hu, B.; Dong, Q.; Hao, Y.; Zhao, Q.; Shen, J.; Zheng, F. Effective Brain Network Analysis with Resting-State EEG Data: A Comparison between Heroin Abstinent and Non-Addicted Subjects. *J. Neural Eng.* **2017**, *14*, 046002. [CrossRef]
18. Valizadeh, S.A.; Riener, R.; Elmer, S.; Jäncke, L. Decrypting the Electrophysiological Individuality of the Human Brain: Identification of Individuals Based on Resting-State EEG Activity. *Neuroimage* **2019**, *197*, 470–481. [CrossRef]
19. Lee, M.; Yoon, J.-G.; Lee, S.-W. Predicting Motor Imagery Performance From Resting-State EEG Using Dynamic Causal Modeling. *Front. Hum. Neurosci.* **2020**, *14*, 321. [CrossRef]
20. Tafreshi, T.F.; Daliri, M.R.; Ghodousi, M. Functional and Effective Connectivity Based Features of EEG Signals for Object Recognition. *Cogn. Neurodynamics* **2019**, *13*, 555–566. [CrossRef]
21. Shahabi, H.; Moghimi, S. Toward Automatic Detection of Brain Responses to Emotional Music through Analysis of EEG Effective Connectivity. *Comput. Hum. Behav.* **2016**, *58*, 231–239. [CrossRef]
22. Olejarczyk, E.; Marzetti, L.; Pizzella, V.; Zappasodi, F. Comparison of Connectivity Analyses for Resting State EEG Data. *J. Neural Eng.* **2017**, *14*, 036017. [CrossRef]
23. Chen, J.L.; Ros, T.; Gruzelier, J.H. Dynamic Changes of ICA-Derived EEG Functional Connectivity in the Resting State. *Hum. Brain Mapp.* **2013**, *34*, 852–868. [CrossRef]
24. Tan, B.; Kong, X.; Yang, P.; Jin, Z.; Li, L. The Difference of Brain Functional Connectivity between Eyes-Closed and Eyes-Open Using Graph Theoretical Analysis. *Comput. Math. Methods Med.* **2013**, *2013*, 976365. [CrossRef]
25. Miraglia, F.; Vecchio, F.; Bramanti, P.; Rossini, P.M. EEG Characteristics in "Eyes-Open" versus "Eyes-Closed" Conditions: Small-World Network Architecture in Healthy Aging and Age-Related Brain Degeneration. *Clin. Neurophysiol.* **2016**, *127*, 1261–1268. [CrossRef]
26. Johnson, J.S.; Kundu, B.; Casali, A.G.; Postle, B.R. Task-Dependent Changes in Cortical Excitability and Effective Connectivity: A Combined TMS-EEG Study. *J. Neurophysiol.* **2012**, *107*, 2383–2392. [CrossRef]
27. Trujillo, P.; Mastropietro, A.; Scano, A.; Chiavenna, A.; Mrakic-Sposta, S.; Caimmi, M.; Molteni, F.; Rizzo, G. Quantitative EEG for Predicting Upper Limb Motor Recovery in Chronic Stroke Robot-Assisted Rehabilitation. *IEEE Trans. Neural Syst. Rehabil. Eng.* **2017**, *27*, 1058–1067. [CrossRef]
28. Mastropietro, A.; Arlati, S.; Mrakic-Sposta, S.; Fontana, L.; Franchin, C.; Malosio, M.; Pittaccio, S.; Gramigna, C.; Molteni, F.; Sacco, M.; et al. Quantitative EEG and Virtual Reality to Support Post-Stroke Rehabilitation at Home. In *Smart Innovation, Systems and Technologies, Proceedings of the International Conference on Innovation in Medicine and Healthcare, Puerto de la Cruz, Spain, 15–17 June 2016*; Springer Science and Business Media Deutschland GmbH: Berlin/Heidelberg, Germany, 2016; Volume 60, pp. 147–157.

29. Belfatto, A.; Scano, A.; Chiavenna, A.; Mastropietro, A.; Mrakic-Sposta, S.; Pittaccio, S.; Tosatti, L.M.; Molteni, F.; Rizzo, G. A Multiparameter Approach to Evaluate Post-Stroke Patients: An Application on Robotic Rehabilitation. *Appl. Sci.* **2018**, *8*, 2248. [CrossRef]
30. Delorme, A.; Makeig, S. EEGLAB: An Open Source Toolbox for Analysis of Single-Trial EEG Dynamics Including Independent Component Analysis. *J. Neurosci. Methods* **2004**, *134*, 9–21. [CrossRef]
31. Andreas Edgar Kothe, C. Artifact Removal Techniques with Signal Reconstruction. U.S. Patent Application 14/895,440, 28 April 2016.
32. Winkler, I.; Haufe, S.; Tangermann, M. Automatic Classification of Artifactual ICA-Components for Artifact Removal in EEG Signals. *Behav. Brain Funct.* **2011**, *7*, 30. [CrossRef]
33. Chang, C.Y.; Hsu, S.H.; Pion-Tonachini, L.; Jung, T.P. Evaluation of Artifact Subspace Reconstruction for Automatic Artifact Components Removal in Multi-Channel EEG Recordings. *IEEE Trans. Biomed. Eng.* **2020**, *67*, 1114–1121. [CrossRef]
34. Sameshima, K.; Baccalá, L.A. *Methods in Brain Connectivity Inference through Multivariate Time Series Analysis*; CRC Press: Boca Raton, FL, USA, 2014; ISBN 9781439845738.
35. Ding, L.; Duan, W.; Wang, Y.; Lei, X. Test-Retest Reproducibility Comparison in Resting and the Mental Task States: A Sensor and Source-Level EEG Spectral Analysis. *Int. J. Psychophysiol.* **2022**, *173*, 20–28. [CrossRef]
36. Zink, N.; Mückschel, M.; Beste, C. Resting-State EEG Dynamics Reveals Differences in Network Organization and Its Fluctuation between Frequency Bands. *Neuroscience* **2021**, *453*, 43–56. [CrossRef] [PubMed]
37. Sanei, S.; Chambers, J.A. *EEG Signal Processing*; John Wiley & Sons Ltd.: West Sussex, UK, 2013; ISBN 9780470025819.
38. Kuś, R.; Kamiński, M.; Blinowska, K.J. Determination of EEG Activity Propagation: Pair-Wise versus Multichannel Estimate. *IEEE Trans. Biomed. Eng.* **2004**, *51*, 1501–1510. [CrossRef] [PubMed]
39. Boccaletti, S.; Latora, V.; Moreno, Y.; Chavez, M.; Hwang, D.U. Complex Networks: Structure and Dynamics. *Phys. Rep.* **2006**, *424*, 175–308. [CrossRef]
40. Kamiński, M.; Ding, M.; Truccolo, W.A.; Bressler, S.L. Evaluating Causal Relations in Neural Systems: Granger Causality, Directed Transfer Function and Statistical Assessment of Significance. *Biol. Cybern.* **2001**, *85*, 145–157. [CrossRef] [PubMed]
41. Kaminski, M.J.; Blinowska, K.J. A New Method of the Description of the Information Flow in the Brain Structures. *Biol. Cybern.* **1991**, *65*, 203–210. [CrossRef] [PubMed]
42. Mullen, T.R.; de Sa, V.; Hollan, J.; Jernigan, T.; Kreutz-Delgado, K.; Makeig, S.; Thompson, W. *The Dynamic Brain: Modeling Neural Dynamics and Interactions From Human Electrophysiological Recordings*; University of California: San Diego, CA, USA, 2014.
43. Blinowska, K.J.; Kaminski, M. Functional Brain Networks: Random, "Small World" or Deterministic? *PLoS ONE* **2013**, *8*, e78763. [CrossRef]
44. van de Steen, F.; Almgren, H.; Razi, A.; Friston, K.; Marinazzo, D. Dynamic Causal Modelling of Fluctuating Connectivity in Resting-State EEG. *Neuroimage* **2019**, *189*, 476–484. [CrossRef]
45. Sakkalis, V. Review of Advanced Techniques for the Estimation of Brain Connectivity Measured with EEG/MEG. *Comput. Biol. Med.* **2011**, *41*, 1110–1117. [CrossRef]
46. Erhardt, E.B.; Rachakonda, S.; Bedrick, E.J.; Allen, E.A.; Adali, T.; Calhoun, V.D. Comparison of Multi-subject ICA Methods for Analysis of FMRI Data. *Hum. Brain Mapp.* **2011**, *32*, 2075. [CrossRef]
47. Hyvärinen, A. Testing the ICA Mixing Matrix Based on Inter-Subject or Inter-Session Consistency. *Neuroimage* **2011**, *58*, 122–136. [CrossRef]
48. von Bünau, P.; Meinecke, F.C.; Király, F.C.; Müller, K.R. Finding Stationary Subspaces in Multivariate Time Series. *Phys. Rev. Lett.* **2009**, *103*, 214101. [CrossRef] [PubMed]
49. Sundararajan, R.R.; Pourahmadi, M. Stationary Subspace Analysis of Nonstationary Processes. *J. Time Ser. Anal.* **2018**, *39*, 338–355. [CrossRef]

sensors

MDPI

Article

Multi-Scale Evaluation of Sleep Quality Based on Motion Signal from Unobtrusive Device

Davide Coluzzi [1,*], Giuseppe Baselli [1,*], Anna Maria Bianchi [1], Guillermina Guerrero-Mora [2], Juha M. Kortelainen [3], Mirja L. Tenhunen [4,5] and Martin O. Mendez [1,6]

1 Dipartimento di Elettronica, Informazione e Bioingegneria Politecnico di Milano, 20133 Milano, Italy; annamaria.bianchi@polimi.it (A.M.B.); martin.mendez@uaslp.mx (M.O.M.)
2 Unidad Académica Multidisciplinaria Zona Media, Universidad Autónoma de San Luis Potosí, San Luis Potosí 79615, Mexico; guillermina.guerrero@uaslp.mx
3 VTT Technical Research Center of Finland, Tampere, Finland; juha.m.kortelainen@vtt.fi
4 Department of Clinical Neurophysiology, Medical Imaging Centre,Pirkanmaa Hospital District, Tampere, Finland; mirja.tenhunen@pshp.fi
5 Department of Medical Physics, Tampere University Hospital, Medical Imaging Centre, Pirkanmaa Hospital District, Tampere, Finland
6 Laboratorio Nacional—Centro de Investigación, Instrumentación e Imagenología Médica, Facultad de Ciencias, Universidad Autónoma de San Luis Potosí, San Luis Potosí 78210, Mexico
* Correspondence: davide.coluzzi@polimi.it (D.C.); giuseppe.baselli@polimi.it (G.B.)

Abstract: Sleep disorders are a growing threat nowadays as they are linked to neurological, cardiovascular and metabolic diseases. The gold standard methodology for sleep study is polysomnography (PSG), an intrusive and onerous technique that can disrupt normal routines. In this perspective, m-Health technologies offer an unobtrusive and rapid solution for home monitoring. We developed a multi-scale method based on motion signal extracted from an unobtrusive device to evaluate sleep behavior. Data used in this study were collected during two different acquisition campaigns by using a Pressure Bed Sensor (PBS). The first one was carried out with 22 subjects for sleep problems, and the second one comprises 11 healthy shift workers. All underwent full PSG and PBS recordings. The algorithm consists of extracting sleep quality and fragmentation indexes correlating to clinical metrics. In particular, the method classifies sleep windows of 1-s of the motion signal into: displacement (DI), quiet sleep (QS), disrupted sleep (DS) and absence from the bed (ABS). QS proved to be positively correlated (0.72 ± 0.014) to Sleep Efficiency (SE) and DS/DI positively correlated (0.85 ± 0.007) to the Apnea-Hypopnea Index (AHI). The work proved to be potentially helpful in the early investigation of sleep in the home environment. The minimized intrusiveness of the device together with a low complexity and good performance might provide valuable indications for the home monitoring of sleep disorders and for subjects' awareness.

Keywords: sleep monitoring; pressure bed sensor (PBS); unobtrusive measure; multi-scale analysis; sleep apnea–hypopnea syndrome (SAHS); shift-working

Citation: Coluzzi, D.; Baselli, G.; Bianchi, A.M.; Guerrero-Mora, G.; Kortelainen, J.M.; Tenhunen, M.L.; Mendez, M.O. Multi-Scale Evaluation of Sleep Quality Based on Motion Signal from Unobtrusive Device. *Sensors* **2022**, *22*, 5295. https://doi.org/10.3390/s22145295

Academic Editor: Susanna Spinsante

Received: 30 May 2022
Accepted: 4 July 2022
Published: 15 July 2022

1. Introduction

Sleep is a biological process intrinsic to life and essential for optimal health as it plays a critical role in brain function and systemic physiology. However, sleep complications and disorders are a growing threat nowadays, affecting up to 70 million people in the United States and approximately 45 million in Europe [1]. Sleep disturbances can involve sleep deprivation and fragmentation [2], occurring when the necessary amount and quality of sleep is not achieved and when there is difficulty in falling asleep [3] or maintaining continuous pattern of sleep [4]. On the other hand, sleep can be affected by other disorder events such as respiratory or motor ones [3].

In this regard, one of the most common and alarming conditions of sleep breathing disorders is Sleep Apnea-Hypopnea Syndrome (SAHS). It affects more adult males with respect to adult females and it is associated with many factors such as overweight and obesity, alcohol, smoking, nasal congestion, and estrogen depletion in menopause, but the only intervention strategy currently supported with enough evidence is weight loss [5,6]. The sleep of subjects suffering of SAHS is characterized by cessations (apnea) or considerable reductions (hypopnea) in respiratory flow. These abnormal episodes are recurrent during the night and can last from a few seconds to minutes [7]. It follows that sleep results are strongly fragmented, whereas other symptoms are excessive sleepiness, decreased cognitive performance, fatigue and also depression [8].

Thus, fragmented sleep can affect the capabilities of memorization, learning and concentration, but also mood and behavior. Due to bad sleep quality, social problems are also frequent such as reduced working efficiency and increased risk in traffic accidents. Importantly, it is also well-known that when the poor sleep condition is prolonged for a long time the risk of developing cardiovascular pathologies such as hypertension increases. For these reasons and the current increase in the number of jobs requiring changing and prolonged shifts, such as nursery, the sleep fragmentation assessment represents a main topic [9,10].

Polysomnography (PSG) is currently the primary method for sleep analysis and is considered the gold standard for sleep monitoring. However, it is an onerous and intrusive technique that can disrupt normal routines. In addition, single nightly measurements of patients, are insufficient to study intrinsic patterns of variability or to correlate sleep with the timing of other activities [11].

With the perspective of minimizing the intrusiveness, m-Health technologies have been developed lately, offering a rapid, customized, and synergistic solution through the use of unobtrusive wearable or home automation devices to monitor vital signs during daily activities [12]. In spite of the fact that great diffusion only occurred in recent years, these devices have found applications in a wide range of scenarios [13] such as fitness or sport [14], rehabilitation [15], health monitoring [16,17] and sleep analysis [18,19] for the aims of prolonged monitoring and preventive interventions.

Different technologies were widely employed for different goals related to sleep analysis such as extracting quality indexes [20], evaluating fragmentation [21] or detecting disorders episodes [22] and sleep phases [23,24]. Methods can be divided according to the devices used, such as electrocardiogram-based [25], actigraphy [26], smartphones [27], smartwatches and complete IMUs [28] or contactless devices, such as bed pressure sensors [7,10,29]. The latter are one of the latest technologies having the advantage of not generating any discomfort. Indeed, these kinds of sensors do not need direct contact with the subject's body, but they can be integrated into the home environment. Furthermore, the position where the devices are located (smartwatches on the wrists, contactless devices embedded in the bed, near chest or under the mattress) was also evaluated in different studies [24,30,31].

Computational methods used to extract valuable information for screening purposes are mainly based on signal processing and Artificial Intelligence (AI). Common features extracted are averages, ranges, angles, skewness, kurtosis and Wavelet coefficients [32,33], whereas classifiers used are K-Nearest Neighbor (KNN) [34], Decision Tree, Random Forest, Support Vector Machine [10,24,34–36] and Hidden Markov Models (HMMs) [37].

In this work, we developed a multi-scale method based on motion signal extracted from an unobtrusive Pressure Bed Sensor (PBS) to evaluate sleep behavior. The contributions of the study are:

- The implementation of a visualization tool for sleep fragmentation as a function of the activity level;
- The evaluation of the sleep activity level dynamics from the multi-scale perspective;
- The sleep quality indexes extracted from the visualization tool and multi-scale analysis which were compared to clinical metrics, such as Sleep Efficiency (SE) and Apnea-Hypopnea Index (AHI);

- The analysis on motion signal from two different datasets composed of shift-working nurses and people with suspicions of sleep apnea;
- An easy tool useful for non-invasive devices based on the only motion signal suitable for home monitoring.

2. Materials and Methods

2.1. Data Acquisition and Study Population

Data used in this study were collected by means of two different acquisition campaigns performed by using the same device, already employed in [7,10,29]. Ethical approval and informed consent details are reported in the cited works.

The PBS device was designed with eight electrodes, located in two columns and four rows, to acquire the measurement of pressure change generated by the sleeping subject. PBS covers a measurement area of 64 cm × 64 cm and it was placed under the mattress at the middle of the sleeping subject's body. A deepened description of the setup and more details of the device are reported in [7,10,29]. The device was used to acquire:

Dataset 1: includes 22 subjects (11 males and 11 females, age: 48–63 years) that underwent full PSG and PBS recording at the laboratory of the Sleep Centre of Tampere University Hospital (TaUH, Tampere, Finland) for suspected sleep apnea. PSG measured cardiac (ECG), neuronal (EEG), and muscular (EMG) activity. In addition to two elastic bands for Respiratory Inductive Plethysmogram on the thorax and abdomen position, a pulse oximeter for oxygen saturation in blood, thermistor, and nasal cannula for airflow measurement were used during the recording. The Respiratory Event (RE) scoring was performed through an automatic procedure (Rem-Logic software - Embla Systems limited liability company) that detects abnormal events from the nasal airflow signal. For example, apneas are detected as a reduction greater or equal to 90% from the baseline. After the evaluation of the thoracic and abdominal respiratory effort for the classification of the REs, an expert clinician made manual corrections (e.g., false positive/negative REs), if necessary. Each RE present in the recordings was labeled according to four different classes corresponding to the type of RE: (1) Obstructive Sleep Apnea (OSA); (2) Central Apnea; (3) Hypopnea; and (4) Mixed Apnea [7].

Dataset 2: comprises 11 healthy females (age: 20–54 years) that underwent standard PSG and PBS recording at the sleep laboratory of the Finnish Institute of Occupational Health (FIOH, Helsinki, Finland) measuring night or day time sleep for shift workers. Two different recordings, one during daytime sleep after a night shift of work and one during nighttime sleep, were obtained from each subject. The hypnograms of the resulting 22 recordings were then scored by medical specialists following a standard procedure. Each sleep phase was labeled according to the 7 possible classes: (1) Stage 1; (2) Stage 2; (3) Stage 3; (4) Stage 4; (5) REM; (6) Wake with lights off and; (7) Wake with lights on [10].

PBS recording data files gathered were written into a memory card and synchronized with the reference PSG for the analysis. Information about all recordings from both datasets are summarized in Table 1.

2.2. Data Conditioning

A signal reflecting the motion and displacement activity occurring during sleep is possible to be captured from the different channels acquired through the PBS.

In Dataset 1, the motion signal was extracted computing the standard deviation for each measurement channel with a sliding raised cosine 4-s window. Then, the average value between channel-wise standard deviations was taken [7]. On the other hand, in Dataset 2, the motion signal was obtained from Principal Component Analysis (PCA) [10]. For both datasets the normalization for the maximum value of the recording was performed.

Table 1. Characteristics of the datasets.

1. Apnea Dataset						2. Shift-Work Dataset				
Rec.	Subj.	ST (h)	SE	TNE	AHI	Rec.	Subj.	ST (h)	SE	Timetable
1 *	S1	6.01	0.72	21	3.49	23	S23	4.34	0.95	D
2	S2	9.66	0.77	145	15.01	24	S23	9.00	0.83	N
3 *	S3	8.98	0.95	368	40.99	25	S24	3.90	0.84	D
4	S4	8.74	0.42	2	0.23	26	S24	9.83	0.85	N
5	S5	7.64	0.44	1	0.13	27	S25	4.94	0.85	D
6	S6	8.87	0.66	454	50.63	28	S25	8.36	0.69	N
7	S7	7.22	0.63	13	1.80	29	S26	4.06	0.69	D
8	S8	8.34	0.59	6	0.72	30	S26	8.37	0.89	N
9	S9	9.65	0.68	5	0.52	31	S27	4.89	0.86	D
10	S10	6.18	0.46	196	31.74	32	S27	9.05	0.83	N
11	S11	6.61	0.61	345	52.21	33	S28	5.54	0.94	D
12	S12	6.49	0.53	180	27.75	34	S28	8.46	0.95	N
13 *	S13	7.69	0.58	99	12.87	35	S29	5.25	0.93	D
14	S14	9.05	0.68	162	17.90	36	S29	8.68	0.75	N
15 *	S15	7.32	0.63	161	22.00	37	S30	4.13	0.93	D
16	S16	11.17	0.64	109	9.76	38	S30	8.09	0.90	N
17	S17	6.79	0.38	319	46.97	39	S31	4.60	0.86	D
18	S18	8.56	0.90	39	4.56	40	S31	9.23	0.85	N
19	S19	8.18	0.87	27	3.30	41	S32	4.80	0.79	D
20	S20	7.02	0.77	161	22.92	42	S32	7.86	0.92	N
21	S21	8.40	0.91	1	0.12	43	S33	5.21	0.47	D
22	S22	5.73	0.80	34	5.93	44	S33	9.52	0.71	N

ST: Sleep Time in hours; SE: Sleep Efficiency; TNE: Total Number of Events; AHI: Apnea-Hypopnea Index; The recordings marked with "*" symbol are the recordings considered uncertain (see the Section 2.6 for the selection of the uncertain recordings).

2.3. Pipeline Overview

A multi-scale algorithm using motion signal was designed to assess the sleep quality on the two different datasets. The pipeline can be divided into different steps with the purpose of identifying different states during sleep and analyze their trends at different time scales. After the extraction of the motion signal and the pre-conditioning, the thresholding method is applied to recognize different kinds of activity in various scenarios. Specifically: TH_{ABS} represents the threshold below which subject's absence from the bed is identified and TH_{DI} is the threshold above which displacements due to subject movements are detected. Afterwards, a multi-scale analysis based on the cumulative histogram of quiet sleep periods is performed to analyze sleep fragmentation to recognize quiet and disrupted sleep. The evaluation is based on prolonged periods of absence of displacements, identified through $minQS$ that represents the minimum duration considered for a quiet sleep interval. A summary of the pipeline is shown in Figure 1.

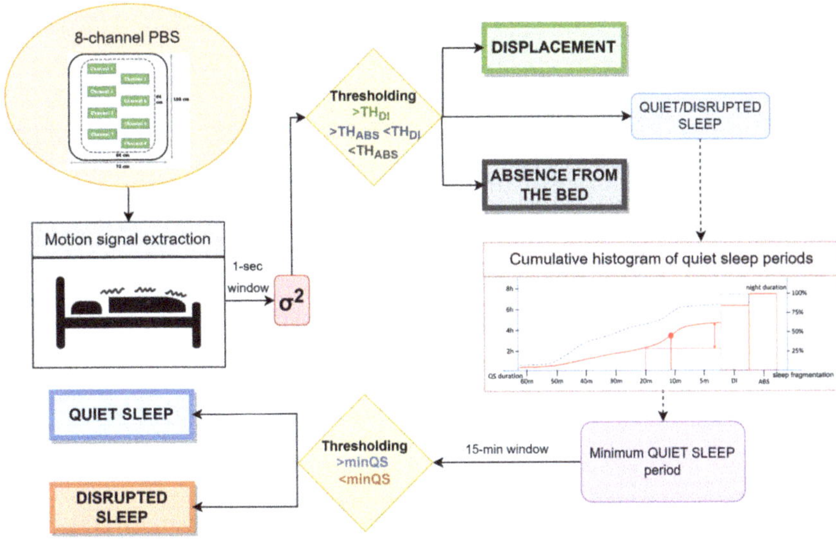

Figure 1. Complete pipeline of the designed algorithm.

2.3.1. Motion Detection

The power of the motion signal varies according to the different types of noise that may arise in the environment. Three types of noise identify three situations of interest to be monitored during sleep:

- External noise: due to the characteristics of the surrounding environment (e.g., traffic). When only this noise is present ($\sigma^2 < TH_{ABS}$) absence from the bed can be assumed (hereafter called ABS);
- Physiological noise: due to the natural physiological activity (e.g., breathing) of the subject. If detected ($TH_{ABS} < \sigma^2 < TH_{DI}$), presence in the bed with no sleep disturbs or movements can be assumed (hereafter called quiet sleep—QS);
- Displacement: due to physiological movements ($\sigma^2 > TH_{DI}$) during sleep cycle or abnormal ones (hereafter called DI).
  Body movements cause the strongest components in the signal, sometimes even saturating the sensor signal, being many orders higher than the other possible components generated by the different noise sources. It is well-known that in typical adult sleep behavior transitions from REM to almost-awake moments generate body movements each 1.5 h that last a few seconds in physiological sleep [38,39]. On the other hand, displacements may also be related to other kind of conditions and scenarios. In particular, the presence of disturbed breathing events (i.e., all thoracic movements stronger than normal physiological activity such as apnea) or abnormal movements (such as myclonias) induce strong fluctuations in the motion signal.
  The major difference between these cases can be identified through the different duration and periodicity of the events. The abnormal ones are, indeed, more frequent and closer to each other, resulting in shorter periods of disrupted sleep (hereafter called DS). An example of signal highlighting apnea events is shown Figure 2 (box 1).

Therefore, due to the huge differences in the power of the motion signal, the first phase of the algorithm consists of detecting the three main states through the thresholding method. In Figure 2 (box 2), a motion signal showing the differences in power during these distinct states (i) ABS; (ii) QS/DS and (iii) DI and the two thresholds that would identify them is reported. In the figure, it is also highlighted that signal intervals between the two thresholds cannot be considered only related to QS, but also to DS, according to the different duration of periods with no displacements.

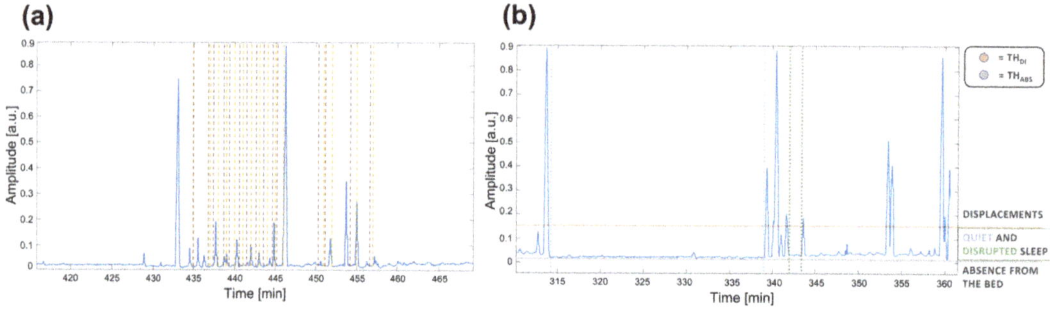

Figure 2. Example of motion signals on time intervals of about one hour of the rec. 2. In (**a**) the labeled apnea events are shown. Brown dashed lines represent the event starting, while yellow dashed lines the ending. In (**b**) the two thresholds are shown to highlight the different sources of noise. In particular, TH_{DI} (horizontal line in red) is the threshold above considering displacements, whereas TH_{ABS} (horizontal line in gray) is the threshold below considering absence from the bed because of the reduced activity due to the only external noise. The activity between the two thresholds highlights the period spent lying on the bed that can identify QS and DS. Furthermore, a long time interval identifying QS is highlighted between the two dashed blue vertical lines, while a short time interval identifying DS is shown between the two dashed green vertical lines.

2.3.2. Multi-Scale Analysis for Sleep Fragmentation

The only identification of body displacements may suggest potential sleep disorders but in some cases it is fundamental to specifically investigate their characteristics. For this purpose, we introduced the cumulative histogram of QS periods.

The proposed visualization method helps to investigate the duration of these periods, as well as the total amount of QS based on the multi-scale approach. In addition, the disruptions are also easily interpretable and analyzable in their characteristic periodicity. This evaluation of the sleep fragmentation allows to highlight random or specific patterns providing a minimum duration to actually consider a period as QS.

In some cases such as SAHS or myoclonia, threshold-based detection occurs frequently and for a short time. As a consequence, short periods of motion signal below TH_{DI} and between two detected DI events surrounding them (for example, intervals between apnea or abnormal movement events or short stationary periods due to physiological movements) would be correctly detected as DS because of the definition the minimum QS interval. On the other hand, intervals in which the subject is simply lying on the bed would not be considered as QS since they are expected to be characterized by shorter periods of absence of DI. Furthermore, cases in which frequent and long movements occur, not necessarily related to any specific disorder, would be highlighted, identifying a fragmented sleep that may be helpful to be aware of.

Therefore, the exploration of sleep fragmentation through the cumulative histogram of QS periods allows to improve the thresholding-based estimation by accurately identifying real QS ($length(TH_{ABS} < \sigma^2 < TH_{DI}) > minQS$) and DS ($length(TH_{ABS} < \sigma^2 < TH_{DI}) < minQS$). The latter, among all the possible scenarios in which it can occur, is indeed generally related to bad rest periods that it would be crucial to detect and distinguish from QS to correctly monitor the sleep.

In Figure 3, the expected cumulative histograms of QS periods in possible disturbed and healthy good sleep cases are shown. Specifically, it is possible to analyze how much time the subject has spent in periods of QS long at least a certain duration (indicated on the x-axis). Reducing this interval, the cumulative duration increases until it is matched to the recording duration. Indeed, the scale of durations is followed by DI and ABS durations which complete 100% of the cumulative. For this reason, the axis is oriented from long to short periods of QS. It is worth noting that the axis starts from periods of 60 min, because an occasional interruption of a longer period does not affect the estimate. It is worth noting that high slope points are marking the step-up of QS interruption below the given duration. This may be a marker of repeated disturbances (e.g., SAHS events or myoclonus) with a period equal or shorter than the step-up point.

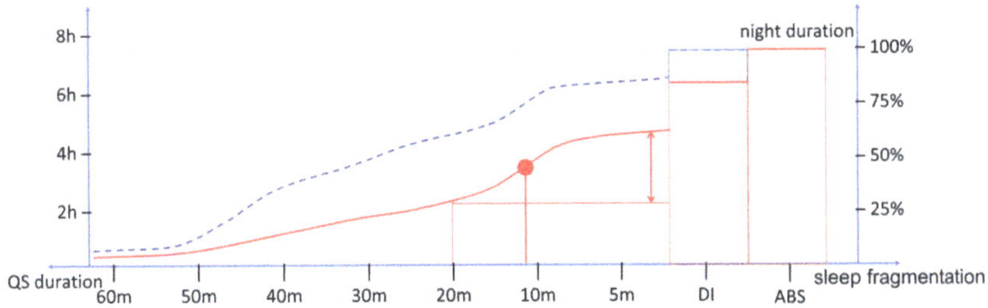

Figure 3. Schema representing the possible cumulative histogram of QS periods in disturbed (red) and healthy good (dashed blue) sleep. The point of maximum slope (red dot) is expected to characterize the dynamics of the fragmented sleep.

The typical sleep pattern is characterized by regular REM/light/deep sleep cycles, thus, it is expected to present no movement other than spontaneous ones occurring during transitions from REM and to result in a modest percentage of fragmented sleep. Conversely, distinct characteristics can be expected and investigated on the cumulative histogram of QS according to the different pathological/disturbed sleep. For example, major percentages of sleep constituted by short periods of QS or significant durations of ABS can be expected in SAHS or insomnia, respectively.

2.4. Displacement Analysis and Parameters Optimization

In order to identify the four states of interest (i) ABS; (ii) QS; (iii) DS (iiii) DI, it is necessary to appropriately tune the parameters of the method.

The first parameter to be set is the window size to be considered to evaluate the power of the signal. For this purpose, the distribution of durations of the characterizing DI events were investigated across the two datasets setting different thresholds at 0.01, 0.05, 0.1, 0.2, 0.3, 0.35 and 0.5 on the normalized signal.

From the Probability Density Functions (PDF) shown in Figure 4, it is possible to notice that the durations of the DI extracted from Dataset 1 are distributed up to 20 s. Conversely, in Dataset 2, the majority of the displacement periods (more than 50% of total duration) are segments of 1- and 2-s.

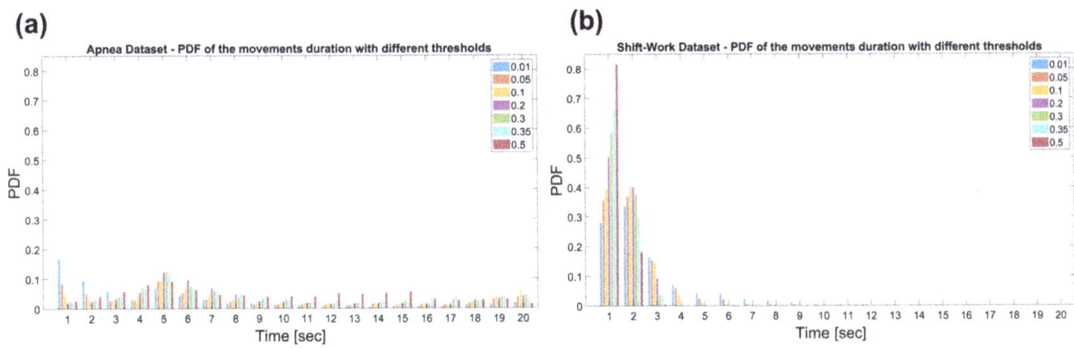

Figure 4. Probability Density Function (PDF) of the displacements duration in the Apnea Dataset (**a**) and the Shift-Work Dataset (**b**). The durations were obtained setting thresholds at 0.01, 0.05, 0.1, 0.2, 0.3, 0.35 and 0.5.

Therefore, the power of the signal was evaluated on 1-s windows as larger intervals would lead to the misdetection of short movements and transitions, which turn out to be very frequent, especially in the Dataset 2.

It is also worth noting that longer DI characterizing Dataset 1 are in agreement with subjects enrolled for sleep problems. Furthermore, the maximum value in the PDF, for almost all thresholds used can be noticed around 5 s, highlighting the typical duration in the order of less than a dozen seconds of the apnea episodes. [6,40]. Conversely, subjects enrolled in Dataset 2 are all healthy resulting in shorter DIs.

Afterwards, the two thresholds and the minimum QS period parameter, described in Section 2.3.1, were set. The optimization was performed using a grid-search based strategy on both datasets. In particular, the best values were found maximizing the correlations between QS and SE and between DS and AHI, when present. In order to have balanced values, both correlation values must be greater or equal than 0.5. Then, the best parameters were obtained maximizing the sum of the two correlations. The resulting values chosen were 0.05 for the threshold to recognize DI and 15 min as the minimum QS period.

2.5. Detrended Fluctuation Analysis

A widely-used multi-scale method is the Detrended Fluctuation Analysis (DFA). DFA is a nonstationary time series technique that allows to recognize long-range correlations. It is widely applied in the biomedical field for a variety of applications, such as [41,42]. DFA calculates the root-mean-square fluctuation of time series, disregarding trends and nonstationarities in the data. It allows the detection of intrinsic self-similarity, and it also avoids the spurious detection of apparent self-similarity.

DFA can be divided into three steps. The first one involves the shifting by the mean and the cumulative sum of the time series. The second one consists of dividing it into epochs (scales) of various size (logarithmically spaced) and considering these different segmentations. In the third step, each epoch e is detrended and locally fit to a polynomial finding the root mean square RMS_e, and then the $RMS_{\Delta s}$:

$$RMS_{\Delta s} = \sqrt{\frac{1}{N}\sum_{i=1}^{N}[y(i) - y_{\Delta s}(i)]^2} \quad (1)$$

where N is the total number of data points, $RMS_{\Delta s}$ is the root mean square obtained for each scale and y is the input signal.

The Hurst exponent H is then estimated by computing the linear fit between log-Δs and log-$RMS_{\Delta s}$ as a function of log-n. H is thus the slope of the line in the range of time scales of interest and can be estimated using linear regression. Through H it is possible to quantify the temporal correlations in the signal scale over different window sizes. In particular, whether:

- $H = 0.5$, the time series is uncorrelated;
- $H > 0.5$, there are larger fluctuations on longer time-scales than expected by chance, thus long-range correlations;
- $H < 0.5$, means that fluctuations are smaller in larger time windows than expected by chance, thus the time series is anti-correlated.

Results of DFA applied to the motion signal extracted were compared to indexes from the cumulative histogram of QS periods. For this reason, we logarithmically selected 15 scales from 1 to 60 min, in agreement to the scales considered by our method (see Section 2.3.2).

2.6. Experimental Evaluation

Pearson's correlation analysis was performed between different indexes in multiple scenarios and conditions to assess the extracted sleep quality.

First, the QS extracted index was correlated to *SE*, which was available for all recordings. *SE* is defined as:

$$SE = \frac{ST}{TIB} \tag{2}$$

where *ST* is the total sleep time and *TIB* the total time spent "in bed".

Second, *AHI*, which is the number of apnea and hypopnea events per hour of sleep, was correlated to DS/DI index. *AHI* is defined as:

$$AHI = \frac{TNE}{TR} \tag{3}$$

where *TNE* is the total number of apnea and hypopnea events and *TR* is total time duration in hours of the recording. The *AHI* values for adults are categorized as:

- Normal (N): $AHI < 5$
- Mild sleep apnea (Mi): $5 \leq AHI < 15$
- Moderate sleep apnea (Mo): $15 \leq AHI < 30$
- Severe sleep apnea (S): $AHI \geq 30$

In the correlation analyses, a recording with high values of SE and AHI (Rec.: 3; SE: 0.95; AHI: 40.99) was marked as uncertain. Furthermore, we also considered as uncertain three recordings where hypopneas composed at least 80% of the total abnormal breathing events (see Table 1). These specific abnormal respiratory events, indeed, do not generate any motion [43], thus resulting in being undetectable by PBS.

Furthermore, the two datasets were split according to the SE. Specifically, the threshold was set to 80%, being considered normal/healthy SE above it [44]. It resulted in 19 recordings with good sleep efficiency (GSE), of which 16 are from the Dataset 2 (8 during day and 8 during night) and 3 from Dataset 1 (all with $AHI < 5$) and 21 bad sleep efficiency (BSE), of which 15 are from the Dataset 1 (4 with $AHI \geq 30$, 4 with $15 \leq AHI < 30$, 2 with $5 \leq AHI < 15$ and 5 with $AHI < 5$) and 6 from the Dataset 2 (3 during day and 3 during night).

Correlation analyses were also separately performed on the two datasets to evaluate the presence and the duration of the displacements (DI state). These durations were statistically evaluated through Mann–Whitney tests between the independent subgroups obtained.

In particular, in Dataset 1, the recordings were analyzed together and divided into the normal and mild sleep apnea (N/Mi) group vs. moderate and severe sleep apnea (Mo/S) group. It resulted in groups of 10 N/Mi and 8 Mo/S recordings. In Dataset 2, the recordings were analyzed together and dividing between the ones acquired during the day (11 recordings) and during the night (11 recordings).

Finally, the multi-scale evaluation performed by using the algorithm was compared to DFA. First, the Hurst exponent, that assesses the self-similarity of the time series, was computed and tested through Mann–Whitney tests to assess statistically significant differences across all groups. Then, SE and AHI were correlated to the Hurst exponent.

3. Results

First, sleep fragmentation was evaluated through the cumulative histogram of QS periods. In general, this visualization revealed a greater area and a lower slope in subjects with high SE and low AHI. In Figure 5, some example cases together with pie charts of the three main states detected are shown.

It can be noticed that, since the percentage of DI is indicated in the cumulative histogram of QS immediately after the value 1 min on *x*-axis, the recordings having more movements (green area) results in steeper slopes between the last QS period and DI percentage (such as subject 17 reported on middle left). Furthermore, it is worth noting that the area of the cumulative histogram increases together with the percentage of QS (blue area). No differences are visible in the last part of the cumulative histogram of QS periods because the ABS state was never detected since no subject ever stood up during the acquisitions.

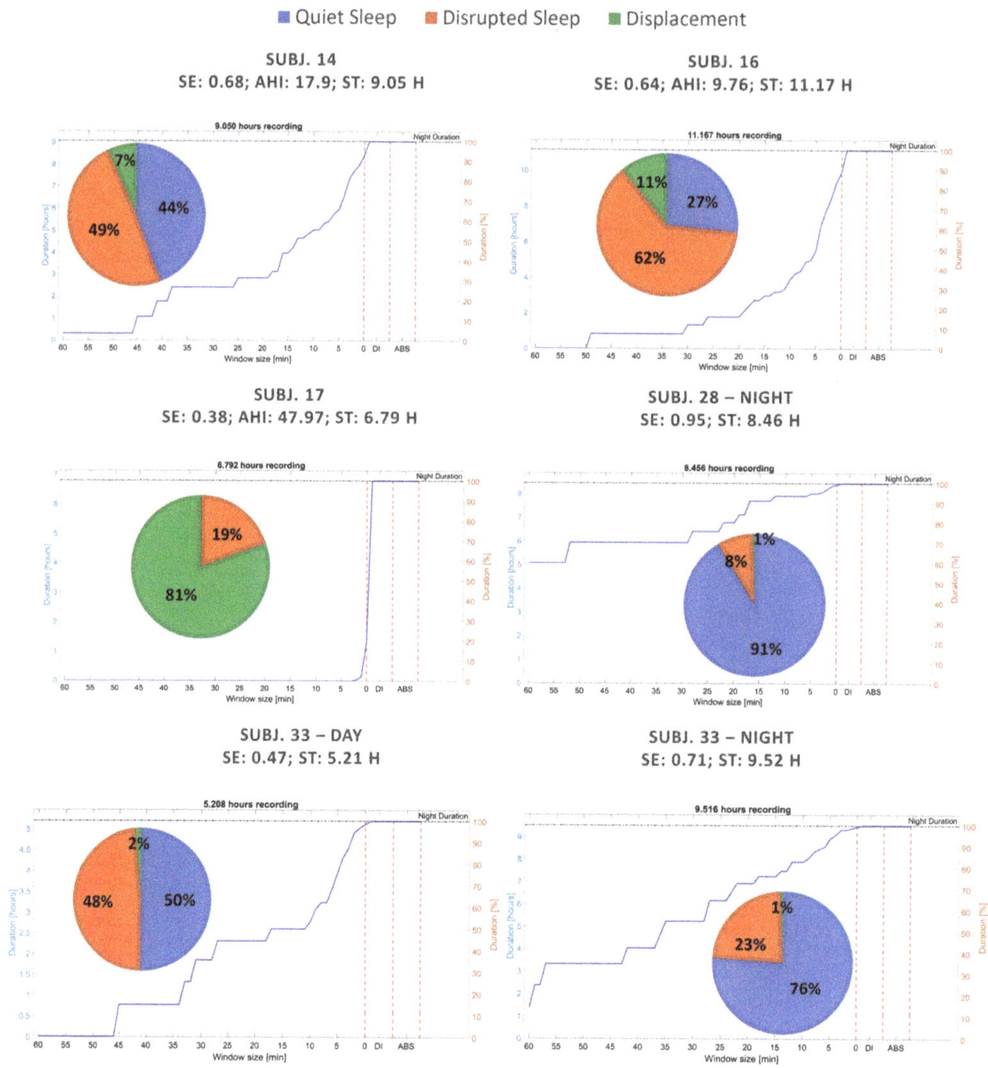

Figure 5. Sleep quality evaluation and fragmentation of recordings from both datasets through pie charts and cumulative histogram of QS periods.

All the results obtained by the algorithm for the three main states to be detected recording by recording are then summarized in Table 2.

Afterwards, the Leave-One-Out Cross-Validation was performed on both datasets evaluating the variability in the correlations between SE and QS detected and between AHI and DS/DI detected. Parameters and results found were in line to those obtained through Grid-Search approach described in the Section 2.4. In particular, the resulting correlations between all SE and QS obtained (0.7162 ± 0.0143) and AHI, when available, and DS/DI obtained (0.8537 ± 0.0073) remain stable across all folds.

Table 2. Sleep quality indexes detected by the proposed algorithm for each recording of the two datasets.

1. Apnea Dataset					2. Shift-Work Dataset				
Rec.	Subj.	QS (%)	DS (%)	DI (%)	Rec.	Subj.	QS (%)	DS (%)	DI (%)
1 *	S1	12.02	51.46	36.52	23	S23	90.42	8.92	0.66
2	S2	35.23	49.81	14.96	24	S23	71.37	27.34	1.29
3 *	S3	12.14	81.02	6.84	25	S24	93.64	5.43	0.93
4	S4	42.05	47.10	10.85	26	S24	79.78	19.66	0.56
5	S5	48.26	41.38	10.36	27	S25	81.31	17.99	0.70
6	S6	10.15	72.20	17.65	28	S25	68.34	30.61	1.05
7	S7	62.73	32.11	5.16	29	S26	73.18	25.99	0.82
8	S8	58.72	36.49	4.79	30	S26	79.03	20.08	0.89
9	S9	44.06	45.47	10.47	31	S27	81.28	18.13	0.59
10	S10	2.30	84.54	13.16	32	S27	68.44	30.76	0.80
11	S11	6.69	7.23	86.08	33	S28	92.63	6.94	0.43
12	S12	24.62	56.64	18.74	34	S28	91.23	8.34	0.43
13 *	S13	0.02	82.63	17.35	35	S29	77.15	22.40	0.45
14	S14	44.02	48.89	7.09	36	S29	74.65	24.63	0.72
15 *	S15	0	60.32	39.68	37	S30	84.57	14.87	0.56
16	S16	26.58	62.37	11.05	38	S30	77.02	22.19	0.79
17	S17	0	19.48	80.52	39	S31	68.61	30.76	0.63
18	S18	58.62	37.71	3.67	40	S31	69.66	29.31	1.03
19	S19	71.82	24.86	3.32	41	S32	82.21	17.14	0.65
20	S20	4.26	70.11	25.63	42	S32	92.58	6.87	0.55
21	S21	65.16	30.89	3.95	43	S33	50.16	48.39	1.45
22	S22	49.38	45.14	5.48	44	S33	76.13	23.27	0.60

The recordings marked with "*" symbol are the recordings considered uncertain (see the Section 2.6 for the selection of the uncertain recordings).

Afterwards, the agreement between SE and QS and AHI and DS/DI was specifically evaluated through Bland–Altman plots reported in Figures 6 and 7, where uncertain recordings were also marked.

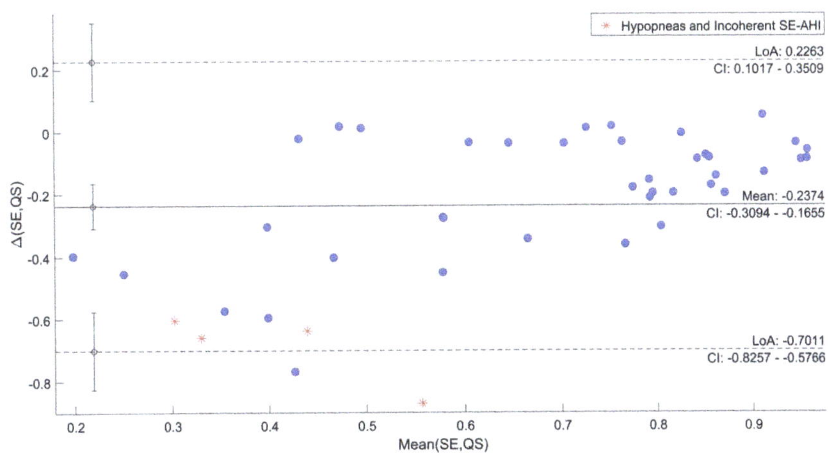

Figure 6. Bland–Altman Plot of SE vs. QS.

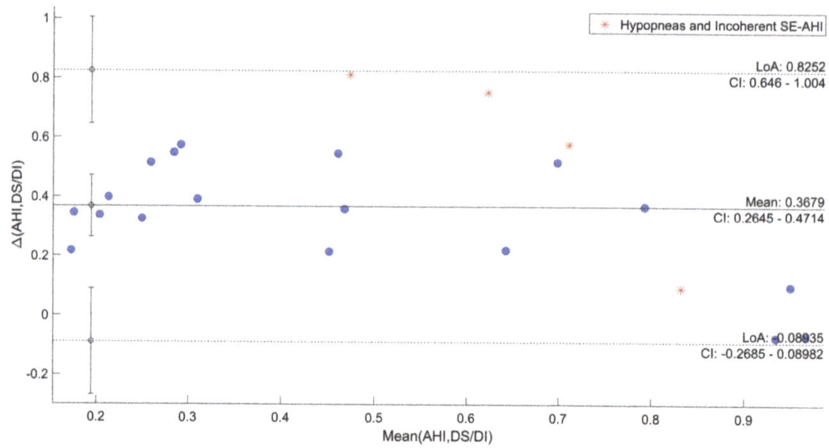

Figure 7. Bland–Altman Plot of AHI vs. DS/DI.

The plots point out differences between each QS and the corresponding SE, and between DS/DI and AHI. In both cases, the mean difference was close to the zero (−0.24 and 0.37, respectively). As regards the comparison between SE and QS, only one recording is out of agreement range (95% range: [−0.70; 0.23]) and it was one of those marked as uncertain. All the other differences resulted in good agreement across a wide range of SE. The other three uncertain recordings were among the recordings that deviated most from the average. Moreover, in the evaluation of AHI in comparison to DS/DI, all differences were within the confidential interval (95% range: [−0.09; 0.82]) with the uncertain recordings among the most deviated ones.

The DI state was then specifically assessed. It is possible to notice that, for the threshold ($th = 0.05$) selected through the procedure described in the Section 2.4, statistically significant differences were found. In particular, the duration of DI differs between Shift-Work and Apnea datasets. Furthermore, this difference was also found in the Apnea Dataset between N/Mi and Mo/S (p-value: <0.05) subgroups and in both datasets between GSE and BSE (p-value: <0.05) subgroups. A similar duration of displacements resulting in no statistical difference was found in day and night recordings of the Shift-Work Dataset. These results, together with the number of DI for each subgroup, are summarized in Table 3. It is worth noting that the duration of displacements is expressed through the mean and rank, at which outliers corresponding to 5% of the displacements in the least group were removed in the pairwise analysis.

Moreover, the Hurst exponent obtained from DFA was evaluated at dataset- and group-level, as described in the Section 2.6. Larger fluctuations resulting in a higher H value in the Apnea Dataset with respect to the Shift-Work Dataset were found. Similar dynamics were found between groups obtained from the same dataset (N/Mi vs. Mo/S, D vs. N; p-value: >0.05), but a statistically significant difference in the self-similarity was found between GSE and BSE groups extracted from both datasets (p-value: <0.05). All the results are summarized in Table 4.

Afterwards, the indexes extracted by the algorithm and Hurst exponent were correlated at dataset- and group-level with SE and AHI. All the results obtained from these correlation analyses are summarized in the Table 5.

The correlation between QS and SE resulted to be positive and strong in almost all groups. The Shift-Work Dataset resulted to be greater (0.82) in the case of recordings acquired during the day than those acquired during the night (0.66). In the whole Apnea Dataset, a minor correlation (0.5) with respect to Shift-Work Dataset (0.76) was noticed. At the same time, in this dataset, a strong correlation was found between DS/DI and AHI (0.85), greater in Mo/S (0.68) than N/Mi (0.44). On the contrary, with respect to QS vs. SE, the two subgroup correlations were comparable (N/Mi: 0.53; Mo/S: 0.48). Considering both datasets, QS and SE strongly correlated as previously mentioned (0.72). Conversely, the two subgroups, divided according to the SE show low and comparable correlation values (GSE: 0.4; BSE: 0.39).

As regards as DFA result evaluation, the Hurst exponent was correlated to SE and AHI, when available. In particular, in the Apnea Dataset, H vs. SE found correlation of −0.6 in N/Mi subgroup, −0.41 in Mo/S subgroup and −0.47 in the whole set. AHI was also compared to the results of the DFA but no correlations were found. On the other hand, in the Shift-Work Dataset, only a high positive correlation of 0.75 in the case of night recording was revealed. In general, good negative correlation (−0.53) between H and SE in both datasets was highlighted, resulting in a great difference between GSE (0.34) and BSE (−0.45) subgroups.

Table 3. Displacements extracted.

Displacements				
1. Apnea Dataset				
Dur mean [rank] (s)	N/Mi (n = 10) 22.93 [1, 63]	Mo/S (n = 8) 26.85 [1, 110]	p <0.05	Wh 25.54 [1, 110]
n. DI	1494	2973		4467
2. Shift-Work Dataset				
Dur mean [rank] (s)	D (n = 11) 2.19 [1, 6]	N (n = 11) 2.17 [1, 5]	p >0.05	Wh 2.17 [1, 6]
n. DI	607	1271		1878
Both				
Dur mean [rank] (s)	GSE (n = 19) 4.65 [1, 21]	BSE (n = 21) 24.12 [1, 111]	p <0.05	Wh 18.62 [1, 111]
n. DI	1791	4554		6345

Dur: duration in seconds; n. DI: number of displacements; Wh: whole dataset; n: number of recordings. Non parametric (Mann–Whitney test). In GSE: 16 are from Dataset 2 — 8 D and 8 N — and 3 from Dataset 1 — all N. In BSE: 15 are from the Dataset 1 — 4 S 4 Mo, 2 Mi and 5 N — and 6 from the Dataset 2 — 3 D and 3 N.

Table 4. Self-similarity through Hurst Exponent (H) computation.

Hurst Exponent				
1. Apnea Dataset				
H mean ± std	N/Mi (n = 10) 0.76 ± 0.07	Mo/S (n = 8) 0.75 ± 0.05	p >0.05	Wh 0.75 ± 0.06
2. Shift-Work Dataset				
H mean ± std	D (n = 11) 0.63 ± 0.06	N (n = 11) 0.65 ± 0.04	p >0.05	Wh 0.64 ± 0.05
Both				
H mean ± std	GSE (n = 19) 0.65 ± 0.05	BSE (n = 21) [6] 0.73 ± 0.08	p <0.05	Wh 0.69 ± 0.08

Wh: whole dataset; n: number of recordings. Non parametric (Mann–Whitney test). In GSE: 16 are from Dataset 2 — 8 D and 8 N — and 3 from Dataset 1 — all N. In BSE: 15 are from the Dataset 1 — 4 S 4 Mo, 2 Mi and 5 N — and 6 from the Dataset 2 — 3 D and 3 N.

Table 5. Correlation analyses.

Correlation Analyses									
	1. Apnea Dataset			**2. Shift-Work Dataset**			**Both**		
	N/Mi (n = 10)	**Mo/S (n = 8)**	**Whole**	**D (n = 11)**	**N (n = 11)**	**Whole**	**GSE (n = 19)**	**BSE (n = 21)**	**Whole**
QS-SE	**0.53**	0.48	**0.50**	**0.82**	**0.66**	**0.76**	0.40	0.39	**0.72**
DS/DI-AHI	0.44	**0.68**	**0.85**	na	na	na	na	na	na
H-SE	**−0.60**	−0.41	−0.47	−0.17	**0.75**	0.07	0.34	−0.45	**−0.53**
H-AHI	−0.21	0.30	−0.02	na	na	na	na	na	na

Pearson's correlation (good correlation in bold). na: unavailable results because of missing AHI; n: number of recordings. In GSE: 16 are from Dataset 2 — 8 D and 8 N — and 3 from Dataset 1 — all N. In BSE: 15 are from the Dataset 1 — 4 S 4 Mo, 2 Mi and 5 N — and 6 from the Dataset 2 — 3 D and 3 N.

4. Discussion

In this work, we proposed a multi-scale method to assess the sleep behavior from motion signals acquired through an unobtrusive device. For this purpose, we computed indexes related to the sleep fragmentation at different temporal scales and evaluated them through the comparison to clinical indexes. The complexity of the method is low, the hardware requirements are low-cost and the four indexes of quality estimated are easily interpretable and informative for users in everyday life.

The multi-scale analysis provided a visualization of sleep fragmentation and a tool to identify states of interest during sleep, with particular attention to the definition of quiet/disrupted sleep (QS/DS). In fact, although numerous valuable indexes are often estimated through objective measures from different devices [20–24], the recognition of real periods of QS is fundamental and may represent an easily interpretable indication for the subject, especially in home monitoring. In different sleep pathologies, multi-scale components of sleep fragmentation are difficult to be recognized and more informative visualizations would be essential in clinics to better interpret the pathology of a specific patient and its characteristics. For example, in the case of a Chronic Obstructive Pulmonary Disease subject, how long is the interval between two apneas? The cumulative histogram of QS periods extracted from the motion signal allowed us to analyze the sleep patterns in comparison to healthy subjects and to visualize differences in sleep fragmentation.

4.1. Sleep Quality Indexes Assessment

From the different examples in Figure 5 some important characteristics were enhanced. As expected, a minimal DI percentage was found in healthy subjects due only to the spontaneous movements before REM phases (brief awakenings), characterizing a typical no-disturbed sleep phase pattern [38,39]. Other general relevant properties observed in these cases were that 50% of sleep is composed of QS periods of more than 30 min and that the movements are exclusively composed of physiological ones that fragment in short periods of QS a modest percentage of sleep. Conversely, during different kinds of disturbed sleep it can be noticed that:

- Total time spent in DI state is greater than in the case of healthy sleep;
- Long periods of QS with an absence of DI constitute a small percentage of the night and fragment a modest percentage of sleep into short periods of QS;
- The point of maximum slope characterizes the dynamics of fragmented sleep.

In particular, the latter represents the minimum QS period to be considered to assure that real QS and DS periods are identified. In the datasets analyzed, 15 min was found to be the best value to distinguish healthy and pathological sleep. However, it is worth noting that, for the pathological cases, this point can significantly change according to the different nature and severity of the disturbance. For example, the maximum slope in the reported examples in Figure 5, although being less than 15 min, varies from being very close to the 0 min in the figure on the middle left (subject 17), to almost 15 min on the top left (subject 14). This value corresponds to the fastest change in the cumulative histogram, thus it is expected to underline the most frequent and characteristic time interval fragmenting the QS of the subject. The present insight points out the necessity of re-calibrating this parameter for the specific sleep disorder to enable an optimal recognition. At the same time, this result confirmed:

- The validity of the cumulative histogram of QS periods as a tool for the qualitative investigation of sleep fragmentation during a night of sleep;
- Its worthiness in longitudinal studies, whatever the chosen period is. In fact, although different sleep disorders can have different and specific dynamics, it is possible to highlight quality trends, showing improvements and worsenings among multiple days.

Furthermore, these findings on the cumulative histogram of QS periods demonstrated that a multi-scale analysis is needed when analyzing sleep from motion signal.

Indeed, for the purpose of assessing QS and DS, we dealt with the problem of the definition of a gold standard. These represent particular states during sleep which are easily interpretable but that, to the best of our knowledge, were not explored through objective measures from contactless devices. In this regard, what is a real period of quiet sleep is not straightforward since in many sleep disorders there are intervals of stillness that may be not actually quiet, as above-mentioned for apparent quiet sleep between abnormal breathing/movement events [40]. Another concern about this finding is about the intrinsic limitation of motion signal. It is indeed impossible to distinguish between real QS periods and intervals in which the patient is completely still but awake. This is an intrinsic limitation of the technology [24] but the identification of a minimum QS period can also improve the robustness of the methods in these cases. On one hand, by correctly setting this value, real QS periods are detected by verifying that they are enough long to be considered undisrupted. On the other hand, it is unlikely that an awake person remains totally still and with completely regular breathing for more than 15 min. Either way, the awareness on this definition of QS must be considered.

To have a direct relation of QS and DS with gold-standard, qualitative and quantitative analyses were carried out to show the agreement. From a first visual exploration of the results (in Table 2) it can be noticed that QS in agreement to SE, shown in Table 1, is higher in Dataset 2 than Dataset 1. On the other hand, the DI state appears to be much less present in Dataset 2, which is consistent with known characteristics of motion signal in SAHS [7]. Indeed, subjects from Dataset 1 were acquired for sleep problems, resulting in numerous members of the group suffering from SAHS and, thus, several abnormal movements. Furthermore, a higher DS/DI tends to be associated with a higher AHI and a reduced SE. A clear example is given by the comparison of recordings 6 and 19. Second, the correlations between SE and QS and DS/DI and AHI of all recordings resulted to be high and with low variability in cross-validation. This result was also confirmed by the Bland–Altman Plot in Figures 6 and 7, where all recordings resulted to be in the agreement range, except for one case, also marked as uncertain. In general, all uncertain recordings were among the most deviated ones. This may suggest a good correlation with the proposed measures, unless unexpected scenarios of SE and AHI and the intrinsic limitation of hypopneas recognition. It is indeed well known that this kind of event can be difficult to be detected by different devices and technologies [45,46], and, especially in motion signals where differences cannot be visualized [43].

The motion signal, indeed, reflects the activity occurring during sleep, capturing all kinds of movement, proving to correlate to wake stage periods [30,47]. The presence of movement was thus tested on the datasets available through DI state to point out possible valuable characteristics of the subgroups. In particular, it can be noticed that in Table 3 differences were found between all subgroups considered. In particular, in the case of splitting through AHI and SE the differences were found to be statistically significant, while dividing by timetable (Day vs. Night) not. Although a slight difference in DI durations resulting in a bit higher variability during day was found (mean $\pm$ std; D: 2.19 $\pm$ 1.47; N: 2.17 $\pm$ 1.35), the number of DI events per hour in the whole of Dataset 2 was higher during the night (D: 10.42; N: 13.84). This seems to confirm the similar results between daytime and nighttime sleep as in [9,10], especially in subjects adapted to the shift-works. In a future perspective, the algorithm may be employed in long-term monitoring at home according to the different shifts and to assess the adaptation to these. Unobtrusive technologies may be of unvaluable interest for the prevention of the well-known risks of occurrence of coronary heart and cardiovascular disease, and beyond that, psychomotor and mood problems [9,10,48].

4.2. Multi-Scale Analyses Comparison

Afterwards, the DFA multi-scale method was applied to investigate differences in the Hurst Exponent. First, it is worth noting that no significant changes were found in Table 4 between subgroups of the same datasets. D and N recordings show similar dynamics in agreement to similar SE values in the two groups (D: 0.83 ± 0.14; N: 0.83 ± 0.09) but also duration and number of DI per hour, as above mentioned. H also did not discriminate N/Mi and Mo/S apnea patients. It is worth noting that 7 of the 10 recordings within N/Mi had healthy AHI, but only 3 of these had healthy SE (≥80%). This result may suggest a reduced sleep quality due to possible other reasons [49], although a low number of apnea episodes occurred. This bias in the results seems to be also confirmed by the analysis performed dividing both datasets between high and low SE. These two subgroups were composed of 19 and 21 subjects, respectively, where the first included the 3 subjects from Dataset 1 considered healthy according to SE and 16 subjects from Dataset 2. This may suggest that self-similarity significantly grows in fragmented sleep, presenting larger fluctuations.

Table 5 showed the correlations found between clinical indexes and computed ones at the group- and subgroup-level. For example, for QS vs. SE a better correlation of daytime recordings was noted, which may be associated to a slightly less variable SE (SE-D: 0.83 ± 0.14; SE-N: 0.83 ± 0.08) due to shorter recordings. For other cases, slight general greater correlation was found in subjects that slept better (N/Mi and GSE), which is probably associated with motion signals from Mo/S and BSE being more variable, in general. For this reason, in cases of bad sleep it is easier to correctly recognize DS/DI, also mirroring the better correlation with AHI for Mo/S subgroup. Furthermore, bad sleep, in general, can be caused by a number of reasons [49]. For example, although cases of recordings with high percentage of hypopneas were excluded, in remaining ones they can still be present and produce false QS periods. To deeply investigate the hypopneas, the abnormal breathing events in the two subgroups of N/Mi and Mo/S were analyzed. Although in Mo/S the number and the duration was clearly greater, the percentage of hypopneas with respect to total duration of abnormal breathing events was less than N/Mi. In particular, 0.39% of the total duration of all breathing events in Mo/S were hypopneas, whereas 0.58% of the duration in N/Mi were hypopneas, hence resulting in a more difficult recognition of DS and higher QS periods identification.

Another interesting finding was that the Hurst exponent resulted to negatively correlate to SE (−0.53). In agreement with previous result, H appears to grow as SE decreases, and it is worth noting that this value is very similar in Mo/S and BSE subgroups (Mo/S: −0.41; BSE: −0.45). The latter indeed contains all recordings of Mo/S (8/21) but also seven from N/MI and six from Dataset 2. These cases appear to not heavily affect the result obtained in Mo/S group and consistency among all unhealthy subjects. In general, this may suggest an auto-affine structure in motion signal of SAHS cases, given by the known periodical pattern [50,51], which is not present instead in low SE cases. On the other hand, a positive good correlation was found between H and SE during night recordings of Shift-Work Dataset (0.75). In D group and whole Dataset 2 this correlation was not found. It is worth noting that the N group comprised the longest acquisitions and with good and the least variable SE across all subjects, resulting to be the most homogeneous group. In this case, a greater agreement between H and SE is suggested and it can be speculated to mirror the less clear auto-affine structure in shorter recordings. Furthermore, we may also speculate that H appears to be prone to great changes according to variable SE values. Furthermore, this could possibly explain the great difference between the values in GSE and BSE, since GSE is mainly composed by these recordings from Dataset 2, joined with three recordings from Dataset 1 containing few apnea episodes. On the other hand, BSE is mainly composed of recordings from Dataset 1, resulting in huge differences in correlations obtained in the use of DFA, probably due to such different apnea case patterns.

4.3. Home Monitoring Perspectives

The results cast a new light on the home sleep assessment measures obtained from unobtrusive devices that may be intuitively monitored by the subject. Recently many studies focused on the problem of minimizing intrusiveness [52], especially during sleep [53]. The problems of intrusiveness and conditioning related to PSG are well-known [54,55], thus the continuous screening through home devices results to be fundamental, especially considering the latest development of these technologies [56]. Moreover, PBS has the advantage of eliminating this problem. Numerous devices for sleep monitoring were successfully developed in recent years, such as smartwatches and waist or chest belts [20,24,27], but with the discomfort of wearing them during the whole sleep night. Although, it may be considered only a small limitation, its continuous use in daily living is discouraged. Conversely, contactless devices do not need direct contact to the patient's body, not generating any discomfort and reaching good performance.

In Table 6, some state-of-the-art studies are reported to compare the proposed work in terms of technologies, methods, datasets used, detected indexes and advantages. In particular, the studies were selected according to the datasets used and to characterize the most widespread and valuable sleep indexes extracted in literature. It is worth noting from the table that the studies based on EEG, ECG and PPG signals [32,36,57–62] can be used to extract valuable information on sleep stages or sleep apnea; however, they need higher computational cost and specialized devices for signal acquisition. On the other hand, other works based on motion signal from accelerometer and PBS [7,10,24,63,64] underline the advantage of causing low or mild discomfort to generally detect sleep and wake phases. However, the proposed work based on PBS allows us to characterize the sleep activity level dynamics from the multi-scale perspective and to provide interpretable indexes for the continuous home monitoring, based only on the motion signal.

It is worth noting that home assessment through these devices must be employed carefully. PSG is the gold-standard for sleep analysis and m-Health technologies may be helpful in raising a first alarm. Indeed, subjects suffering from many sleep disturbances are not often aware of their condition resulting in fatigue, low concentration and memorization [65]. In other cases, there is hope for the clinicians that biomarkers and other indicators will help diagnose presymptomatic signal of diseases. It was indeed found for example that Parkinson's Disease can be associated with Restless Leg Syndrome [66–68]. It follows that its preventive identification would be of great importance.

Furthermore, as above mentioned, the algorithm results to be particularly helpful for longitudinal study and, in general, to have an easy monitoring of personal sleep. It could be helpful, for example, to visualize the sleep fragmentation of specific disturbed nights or analyze the trend of QS/DS in correspondence to the introduction of preventive measures. Examples may be the better care of personal sleep hygiene, such as making sport [69,70], avoiding the use of electronic devices before sleeping [71], or in the worst scenarios, the introduction of sleeping medication.

Table 6. State of the art comparison.

State of the Art						
Reference	**Year**	**Device**	**Method**	**Dataset (n. sub)**	**Detected Indexes**	**Advantages**
Proposed work	2022	PBS	Multi-Scale Signal Processing based method	33 (HC vs. SAHS vs. SW)	ABS, QS, DS, DI	No discomfort, interpretability, model complexity
Hussain et al. [57]	2022	EEG	MLP	154	Sleep stages	Performance, low number of channels, no feature extraction
Yang et al. [58]	2022	ECG	1D-SEResGNet	25 (HC vs. SAHS)	OSA	Embeddable in wearable, no feature extraction
Wu et al. [59]	2021	PPG (wrist)	IBS for fluctuation analysis, RFC	92 (HC vs. SAHS)	AHI	Mild discomfort, interpretability
Banfi et al. [63]	2021	ACC (wrist)	CNN	81	Sleep vs. Wake	Mild discomfort, no feature extraction
Baty et al. [36]	2020	ECG belt	SVM	241 (HC vs. SAHS)	AHI	Mild discomfort, interpretability
Hulsegge et al. [64]	2019	2 ACC (thigh, ankle)	LMM and GEE logistic regression	194 (SW vs. non-SW)	Onset, Offset, TST	Mild discomfort, interpretability, model complexity
Mendez et al. [10]	2017	PBS	SVM	6 SW	Sleep Stages	No discomfort, interpretability, model complexity
Aktaruzzaman et al. [24]	2017	ACC (wrist), HRV	SVM	18 HC	Sleep vs. Wake	Mild discomfort, interpretability, model complexity
Mora et al. [7]	2015	PBS	Signal Processing based method	24 (HC vs. SAHS)	AHI	No discomfort, interpretability, model complexity

EEG: Electroencephalography; ECG: Electrocardiography; PPG: Photoplethysmography; ACC: triaxial accelerometer; HRV: Heart Rate Variability; MLP: Multilayer Perceptron; 1D-SEResGNet: one-dimensional squeeze-and-excitation residual group network; IBS: Information-Based Similarity; RFC: Random Forest Classifier; CNN: Convolutional Neural Network; SVM: Support Vector Machine; LMM: Linear Mixed Models; GEE: Generalized estimation equations; SW: Shift Workers.

4.4. Accelerometer Experimentation and Adaptability

The present study was conceived with the aim of also identifying periods of absence from the bed, which may be particularly helpful in cases of subjects with insomnia disorder or that awakens multiple times during night [72]. Due to the fact that no subject in either dataset ever got out of bed during the recordings, this investigation was not possible on the presented data. This points out the importance of home monitoring, since the acquisition conditions in a controlled environment do not perfectly mirror the real conditions. This state was qualitatively assessed through an experimentation performed on a prototypal device with a triaxial accelerometer, designed to monitor the sleep of subjects during daily living. In particular, the recognition of the state of absence from the bed was achieved through a manual setup, lasting 1 min. The device acquired data for 1 min with and without the subject on the bed and setting the threshold through a ROC analysis. Due to the significant difference between the only external noise, due to traffic and environmental conditions, visible when subject is not lying on the bed, and physiological noise, due to breathing for example, the identification did not result in relevant errors. All the results on the other states representing sleep indexes resulted to be in line with PBS performance, confirming the adaptability of accelerometer data.

5. Conclusions

In this work, we studied the multi-scale behavior of the motion signal extracted from PBS during sleep. The experimentation conducted on two different datasets acquired from shift-working nurses and people with suspicions of sleep apnea was assessed in correlation to clinical indexes and compared to a multi-scale method. The entire pipeline is suitable for online computation on an unobtrusive device dedicated to the described purpose of avoiding any discomfort to the subject. This may provide valuable indications in daily living for a rapid and continuous screening of sleep through a home device.

Author Contributions: Conceptualization, D.C., G.B. and A.M.B.; methodology, D.C., M.O.M., G.B. and A.M.B.; investigation, D.C. and M.O.M.; data curation, M.O.M., G.G.-M., J.M.K. and M.L.T.; writing—original draft preparation, all authors; writing—review and editing, all authors. All authors have read and agreed to the published version of the manuscript.

Funding: The research was partly funded by the Lombardia project (Announcement POR-FESR 2014-2020) SIDERA^B–Sistema Integrato DomiciliarE e Riabilitazione Assistita al Benessere (https://www.liuc.it/ricerca/ricerca-accademica/progetti/siderab-sistema-integrato-domiciliare-e-riabilitazione-assistita-al-benessere/), a project that aims at implementing a new platform for monitoring and coaching of patients with chronic diseases during home rehabilitation.

Institutional Review Board Statement: The study was conducted according to the guidelines of the Declaration of Helsinki, and approved by the Ethics Committee of the Sleep Centre of Tampere University Hospital (TaUH, Tampere, Finland) and the Finnish Institute of Occupational Health (FIOH, Helsinki, Finland).

Informed Consent Statement: Written informed consent has been obtained from the subjects to publish this paper.

Acknowledgments: The authors are grateful to collaborators from Tenacta Group S.P.A, Bergamo, Italy, which developed a prototypal device for sleep monitoring.

Conflicts of Interest: The authors declare no conflict of interest.

References

1. Medic, G.; Wille, M.; Hemels, M.E. Short- and long-term health consequences of sleep disruption. *Nat. Sci. Sleep* **2017**, *9*, 151–161. [CrossRef] [PubMed]
2. Short, M.A.; Banks, S. The Functional Impact of Sleep Deprivation, Sleep Restriction, and Sleep Fragmentation. In *Sleep Deprivation and Disease: Effects on the Body, Brain and Behavior*; Bianchi, M.T., Ed.; Springer: New York, NY, USA, 2014; pp. 13–26. [CrossRef]
3. Institute of Medicine (US) Committee on Sleep Medicine and Research. *Sleep Disorders and Sleep Deprivation: An Unmet Public Health Problem*; The National Academies Collection: Reports Funded by National Institutes of Health; National Academies Press (US): Washington, DC, USA, 2006.

4. Bonnet, M.H.; Arand, D.L. Clinical effects of sleep fragmentation versus sleep deprivation. *Sleep Med. Rev.* **2003**, *7*, 297–310. [CrossRef] [PubMed]
5. Young, T.; Peppard, P.E.; Gottlieb, D.J. Epidemiology of Obstructive Sleep Apnea. *Am. J. Respir. Crit. Care Med.* **2002**, *165*, 1217–1239.
6. Kim, N.H.; Lee, S.K.; Eun, C.R.; Seo, J.A.; Kim, S.G.; Choi, K.M.; Baik, S.H.; Choi, D.S.; Yun, C.H.; Kim, N.H.; et al. Short Sleep Duration Combined with Obstructive Sleep Apnea is Associated with Visceral Obesity in Korean Adults. *Sleep* **2013**, *36*, 723–729. [CrossRef]
7. Mora, G.G.; Kortelainen, J.M.; Hernández, E.R.P.; Tenhunen, M.; Bianchi, A.M.; Méndez, M.O. Evaluation of Pressure Bed Sensor for Automatic SAHS Screening. *IEEE Trans. Instrum. Meas.* **2015**, *64*, 1935–1943.
8. Sleep-related breathing disorders in adults: Recommendations for syndrome definition and measurement techniques in clinical research. The Report of an American Academy of Sleep Medicine Task Force. *Sleep* **1999**, *22*, 667–689. [CrossRef]
9. Boudreau, P.; Dumont, G.A.; Boivin, D.B. Circadian Adaptation to Night Shift Work Influences Sleep, Performance, Mood and the Autonomic Modulation of the Heart. *PLoS ONE* **2013**, *8*, e70813.
10. Mendez, M.O.; Palacios-Hernandez, E.R.; Alba, A.; Kortelainen, J.M.; Tenhunen, M.L.; Bianchi, A.M. Detection of the Sleep Stages Throughout Non-Obtrusive Measures of Inter-Beat Fluctuations and Motion: Night and Day Sleep of Female Shift Workers. *Fluct. Noise Lett.* **2017**, *16*, 1750033.
11. Kelly, J.M.; Strecker, R.E.; Bianchi, M.T. Recent Developments in Home Sleep-Monitoring Devices. *ISRN Neurol.* **2012**, *2012*, 768794. [CrossRef]
12. Isetta, V.; Torres, M.; González, K.; Ruiz, C.; Dalmases, M.; Embid, C.; Navajas, D.; Farré, R.; Montserrat, J.M. A New mHealth application to support treatment of sleep apnoea patients. *J. Telemed. Telecare* **2017**, *23*, 14–18.
13. Lara, O.D.; Labrador, M.A. A Survey on Human Activity Recognition using Wearable Sensors. *IEEE Commun. Surv. Tutor.* **2013**, *15*, 1192–1209.
14. Li, R.T.; Kling, S.R.; Salata, M.J.; Cupp, S.A.; Sheehan, J.; Voos, J.E. Wearable Performance Devices in Sports Medicine. *Sports Health* **2016**, *8*, 74–78.
15. Patel, S.; Park, H.; Bonato, P.; Chan, L.; Rodgers, M. A review of wearable sensors and systems with application in rehabilitation. *J. Neuroeng. Rehabil.* **2012**, *9*, 21. [CrossRef]
16. Rivolta, M.W.; Aktaruzzaman, M.; Rizzo, G.; Lafortuna, C.L.; Ferrarin, M.; Bovi, G.; Bonardi, D.R.; Caspani, A.; Sassi, R. Evaluation of the Tinetti score and fall risk assessment via accelerometry-based movement analysis. *Artif. Intell. Med.* **2019**, *95*, 38–47. [CrossRef]
17. Coluzzi, D.; Rivolta, M.W.; Mastropietro, A.; Porcelli, S.; Mauri, M.L.; Civiello, M.T.L.; Denna, E.; Rizzo, G.; Sassi, R. Design and Validation of a Minimal Complexity Algorithm for Stair Step Counting. *Computers* **2020**, *9*, 31.
18. Ibáñez, V.; Silva, J.; Cauli, O. A survey on sleep assessment methods. *PeerJ* **2018**, *6*, e4849.
19. de Zambotti, M.; Cellini, N.; Goldstone, A.; Colrain, I.M.; Baker, F.C. Wearable Sleep Technology in Clinical and Research Settings. *Med. Sci. Sport. Exerc.* **2019**, *51*, 1538–1557. [CrossRef]
20. Hees, V.T.v.; Sabia, S.; Anderson, K.N.; Denton, S.J.; Oliver, J.; Catt, M.; Abell, J.G.; Kivimäki, M.; Trenell, M.I.; Singh-Manoux, A. A Novel, Open Access Method to Assess Sleep Duration Using a Wrist-Worn Accelerometer. *PLoS ONE* **2015**, *10*, e0142533.
21. Cerutti, S.; Bianchi, A.M.; Reiter, H. Analysis of sleep and stress profiles from biomedical signal processing in wearable devices. In Proceedings of the 2006 International Conference of the IEEE Engineering in Medicine and Biology Society, New York, NY, USA, 30 August–3 September 2006; pp. 6530–6532.
22. Mendez, M.O.; Ruini, D.D.; Villantieri, O.P.; Matteucci, M.; Penzel, T.; Cerutti, S.; Bianchi, A.M. Detection of Sleep Apnea from surface ECG based on features extracted by an Autoregressive Model. In Proceedings of the 2007 29th Annual International Conference of the IEEE Engineering in Medicine and Biology Society, Lyon, France, 22–26 August 2007; pp. 6105–6108.
23. Kortelainen, J.M.; Mendez, M.O.; Bianchi, A.M.; Matteucci, M.; Cerutti, S. Sleep Staging Based on Signals Acquired Through Bed Sensor. *IEEE Trans. Inf. Technol. B* **2010**, *14*, 776–785.
24. Aktaruzzaman, M.; Rivolta, M.W.; Karmacharya, R.; Scarabottolo, N.; Pugnetti, L.; Garegnani, M.; Bovi, G.; Scalera, G.; Ferrarin, M.; Sassi, R. Performance comparison between wrist and chest actigraphy in combination with heart rate variability for sleep classification. *Comput. Biol. Med.* **2017**, *89*, 212–221. [CrossRef]
25. Kesper, K.; Canisius, S.; Penzel, T.; Ploch, T.; Cassel, W. ECG signal analysis for the assessment of sleep-disordered breathing and sleep pattern. *Med. Biol. Eng. Comput.* **2012**, *50*, 135–144. [CrossRef]
26. Smith, M.T.; McCrae, C.S.; Cheung, J.; Martin, J.L.; Harrod, C.G.; Heald, J.L.; Carden, K.A. Use of Actigraphy for the Evaluation of Sleep Disorders and Circadian Rhythm Sleep-Wake Disorders: An American Academy of Sleep Medicine Clinical Practice Guideline. *J. Clin. Sleep Med.* **2018**, *14*, 1231–1237.
27. Reimer, U.; Emmenegger, S.; Maier, E.; Zhang, Z.; Khatami, R. Recognizing sleep stages with wearable sensors in everyday settings. In Proceedings of the 3rd International Conference on Information and Communication Technologies for Ageing Well and e-Health (ICT4AWE 2017), Porto, Portugal, 1 January 2017; pp. 172–179. [CrossRef]
28. Kalkbrenner, C.; Stark, P.; Kouemou, G.; Algorri, M.E.; Brucher, R. Sleep monitoring using body sounds and motion tracking. In Proceedings of the 2014 36th Annual International Conference of the IEEE Engineering in Medicine and Biology Society, Chicago, IL, USA, 26–30 August 2014; pp. 6941–6944.

29. Kortelainen, J.M.; Gils, M.v.; Pärkkä, J. Multichannel bed pressure sensor for sleep monitoring. In Proceedings of the 2012 Computing in Cardiology, Krakow, Poland, 9–12 September 2012; pp. 313–316.
30. Sadeh, A.; Sharkey, M.; Carskadon, M.A. Activity-Based Sleep-Wake Identification: An Empirical Test of Methodological Issues. *Sleep* **1994**, *17*, 201–207. [CrossRef]
31. Lamprecht, M.L.; Bradley, A.P.; Tran, T.; Boynton, A.; Terrill, P.I. Multisite accelerometry for sleep and wake classification in children. *Physiol. Meas.* **2014**, *36*, 133–147.
32. Mendez, M.O.; Corthout, J.; Huffel, S.V.; Matteucci, M.; Penzel, T.; Cerutti, S.; Bianchi, A.M. Automatic screening of obstructive sleep apnea from the ECG based on empirical mode decomposition and wavelet analysis. *Physiol. Meas.* **2010**, *31*, 273–289.
33. Migliorini, M.; Bianchi, A.M.; Nisticò, D.; Kortelainen, J.; Arce-Santana, E.; Cerutti, S.; Mendez, M.O. Automatic sleep staging based on ballistocardiographic signals recorded through bed sensors. In Proceedings of the 2010 Annual International Conference of the IEEE Engineering in Medicine and Biology, Buenos Aires, Argentina, 31 August–4 September 2010; pp. 3273–3276.
34. Camcı, B.; Kahveci, A.Y.; Arnrich, B.; Ersoy, C. Sleep apnea detection via smart phones. In Proceedings of the 2017 25th Signal Processing and Communications Applications Conference (SIU), Antalya, Turkey, 15–18 May 2017; pp. 1–4. [CrossRef]
35. Sundararajan, K.; Georgievska, S.; te Lindert, B.H.W.; Gehrman, P.R.; Ramautar, J.; Mazzotti, D.R.; Sabia, S.; Weedon, M.N.; van Someren, E.J.W.; Ridder, L.; et al. Sleep classification from wrist-worn accelerometer data using random forests. *Sci. Rep.* **2021**, *11*, 24.
36. Baty, F.; Boesch, M.; Widmer, S.; Annaheim, S.; Fontana, P.; Camenzind, M.; Rossi, R.M.; Schoch, O.D.; Brutsche, M.H. Classification of Sleep Apnea Severity by Electrocardiogram Monitoring Using a Novel Wearable Device. *Sensors* **2020**, *20*, 286.
37. Trevenen, M.L.; Turlach, B.A.; Eastwood, P.R.; Straker, L.M.; Murray, K. Using hidden Markov models with raw, triaxial wrist accelerometry data to determine sleep stages. *Aust. N. Z. J. Stat.* **2019**, *61*, 273–298.
38. El Shakankiry, H.M. Sleep physiology and sleep disorders in childhood. *Nat. Sci. Sleep* **2011**, *3*, 101–114. [CrossRef]
39. Mong, J.A.; Cusmano, D.M. Sex differences in sleep: Impact of biological sex and sex steroids. *Philos. Trans. R. Soc. Lond. B Biol. Sci.* **2016**, *371*, 20150110. [CrossRef]
40. Penzel, T.; McNames, J.; de Chazal, P.; Raymond, B.; Murray, A.; Moody, G. Systematic comparison of different algorithms for apnoea detection based on electrocardiogram recordings. *Med. Biol. Eng. Comput.* **2002**, *40*, 402–407. [CrossRef] [PubMed]
41. Castiglioni, P.; Parati, G.; Civijian, A.; Quintin, L.; Rienzo, M.D. Local Scale Exponents of Blood Pressure and Heart Rate Variability by Detrended Fluctuation Analysis: Effects of Posture, Exercise, and Aging. *IEEE Trans. Bio-Med. Eng.* **2009**, *56*, 675–684.
42. Márton, L.F.; Brassai, S.T.; Bakó, L.; Losonczi, L. Detrended Fluctuation Analysis of EEG Signals. *Proc. Technol.* **2014**, *12*, 125–132. [CrossRef]
43. Trang, H.; Leske, V.; Gaultier, C. Use of Nasal Cannula for Detecting Sleep Apneas and Hypopneas in Infants and Children. *Am. J. Respir. Crit. Care Med.* **2002**, *166*, 464–468.
44. Desjardins, S.; Lapierre, S.; Hudon, C.; Desgagné, A. Factors involved in sleep efficiency: A population-based study of community-dwelling elderly persons. *Sleep* **2019**, *42*, zsz038. [CrossRef]
45. Varady, P.; Micsik, T.; Benedek, S.; Benyo, Z. A novel method for the detection of apnea and hypopnea events in respiration signals. *IEEE Trans. Biomed. Eng.* **2002**, *49*, 936–942.
46. Schlüter, T.; Conrad, S. An approach for automatic sleep stage scoring and apnea-hypopnea detection. *Front. Comput. Sci.* **2012**, *6*, 230–241. [CrossRef]
47. Sadek, I.; Mohktari, M. Nonintrusive Remote Monitoring of Sleep in Home-Based Situation. *J. Med. Syst.* **2018**, *42*, 64. [CrossRef]
48. Murata, K.; Yano, E.; Hashimoto, H.; Karita, K.; Dakeishi, M. Effects of shift work on QTc interval and blood pressure in relation to heart rate variability. *Int. Arch. Occup. Environ. Health* **2005**, *78*, 287–292. [CrossRef]
49. Panossian, L.A.; Avidan, A.Y. Review of Sleep Disorders. *Med. Clin. N. Am.* **2009**, *93*, 407–425. [CrossRef]
50. Hedner, J.; Pillar, G.; Pittman, S.D.; Zou, D.; Grote, L.; White, D.P. A Novel Adaptive Wrist Actigraphy Algorithm for Sleep-Wake Assessment in Sleep Apnea Patients. *Sleep* **2004**, *27*, 1560–1566. [CrossRef]
51. Nandakumar, R.; Gollakota, S.; Watson, N. Contactless Sleep Apnea Detection on Smartphones. In Proceedings of the 13th Annual International Conference on Mobile Systems, Applications, and Services (MobiSys'15), Florence, Italy, 18–22 May 2015; Association for Computing Machinery: New York, NY, USA, 2015; pp. 45–57. [CrossRef]
52. Zheng, Y.L.; Ding, X.R.; Poon, C.C.Y.; Lo, B.P.L.; Zhang, H.; Zhou, X.L.; Yang, G.Z.; Zhao, N.; Zhang, Y.T. Unobtrusive Sensing and Wearable Devices for Health Informatics. *IEEE Trans. Bio-Med. Eng.* **2014**, *61*, 1538–1554.
53. Chen, Z.; Lin, M.; Chen, F.; Lane, N.D.; Cardone, G.; Wang, R.; Li, T.; Chen, Y.; Choudhury, T.; Campbell, A.T. Unobtrusive sleep monitoring using smartphones. In Proceedings of the 2013 7th International Conference on Pervasive Computing Technologies for Healthcare and Workshops, Venice, Italy, 5–8 May 2013; pp. 145–152.
54. Hutchison, K.N.; Song, Y.; Wang, L.; Malow, B.A. Analysis of Sleep Parameters in Patients with Obstructive Sleep Apnea Studied in a Hospital vs. a Hotel-Based Sleep Center. *J. Clin. Sleep Med.* **2008**, *4*, 119–122. [CrossRef]
55. Church, G.D. The Role of Polysomnography in Diagnosing and Treating Obstructive Sleep Apnea in Pediatric Patients. *Curr. Probl. Pediatr. Adolesc. Health Care* **2012**, *42*, 2–25. [CrossRef]
56. Ometov, A.; Shubina, V.; Klus, L.; Skibińska, J.; Saafi, S.; Pascacio, P.; Flueratoru, L.; Gaibor, D.Q.; Chukhno, N.; Chukhno, O.; et al. A Survey on Wearable Technology: History, State-of-the-Art and Current Challenges. *Comput. Netw.* **2021**, *193*, 108074. [CrossRef]
57. Hussain, I.; Hossain, M.A.; Jany, R.; Bari, M.A.; Uddin, M.; Kamal, A.R.M.; Ku, Y.; Kim, J.S. Quantitative Evaluation of EEG-Biomarkers for Prediction of Sleep Stages. *Sensors* **2022**, *22*, 3079.

58. Yang, Q.; Zou, L.; Wei, K.; Liu, G. Obstructive sleep apnea detection from single-lead electrocardiogram signals using one-dimensional squeeze-and-excitation residual group network. *Comput. Biol. Med.* **2022**, *140*, 105124. [CrossRef]
59. Wu, S.; Chen, M.; Wei, K.; Liu, G. Sleep apnea screening based on Photoplethysmography data from wearable bracelets using an information-based similarity approach. *Comput. Methods Programs Biomed.* **2021**, *211*, 106442. [CrossRef]
60. Hussain, I.; Park, S.J. HealthSOS: Real-Time Health Monitoring System for Stroke Prognostics. *IEEE Access* **2020**, *8*, 213574–213586.
61. Qin, H.; Liu, G. A dual-model deep learning method for sleep apnea detection based on representation learning and temporal dependence. *Neurocomputing* **2022**, *473*, 24–36. [CrossRef]
62. Tang, L.; Liu, G. The novel approach of temporal dependency complexity analysis of heart rate variability in obstructive sleep apnea. *Comput. Biol. Med.* **2021**, *135*, 104632. [CrossRef]
63. Banfi, T.; Valigi, N.; di Galante, M.; d'Ascanio, P.; Ciuti, G.; Faraguna, U. Efficient embedded sleep wake classification for open-source actigraphy. *Sci. Rep.* **2021**, *11*, 345.
64. Hulsegge, G.; Loef, B.; Kerkhof, L.W.v.; Roenneberg, T.; Beek, A.J.v.d.; Proper, K.I. Shift work, sleep disturbances and social jetlag in healthcare workers. *J. Sleep Res.* **2019**, *28*, e12802.
65. Stöberl, A.S.; Schwarz, E.I.; Haile, S.R.; Turnbull, C.D.; Rossi, V.A.; Stradling, J.R.; Kohler, M. Night-to-night variability of obstructive sleep apnea. *J. Sleep Res.* **2017**, *26*, 782–788.
66. Ondo, W.G.; Vuong, K.D.; Jankovic, J. Exploring the Relationship Between Parkinson Disease and Restless Legs Syndrome. *Arch. Neurol.* **2002**, *59*, 421–424. [CrossRef]
67. Natarajan, R. Review of periodic limb movement and restless leg syndrome. *J. Postgrad. Med.* **2010**, *56*, 157.
68. Postuma, R.B.; Berg, D. Advances in markers of prodromal Parkinson disease. *Nat. Rev. Neurol.* **2016**, *12*, 622–634.
69. Mendelson, M.; Borowik, A.; Michallet, A.S.; Perrin, C.; Monneret, D.; Faure, P.; Levy, P.; Pépin, J.L.; Wuyam, B.; Flore, P. Sleep quality, sleep duration and physical activity in obese adolescents: Effects of exercise training. *Pediatr. Obes.* **2016**, *11*, 26–32.
70. Vanderlinden, J.; Boen, F.; van Uffelen, J.G.Z. Effects of physical activity programs on sleep outcomes in older adults: A systematic review. *Int. J. Behav. Nutr. Phys. Act.* **2020**, *17*, 11. [CrossRef]
71. Lastella, M.; Rigney, G.; Browne, M.; Sargent, C. Electronic device use in bed reduces sleep duration and quality in adults. *Sleep Biol. Rhythm.* **2020**, *18*, 121–129. [CrossRef]
72. Natale, V.; Léger, D.; Martoni, M.; Bayon, V.; Erbacci, A. The role of actigraphy in the assessment of primary insomnia: A retrospective study. *Sleep Med.* **2014**, *15*, 111–115. [CrossRef] [PubMed]

sensors

MDPI

Article

Towards a Practical Implementation of a Single-Beam All-Optical Non-Zero-Field Magnetic Sensor for Magnetoencephalographic Complexes

Mikhail Petrenko and Anton Vershovskii *

Ioffe Institute, Russian Academy of Sciences, 194021 St. Petersburg, Russia
* Correspondence: antver@mail.ioffe.ru

Abstract: We present a single-beam all-optical two-channel magnetic sensor scheme developed for biological applications such as non-zero-field magnetoencephalography and magnetocardiography. The pumping, excitation and detection of magnetic resonance in two cells are performed using a single laser beam with time-modulated linear polarization: the linear polarization of the beam switches to orthogonal every half-cycle of the Larmor frequency. Light with such characteristics can be transmitted over a single-mode polarization-maintaining fiber without any loss in the quality of the polarization characteristics. We also present an algorithm for calculating optical elements in a sensor scheme, the results of measuring the parametric dependences of magnetic resonance in cells, and the results of direct testing of a sensor in a magnetic shield. We demonstrate sensitivity at the level of 20 fT/$\sqrt{}$Hz in one sensor channel in the frequency range of 80–200 Hz.

Keywords: optically detected magnetic resonance; quantum magnetometer; magnetoencephalography

Citation: Petrenko, M.; Vershovskii, A. Towards a Practical Implementation of a Single-Beam All-Optical Non-Zero-Field Magnetic Sensor for Magnetoencephalographic Complexes. *Sensors* **2022**, *22*, 9862. https://doi.org/10.3390/s22249862

Academic Editors: Alfonso Mastropietro, Alessandro Scano and Massimo W. Rivolta

Received: 22 November 2022
Accepted: 12 December 2022
Published: 15 December 2022

1. Introduction

One of the most notable challenges of our time is the task of investigating ultra-weak magnetic fields of the brain. The set of scientific methods that provide a solution to this problem is called magnetoencephalography (MEG) [1,2]. The avalanche growth of interest in this problem, which has manifested itself over the past ten years, is mainly associated with the advent of compact, optical magnetic field sensors. The principle of operation of these sensors is based on the effect of magnetic resonance (MR) [3–5]. The application of these sensors to MEG problems has shaken the long-term monopoly of superconducting SQUID (superconducting quantum interference device) systems [6,7] and made it possible to overcome their inherent limitations.

The first (and still the most sensitive) optical sensors capable of competing with SQUID systems were sensors based on the SERF (spin exchange relaxation-free) effect [8–15]. These are zero-field since they operate only in a zero magnetic field, that is, in stationary magnetically shielded rooms. After SERF sensors convincingly demonstrated their competitiveness in MEG tasks, a number of research groups began to explore the possibility of adapting non-zero-field sensors to MEG tasks. These sensors are initially characterized by somewhat less sensitivity than SERF sensors. Still, their use would make it possible to drastically reduce the requirements for suppressing the external field and its spatial gradients. This, in turn, would make it possible to replace expensive magnetically shielded rooms with magnetic shields and, in the future, to do without shields at all [16–22]. The possibilities and prospects for the use of scalar non-zero-field optical magnetometers (the class to which the sensor presented in this work belongs) were studied in [20] and partially in [23]. A recent review [24] summarizes the general aspects of optical and magnetic field sensors and the problems associated with applications to biomagnetic measurements.

This paper presents a scheme of such a sensor, a single-beam all-optical non-zero-field two-channel magnetometer, i.e., a magnetometer-gradientometer of a non-zero field. The sensor is built in accordance with the principles we outlined earlier in [25,26]; it meets the MEG requirements for all the main parameters, namely, for sensitivity, speed and ability to function without creating RF interference to adjacent sensors.

2. Materials and Methods

The scheme proposed by us in [25] is extremely simple and compact. This advantage is due to several factors.

First, it uses a single beam with modulated (from partial left circular polarization to linear and then to partial right circular polarization) ellipticity for pumping, excitation, and detection of the MR. This scheme differs from numerous single-beam schemes proposed earlier [27–29] by the absence of sensitivity-reducing compromises. Pumping by the circularly polarized component and detection by the linearly polarized radiation component are separated in time, according to the phases of the Larmor precession. The ellipticity of the output beam changes its sign during the modulation period and acquires the maximum absolute value twice during the period $T_M = 2\pi/\omega_M$ [25]. The optimal values of maximum ellipticity lie in the range of 15–20°, meaning that the linear component is always present in the beam. Twice per period, the polarization becomes purely linear (π), with the polarization azimuth corresponding to the polarization azimuth of the incoming beam. For the purposes of the following discussion, radiation can be considered as the sum of two components, purely linear (π) and purely circular ($\sigma^{\pm}$), characterized by time-modulated intensities. This type of modulation is achieved using an electro-optical modulator (EOM). This allows for pumping and detection to be carried out with the highest possible efficiency.

Second, we use combined (hyperfine + Zeeman) pumping, first proposed in [30] and theoretically justified in [31]. The frequency of the beam is tuned to the D_1 optical line of the alkali metal line; it links the hyperfine level $F = I - \frac{1}{2}$ of the ground state $S_{1/2}$ of the atom with levels $F' = I \pm \frac{1}{2}$ of the nearest excited state $P_{1/2}$ [30,31]. The effective Zeeman pumping of the $F = I + \frac{1}{2}$, $m_F = F$ sublevel is due to the partial conservation of momentum in the excited state: the electronic part of the momentum is completely destroyed in collisions with the buffer gas, but the nuclear component is predominantly preserved [31].

Third, we use a modification of the M_x design, known as the Bell–Bloom scheme [4,32]. In this modification, the excitation of the MR is carried out by modulating the circular component of the pumping light at the Larmor frequency. This makes it possible to perform the excitation without a resonant radio-frequency field and, as a result, eliminate the interference such a field creates.

Fourth, we use strong optical pumping, which allows us to collect most of the atoms at the level $F = I + \frac{1}{2}$, $m_F = F$. William Happer called this state "end-state" or "stretched", and showed [33] that the spin-exchange rate in this state can decrease significantly. Indeed, as the pump intensity increases, the broadening of the magnetic resonance is preceded by its narrowing [34], which makes it possible to bring the sensitivity of the nonzero field sensor closer to that of the SERF sensor to some extent.

Finally, we detect MR at the transition $F = I + 1/2$, $m_F = F \leftrightarrow F - 1$ of the ground state by rotating the polarization angle of the linearly polarized (π) radiation component [35,36]. Therefore, the π-component of the beam is detuned in frequency from the interrogated optical transition by the hyperfine splitting of the ground state (for Cs, this is 9.192 GHz). Thus, the conditions for quantum non-demolition measurement (QND) are realized.

Thus, we simultaneously achieve near-optimal conditions for both optical pumping and MR excitation and detection. However, adapting the scheme [25] for application in MEG sensors is associated with certain difficulties. Since light contains both linearly and circularly polarized components, it cannot be transmitted through an optical fiber [37] without deteriorating its polarization characteristics. The obvious solution is to use a separate EOM in each sensor, which can significantly increase the cost of a multichannel MEG complex. On the contrary, the use of a common (sufficiently powerful compared to VCSEL lasers used in SERF zero-field sensors) pump source with a common EOM for several sensors would not only significantly simplify and reduce the cost of the MEG complex but also reduce technical noise by suppressing the common light noise.

In [38], we proposed a modification of the scheme, which will subsequently allow using a standard single-mode polarization-maintaining (SM-PM) optical fiber to solve this problem. Such a fiber has two eigenmodes characterized by orthogonal (s and p) polarizations propagating along the fiber's axis [39]. The phase delay between the modes is not fixed and can change when the fiber is bent, preventing radiation transmission with elliptical polarization. Nothing, however, prevents the transmission of *linearly polarized radiation with modulated azimuth* over the SM-PM fiber. The azimuth of the polarization is modulated as follows: s-polarization is transmitted through the fiber during the first half-cycle of the Larmor frequency, while p-polarization is transmitted during the second half-cycle (note that we do not impose any requirements on the stability of the phase delay between these two half-cycles). Now the problem is reduced to ensuring that this radiation can be converted into radiation containing π and $\sigma\pm$ components, properly modulated in intensity. As will be shown below, such a conversion can be achieved using a combination of a quarter-wave plate (QWP) and a regulated linear polarizer.

This paper presents a scheme of a single-beam all-optical non-zero field two-channel magnetometer-gradientometer (Figure 1) with two channels pumped and interrogated by one common beam; we also present a general algorithm for calculating the optical scheme of the sensor and the results of a study of its characteristics.

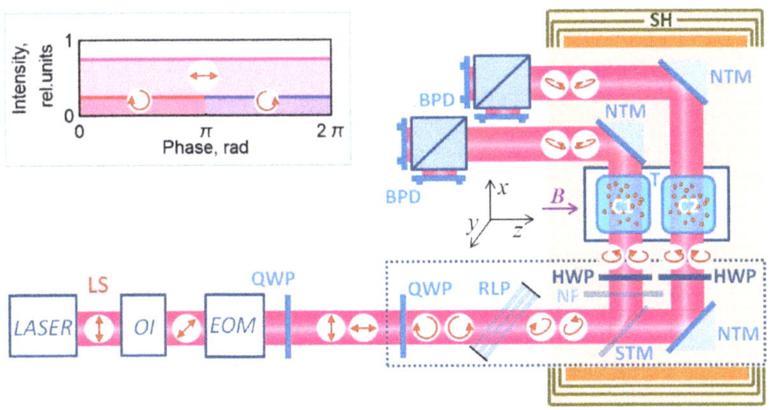

Figure 1. Simplified scheme of the experiment: LS—radiation source, OI—optical isolator, EOM—electro-optical polarization modulator, QWP—quarter-wave plate, RLP—regulated linear polarizer, NF—neutral filter, HWP—half-wave plates, C1, C2—gas cells with Cs vapors, STM is a semitransparent mirror, NTM is a non-transparent (opaque) mirror, BPD are balanced photodetectors, T is a thermostat, SH is a magnetic shield with a solenoid. Arrows indicate beam polarization states corresponding to two modulation half-cycles. Inset: time diagram of the polarization composition of the beam during one modulation period.

The measurements were carried out on the setup described in [23,25,40] and modified in accordance with the task of the experiment. The light source (LS) consisted of an external cavity diode laser (VitaWave ECDL 895R) generating about 25 mW at a wavelength of 894.592 nm, an optical isolator, and an electro-optical modulator (Thorlabs EO-AM-NR-C1). The control voltage at the EOM, modulated at a frequency of ~42 kHz with an amplitude of 200 V, provided a phase shift of ±45° between the components of the light decomposed along the EOM's own axes. An additional QWP (quarter-wave plate) provided linearly polarized radiation with modulated azimuth generation at the output of the radiation source.

The sensitive elements of the gradiometric sensor were cubic cells 8 × 8 × 8 mm^3 in size, containing saturated cesium vapor and nitrogen at a pressure of ~100 torr. A thermostat with cells and a heater was placed in the central region of a multilayer magnetic shield. A magnetic field induction of ~12 μT was maintained in the shield. A quarter-wave plate (QWP) installed at the sensor input converts linearly polarized radiation with modulated azimuth into radiation with switchable (from left to right and vice versa) circular polarization (the angle between the QWP axes and the fiber's own axes is 45°). Further, the regulated linear polarizer converts the circular polarization into an elliptical one, and the linear component necessary for detection appears in the beam. The linear polarizer used in our experiment is a stack of plane-parallel glass plates fixed at a Brewster angle to the beam direction in a common frame. The polarizer is adjusted by changing the number of plates. The angle of rotation of the frame around the beam determines the polarization azimuth of the π component. Unfortunately, in our experiment, the power of the laser source (taking into account the losses introduced by additional optical elements) turned out to be insufficient to ensure the optimal light intensity for pumping and interrogating two channels of the gradiometer. This prevented us from using SM-PM fiber. Instead, we had to confine ourselves to a model experiment, i.e., to reproduce at the output of the light source those characteristics that can certainly be obtained at the output of an ideal SM-PM fiber.

Half-wave plates (HWP) are installed in such a way as to ensure the optimal azimuth of the π-component of radiation in the cells with respect to the direction of the magnetic field vector. In our experiment, the linear polarizer was positioned in such a way that the electric vector $\boldsymbol{E}$ of the linear radiation component was parallel to the field vector $\boldsymbol{B}$. When D_1 line is used for the pump, the above makes it possible to minimize the broadening of the MR by the linear radiation component by eliminating its destructive interaction with the most populated (as a result of optical pumping) levels $F = I + 1/2$, $m_F = \pm F$. The sensor axis passes through the centers of cells C1 and C2 in the direction of light propagation—along the x-axis in Figure 1. When the sensor is rotated around its axis, the parallelism of vectors E and B can be ensured by choosing the direction of the HWP axis. This will make it possible to rotate the sensor around its axis by 360° without degrading its parameters, which should be considered an additional advantage of the proposed scheme.

The block of the optical scheme, which requires preliminary calculation, is enclosed in a dotted rectangle in Figure 1. Two problems were solved: (1) conversion of the input linearly polarized light with modulated azimuth into the light with the required polarization parameters, and (2) preservation of the polarization parameters of the light when the beam is split into two beams necessary for pumping and interrogating two cells. The ultimate goal of optimization was to ensure identical characteristics of the beams in the two cells in all phases of modulation.

The second task turned out to be non-trivial since any beam-splitting mirror, as well as any interference beam splitter, either changes the ratio of the intensities of the s and p radiation components or introduces a significant phase delay between them. Of the possible solutions, we chose the most compact one: rotating the beam polarization azimuth in front of the beam-splitting mirror and introducing a neutral filter into one of the channels. The rotation is carried out by rotating the linear polarizer frame; after passing through the beam-splitting unit, it has to be compensated by additional HWPs.

To calculate the optical scheme, we used the formalism of Mueller matrices [41]. The Stokes vector of radiation that has passed through a number of optical elements is described by successive multiplication by matrices corresponding to these elements. Thus, the Stokes vectors in two cells can be described by the expressions:

$$\begin{aligned} S_1 &= M_{HWP} M_{NTM} M_{STM-T} M_{RLP} M_{QWP} S_0; \\ S_2 &= M_{HWP} M_{NF} M_{STM-R} M_{RLP} M_{QWP} S_0, \end{aligned} \tag{1}$$

where S_0 is the Stokes vector of the input beam, M_{NTM} is the non-transparent mirror matrix, $M_{STM\text{-}T}$ is the semitransparent mirror matrix for the transmitted beam, $M_{STM\text{-}R}$ is the semitransparent mirror matrix for the reflected beam, M_{NF} is the neutral density filter matrix, M_{RLP} is the variable linear polarizer array, M_{QWP} is quarter-wave plate matrix, M_{HWP}—half-wave plate matrix. A stack of N plane-parallel glass plates located at the Brewster angle ($M_{RLP} = M_G{}^N$, one glass is described by the M_G matrix [42]) was used as a regulated linear polarizer.

The Mueller matrices used in our calculations are given in Appendix A. During the optimization, the following parameters varied: α, the RLR rotation angle, and T_{NF}, which is the density of the neutral filter.

Figure 2a shows the calculation result for the optical elements used in our experiment. The reflection and transmission coefficients of the beam-splitting mirror for the s component are $R_s = 0.72$ and $T_s = 0.28$, respectively, and for the p component, $R_p = 0.37$ and $T_p = 0.63$. The reflection coefficients for an opaque silver mirror for the s and p components are $R_s = 0.997$ and $R_p = 0.976$, respectively. Equalization of radiation parameters in two cells is achieved at $\alpha = 46°$ and $T_{NF} = 0.82$.

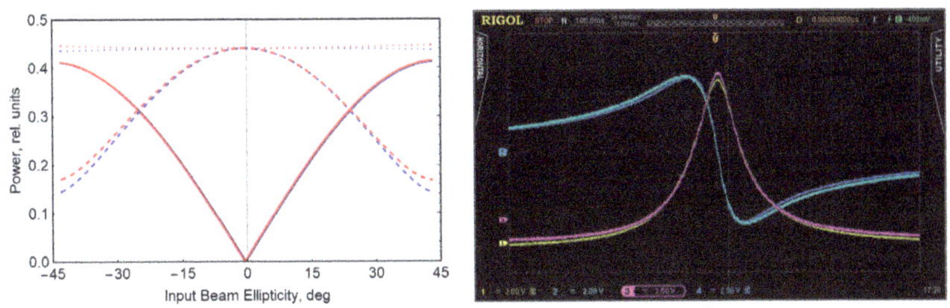

(**a**) (**b**)

Figure 2. (**a**) Example of calculation results: red lines are the light intensity at the input to cell C1, and blue lines are the light intensity at the input to cell C2. Solid lines are the circular component; dashed lines are the linear component; dotted lines are the total intensity. (**b**) Oscillograms of the magnetic resonance signals in cells C1 and C2 after synchronous detection (one component and MR signal module are shown).

Oscillograms of MR signals in two cells after synchronous detection (one component and MR signal module) are also shown (Figure 2b). As Figure 2b illustrates, the amplitudes and widths of the resonances in the cells are approximately the same, and there is no frequency shift between the resonances, which indicates a good balance of the light parameters in the two cells.

3. Results

Differences in the radiation characteristics in the proposed scheme from those required in [25] are reduced to the fact that the ellipticity modulation is carried out to a rectangular law (Figure 1) instead of a sinusoidal one. Thus, both the circular and linear components are characterized by constant intensities, and the phases of MR signal detection are not separated in time from the pump phases. The influence of the modulation shape in the standard two-beam Bell–Bloom scheme was studied in [40], and it was shown that although rectangular modulation leads to a slight broadening of the MR signal, it nevertheless allows values close to the ultimate sensitivity to be reached; however, the assumption that this is also true for the single-beam scheme requires proof. Therefore, we simulated the pumping conditions during the light transmission by the method described above and studied the MR parameters. The measurement results are shown in Figure 3. As in [25], we estimated the ultimate short-term sensitivity by calculating the ratio of the measured resonance amplitude to its measured width and to the calculated spectral density of the photocurrent shot noise.

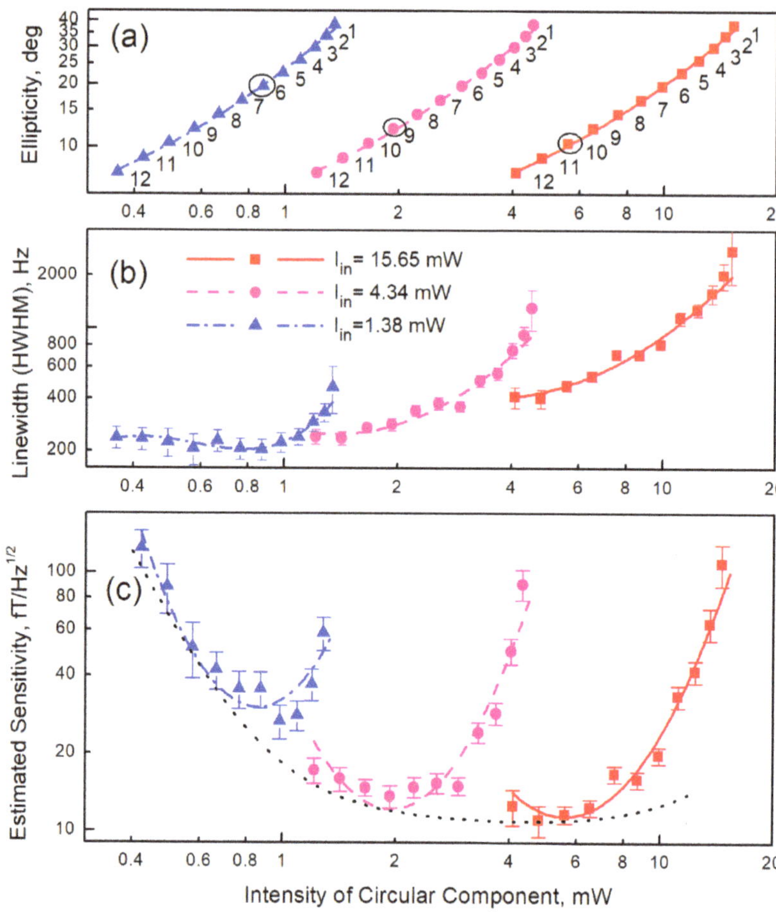

Figure 3. Dependence of the parameters of magnetic resonance when pumped with light with modulated ellipticity on the light intensity at the input of the cell: (**a**) ellipticity for different numbers (indicated by numbers in the graph field) of glass plates in a linear polarizer; the black circles indicate the optimal ellipticity values for this series, (**b**) magnetic resonance half-width, (**c**) estimation of the ultimate (limited by calculated shot noise) sensitivity. Connecting lines are guides to the eye.

In accordance with the results presented in Figure 3, the required value of ellipticity (Figure 3a) was chosen according to the criterion of maximum sensitivity (Figure 3c), based on the available intensity of laser light and the value of losses on the elements of the optical scheme. As a consequence, the parameters of the linear polarizer (the number of glass plates in a stack) and the light intensity in each cell were determined (see Section 4).

Next, we measured the gradiometric sensitivity of the proposed scheme when pumped with linearly polarized radiation with modulated azimuth. To do this, a magnetic coil was mounted on the frontal plane of the thermostat. The field generated by the coil in each of the cells was measured by the displacement of the magnetic resonance line. Based on the response to the same field, the frequency band of the sensor was determined: $f_0 = \Gamma/(2\pi) \approx 315$ Hz. Further, in the experiment, the response speed was additionally limited by the time constant of the synchronous detector (τ = 0.3 ms, 18 dB/octave). The measurement results are shown in Figure 4.

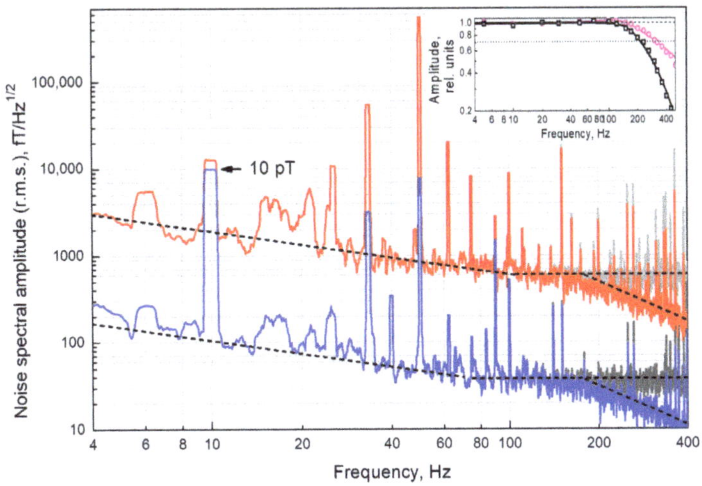

Figure 4. Noise spectrum of the magnetic resonance signal in cell C2 (red line), and the difference signal of magnetic resonances in cells C1 and C2 (blue line)—moving r.m.s. average in 1 Hz band. Gray lines are the spectra corrected for the frequency response of the sensor. The peak at a frequency of 10 Hz (marked with an arrow) is a calibration signal with an amplitude of 10 pT r.m.s. The peak at a frequency of 50 Hz is the interference from the main currents. The dashed lines are the noise floors of the signal in cell C2 and of the difference signal, respectively. Inset: the magenta line is the sensor's frequency response (cutoff frequency f_0 = 315 Hz), the black line is the sensor's frequency response, taking into account the time constant of the SR830 synchronous detector (τ = 0.3 ms, 18 dB/octave).

4. Discussion

Let us try to evaluate how the proposed changes in the sensor design affect its ultimate characteristics, the most significant of which are the achievable sensitivity and bandwidth. For this, we compare the MR parameters obtained in this work with the parameters obtained in [25]. According to the evaluation given in [25], the shot-noise-limited sensitivity reached 8.8 fT/$\sqrt{}$Hz at a bandwidth (determined by the MR width) of the order of $\Gamma/(2\pi) \approx 580$ Hz, whereas, according to Figure 3, the shot-noise-limited sensitivity reaches (11.0 $\pm$ 0.7) fT/$\sqrt{}$Hz at a bandwidth of $\Gamma/(2\pi) \approx 430$ Hz. These results show that the proposed scheme can be used in MEG complexes without noticeable deterioration in their parameters.

The difference in sensitivity is explained, in particular, by the additional light loss in the linear polarizer. The optimal value of ellipticity lies in the range of 10–20° (Figure 3a,c), which is fully consistent with the data [25]. With the intensity available to us in one cell (roughly corresponding to the magenta series in Figure 3), the ellipticity of (11 $\pm$ 1)° is optimal. This means that (47.9 $\pm$ 0.3)% of the total intensity is lost in an ideal adjustable linear polarizer. As a polarizer, we used a stack of conventional microscope coverslips. Due to the imperfection of the surfaces and the spread of their installation angles, the loss on a stack of 9–10 glass plates providing the corresponding ellipticity (see Figure A1) amounted to (66.7 $\pm$ 0.9)%. Under the conditions of limited laser power (15.65 mW at the EOM output); this loss forced us to reduce the working cell temperature to ~80 °C compared to 90 °C in [25].

It should be noted that the data in Figure 3 were obtained without using a beam splitter, i.e., all the light intensity was fed into one cell. When we operate with two cells (Figure 4), the power available in our experiment in each channel is ~40% of the maximum (see Figure 2a),—i.e., about 2.1 mW per cell. As a result, the ultimate shot-noise-limited sensitivity deteriorates to the value of (15.1 $\pm$ 0.7) fT/$\sqrt{}$Hz, and a MR half-width Γ is reduced to $2\pi{\cdot}350$ Hz.

The MR width, in addition to the bandwidth, also determines the permissible field inhomogeneity, that is, the maximum difference in magnetic fields at the points of location of individual sensors. Thus, at half-width $\Gamma = 2\pi \cdot 350$ Hz, the maximum allowable deviation of the field from the array-average value for a sensor based on cesium atoms will be approximately $k \cdot \Gamma / \gamma_{Cs} \approx 50$ nT (here, $\gamma_{Cs} \approx 2\pi \cdot 3.5$ Hz/nT is the gyromagnetic ratio Cs, $k \approx 0.5$ is the width of the conditionally linear section on the dispersion contour of the MR, referred to Γ). An array radius of 0.1 m corresponds to an allowable gradient of 1 μT/m.

If we exclude from the spectra in Figure 4 the zones of technical interference and technical noise that dominates at low frequencies (up to 80 Hz), the gradient noise lies in the range of 30–60 fT/$\sqrt{\text{Hz}}$. In terms of one channel of the sensor, this is 20–40 fT/$\sqrt{\text{Hz}}$, and approximately corresponds to the sensitivity limit estimate given earlier in this section. In addition to photon shot noise, the contribution to the white noise recorded at frequencies above 80 Hz can come from both technical factors (white thermal Johnson noise) and fundamental ones (atomic projection noise). The atomic projection noise amplitude with the optimal parameter configuration is comparable to the shot noise amplitude.

According to [43], in our cylindrical shield, in which the radius of the inner shell made of steel is $a = 17$ cm, the thermal noise amplitude should be ~23 fT/$\sqrt{\text{Hz}}$. The noise suppression coefficient in the gradiometric scheme in this shield should be about $1.19 \cdot (d/a)$, where d is the distance between the cells. In our experiment, $d = 1.0$ cm, which corresponds to noise suppression by a factor of 20, down to 1.1 fT/$\sqrt{\text{Hz}}$. The value of the thermal noise component proportional to $f^{-1/2}$ should also not exceed units of fT/$\sqrt{\text{Hz}}$ at a frequency of 1 Hz [43]. Thus, the thermal noise of the shield should not make a significant contribution to our measurements.

The external field's suppression level in the gradiometric scheme can be estimated from the suppression of pickup at a frequency of 50 Hz: it is suppressed approximately 70-fold. We can take the residual pickup level (~1.4%) as an upper bound for the unbalance of the gradiometer parameters.

At the same time, both the $f^{-1/2}$ noise, which dominates at frequencies up to 80 Hz, and white noise, which dominates at frequencies above 80 Hz, are suppressed much less, approximately by a factor of $16\sqrt{2} \approx 23$ (taking into account that two channels contribute to the noise of the difference signal). This can be explained by laser radiation noise, both intrinsic and acoustic, during the transmission of radiation through the air over a distance of ~2 m. Thus, to further improve the scheme, it is necessary, first, to increase the power of laser radiation (taking into account the inevitable losses during input into the SM-PM fiber) and, second, to actively stabilize its parameters.

5. Conclusions

We have shown that the earlier proposed scheme can be modified to exclude the transmission of elliptically polarized radiation from the pump source to the sensor—which makes it possible to use optical fiber for radiation transmission. This eliminates the last fundamental obstacle to constructing a magnetoencephalographic system of a non-zero field based on single-beam optical sensors. A magnetometer-gradientometer based on this principle has demonstrated a limiting sensitivity (estimated from the ratio of signal to linewidth and photon shot noise) at the level of (11.0 ± 0.7) fT/$\sqrt{\text{Hz}}$ at the optimum optical pump intensity and 15–18 fT/$\sqrt{\text{Hz}}$ at the distribution of pump radiation on two sensor channels. Direct measurement of the gradiometric sensitivity of the proposed scheme showed that the sensitivity of one sensor channel in the range of 80–200 Hz reaches 20 fT/$\sqrt{\text{Hz}}$. Further improvement in sensitivity can be achieved by using a more powerful laser pump source with a fiber output and active methods for suppressing laser radiation noise.

Author Contributions: Conceptualization, A.V.; Methodology, M.P. and A.V.; Software, A.V.; Supervision, A.V.; Validation, M.P. and A.V.; Writing, A.V.; funding acquisition, A.V. All authors have read and agreed to the published version of the manuscript.

Funding: The reported study was funded by RFBR, project number 19-29-10004.

Conflicts of Interest: The authors declare no conflict of interest.

Appendix A

The Muller matrix [41] for neutral filter:

$$M_{NF} = T_{NF}\begin{pmatrix} 1 & 0 & 0 & 0 \\ 0 & 1 & 0 & 0 \\ 0 & 0 & 1 & 0 \\ 0 & 0 & 0 & 1 \end{pmatrix}, \tag{A1}$$

where T_{NF} is the transmittance of the neutral filter.

Muller matrix for the phase plate (the expression is used to calculate the matrices M_{QWP}, M_{HWP}):

$$M_{WP} = \begin{pmatrix} 1 & 0 & 0 & 0 \\ 0 & Cos^2(2\Theta) + Sin^2(2\Theta)Cos(\delta) & Cos(2\Theta)Sin(2\Theta)[1 - Cos(\delta)] & Sin(2\Theta)Sin(\delta) \\ 0 & Cos(2\Theta)Sin(2\Theta)[1 - Cos(\delta)] & Cos^2(2\Theta)Cos(\delta) + Sin^2(2\Theta) & -Cos(2\Theta)Sin(\delta) \\ 0 & -Sin(2\Theta)Sin(\delta) & Cos(2\Theta)Sin(\delta) & Cos(\delta) \end{pmatrix}, \tag{A2}$$

where Θ is the angle of rotation of the main axis of the plate, δ is the phase delay angle (equal to $\pi/2$ for QWP, and π for HWP).

Muller matrix for a mirror (the expression is used to calculate the matrices M_{NTM}, $M_{STM\text{-}T}$, $M_{STM\text{-}R}$)):

$$M_M = \begin{pmatrix} \frac{R_p+R_s}{2} & \frac{R_p-R_s}{2} & 0 & 0 \\ \frac{R_p-R_s}{2} & \frac{R_p+R_s}{2} & 0 & 0 \\ 0 & 0 & -\sqrt{R_pR_s}Cos(\delta) & -\sqrt{R_pR_s}Sin(\delta) \\ 0 & 0 & \sqrt{R_pR_s}Sin(\delta) & -\sqrt{R_pR_s}Cos(\delta) \end{pmatrix}, \tag{A3}$$

where R_p and R_s are the reflection coefficients for p and s polarizations, respectively, and δ is the phase delay angle. When calculating the transmission through a semitransparent mirror, the reflection coefficients are replaced by the transmission coefficients $T_p = 1 - R_p$, $T_s = 1 - R_s$, and the sign of δ is inverted.

In general, the normal mirror plane is not parallel to the beam. If the axis of rotation of the mirror does not coincide with the axis of the coordinate system, the Muller matrix of the mirror is transformed using the matrix M_R, which describes the rotation of the polarization plane through the angle Θ:

$$M_{M'} = M_R(\Theta)M_M M_R(-\Theta) = M_R M_M M_R{}^{-1} \tag{A4}$$

where

$$M_R = \begin{pmatrix} 1 & 0 & 0 & 0 \\ 0 & Cos(2\Theta) & Sin(2\Theta) & 0 \\ 0 & -Sin(2\Theta) & Cos(2\Theta) & 0 \\ 0 & 0 & 0 & 1 \end{pmatrix}, \tag{A5}$$

Expression (A5) is also applicable to the calculation of the matrix $M_{SF}(n_1,n_2,\varphi)$, which describes reflection and refraction when light is incident at an angle φ on the boundary of two media with refractive indices n_1 and n_2, and the coefficients $R_p(n_1,n_2,\varphi)$, $R_s(n_1,n_2,\varphi)$, $T_p(n_1,n_2,\varphi)$, $T_s(n_1,n_2,\varphi)$ are described by Fresnel equations [42].

The matrix $M_{GL}(n_1,n_2,\varphi)$ describes the through passage of a beam through two surfaces of a plane-parallel glass plate:

$$M_{GL0}(n_1, n_2, \varphi) = M_{SF}(n_2, n_1, \psi)M_{SF}(n_1, n_2, \varphi). \tag{A6}$$

where ψ is the direction of the refracted beam inside the plate: $\sin(\psi) = (n_1/n_2)\sin(\varphi)$.

If the beam displacement is large compared to the beam diameter, subsequent re-reflections can be neglected and vice versa. If the plate thickness is small compared to the beam diameter, the beam shift due to re-reflections can be neglected. Then, the expression for the transmission of a plane-parallel glass plate $M_{GL}(n_1,n_2,\varphi)$ must be constructed by summing an infinite series describing multiple reflections from two surfaces:

$$M_{GL}(n_1,n_2,\varphi) = M_{SF}(n_2,n_1,\psi)\sum_{i=0}^{\infty}[1 - M_{SF}(n_2,n_1,\psi)]^{2i}M_{SF}(n_1,n_2,\varphi), \quad \text{(A7)}$$

Consequently,

$$M_{GL}(n_1,n_2,\varphi) = [2 - M_{SF}(n_2,n_1,\psi)]^{-1}M_{SF}(n_1,n_2,\varphi). \quad \text{(A8)}$$

Accordingly, a stack of N plates located at an angle φ to the beam direction is described by the matrix:

$$M_{RLP}(n_1,n_2,\varphi,\Theta) = M_R(\Theta)(M_{GL}(n_1,n_2,\varphi))^N M_R(-\Theta). \quad \text{(A9)}$$

Since the RLP must provide partial suppression of one linear component with the maximum transmission of another, the angle φ should be chosen equal to the Brewster angle: $\varphi = \varphi_{Br} = \mathrm{arctg}(n_2/n_1)$. The calculated linear polarizer parameters are shown in Figure A1.

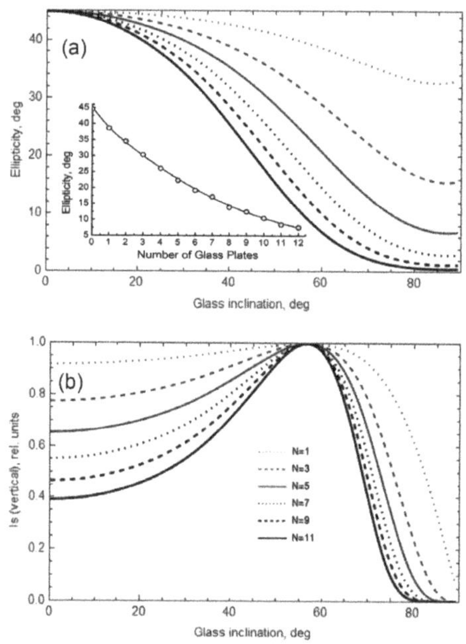

Figure A1. Characteristics of the output radiation of a regulated linear polarizer consisting of N thin glass plates-calculation by Formulas (A5)–(A9); the input light is circularly polarized. (**a**) Dependence of the ellipticity of the output light on the angle of inclination of the glass plates; Inset: ellipticity of the output light as a function of the number N of glass plates set at the Brewster angle. The line is the calculation; the circles are the experiment. (**b**) Dependence of the transmission of the s-component of radiation on the angle of inclination of the glass plates.

References

1. Hämäläinen, M.; Hari, R.; Ilmoniemi, R.J.; Knuutila, J.; Lounasmaa, O.V. Magnetoencephalography—Theory, Instrumentation, and Applications to Noninvasive Studies of the Working Human Brain. *Rev. Mod. Phys.* **1993**, *65*, 413–497. [CrossRef]
2. Liang, A.; Wu, H.; Ma, Y.; Liang, X.; Jia, Y.; Gao, Y.; Ning, X.; Fang, J. Whole-Head Magnetoencephalogram and Its Application in Developmental Communication Disorders Research: A Review. *IEEE Access* **2021**, *9*, 42515–42532. [CrossRef]
3. Dehmelt, H.G. Modulation of a Light Beam by Precessing Absorbing Atoms. *Phys. Rev.* **1957**, *105*, 1924–1925. [CrossRef]
4. Bell, W.E.; Bloom, A.L. Optically Driven Spin Precession. *Phys. Rev. Lett.* **1961**, *6*, 280–281. [CrossRef]
5. Budker, D.; Romalis, M. Optical Magnetometry. *Nat. Phys.* **2007**, *3*, 227. [CrossRef]
6. Cohen, D. Magnetoencephalography: Detection of the Brain's Electrical Activity with a Superconducting Magnetometer. *Science* **1972**, *175*, 664–666. [CrossRef]
7. Marhl, U.; Jodko-Wladzińska, A.; Brühl, R.; Sander, T.; Jazbinšek, V. Transforming and Comparing Data between Standard SQUID and OPM-MEG Systems. *PLoS ONE* **2022**, *17*, e0262669. [CrossRef]
8. Kominis, I.K.; Kornack, T.W.; Allred, J.C.; Romalis, M.V. A Subfemtotesla Multichannel Atomic Magnetometer. *Nature* **2003**, *422*, 596–599. [CrossRef]
9. Iivanainen, J.; Stenroos, M.; Parkkonen, L. Measuring MEG Closer to the Brain: Performance of on-Scalp Sensor Arrays. *NeuroImage* **2017**, *147*, 542–553. [CrossRef]
10. Boto, E.; Meyer, S.S.; Shah, V.; Alem, O.; Knappe, S.; Kruger, P.; Fromhold, T.M.; Lim, M.; Glover, P.M.; Morris, P.G.; et al. A New Generation of Magnetoencephalography: Room Temperature Measurements Using Optically-Pumped Magnetometers. *NeuroImage* **2017**, *149*, 404–414. [CrossRef]
11. Tierney, T.M.; Holmes, N.; Mellor, S.; López, J.D.; Roberts, G.; Hill, R.M.; Boto, E.; Leggett, J.; Shah, V.; Brookes, M.J.; et al. Optically Pumped Magnetometers: From Quantum Origins to Multi-Channel Magnetoencephalography. *NeuroImage* **2019**, *199*, 598–608. [CrossRef]
12. Boto, E.; Holmes, N.; Leggett, J.; Roberts, G.; Shah, V.; Meyer, S.S.; Muñoz, L.D.; Mullinger, K.J.; Tierney, T.M.; Bestmann, S.; et al. Moving Magnetoencephalography towards Real-World Applications with a Wearable System. *Nature* **2018**, *555*, 657. [CrossRef] [PubMed]
13. Iivanainen, J.; Zetter, R.; Parkkonen, L. Potential of On-Scalp MEG: Robust Detection of Human Visual Gamma-Band Responses. *Hum. Brain Mapp.* **2020**, *41*, 150–161. [CrossRef] [PubMed]
14. Nardelli, N.V.; Perry, A.R.; Krzyzewski, S.P.; Knappe, S.A. A Conformal Array of Microfabricated Optically-Pumped First-Order Gradiometers for Magnetoencephalography. *EPJ Quantum Technol.* **2020**, *7*, 11. [CrossRef]
15. Rea, M.; Boto, E.; Holmes, N.; Hill, R.; Osborne, J.; Rhodes, N.; Leggett, J.; Rier, L.; Bowtell, R.; Shah, V.; et al. A 90-Channel Triaxial Magnetoencephalography System Using Optically Pumped Magnetometers. *Ann. N. Y. Acad. Sci.* **2022**, *1517*, 107–124. [CrossRef]
16. Limes, M.E.; Foley, E.L.; Kornack, T.W.; Caliga, S.; McBride, S.; Braun, A.; Lee, W.; Lucivero, V.G.; Romalis, M.V. Portable Magnetometry for Detection of Biomagnetism in Ambient Environments. *Phys. Rev. Appl.* **2020**, *14*, 011002. [CrossRef]
17. Guo, Y.; Wan, S.; Sun, X.; Qin, J. Compact, High-Sensitivity Atomic Magnetometer Utilizing the Light-Narrowing Effect and in-Phase Excitation. *Appl. Opt.* **2019**, *58*, 734–738. [CrossRef]
18. Zhang, R.; Xiao, W.; Ding, Y.; Feng, Y.; Peng, X.; Shen, L.; Sun, C.; Wu, T.; Wu, Y.; Yang, Y.; et al. Recording Brain Activities in Unshielded Earth's Field with Optically Pumped Atomic Magnetometers. *Sci. Adv.* **2020**, *6*, eaba8792. [CrossRef]
19. Perry, A.; Bulatowicz, M.; Larsen, M.; Walker, T.; Wyllie, R. All-Optical Intrinsic Atomic Gradiometer with Sub-20 FT/Cm/$\sqrt{}$Hz Sensitivity in a 22 MT Earth-Scale Magnetic Field. *Opt. Express* **2020**, *28*, 36696–36705. [CrossRef]
20. Clancy, R.J.; Gerginov, V.; Alem, O.; Becker, S.; Knappe, S. A Study of Scalar Optically-Pumped Magnetometers for Use in Magnetoencephalography without Shielding. *Phys. Med. Biol.* **2021**, *66*, 175030. [CrossRef]
21. Lucivero, V.; Lee, W.; Kornack, T.; Limes, M.; Foley, E.; Romalis, M. Femtotesla Nearly-Quantum-Noise-Limited Pulsed Gradiometer at Earth-Scale Fields. *Phys. Rev. Appl.* **2022**, *18*, L021001. [CrossRef]
22. Fu, K.-M.C.; Iwata, G.Z.; Wickenbrock, A.; Budker, D. Sensitive Magnetometry in Challenging Environments. *AVS Quantum Sci.* **2020**, *2*, 044702. [CrossRef]
23. Petrenko, M.V.; Dmitriev, S.P.; Pazgalev, A.S.; Ossadtchi, A.E.; Vershovskii, A.K. Towards the Non-Zero Field Cesium Magnetic Sensor Array for Magnetoencephalography. *IEEE Sens. J.* **2021**, *21*, 18626–18632. [CrossRef]
24. Romalis, M.V. Optically Pumped Magnetometers for Biomagnetic Measurements. In *Flexible High Performance Magnetic Field Sensors*; Springer: Berlin/Heidelberg, Germany, 2022; pp. 3–15.
25. Petrenko, M.V.; Pazgalev, A.S.; Vershovskii, A.K. Single-Beam All-Optical Non-Zero Field Magnetometric Sensor for Magnetoencephalography Applications. *Phys. Rev. Appl.* **2021**, *15*, 064072. [CrossRef]
26. Petrenko, M.V.; Pazgalev, A.S.; Vershovskii, A.K. Ultimate Parameters of the All-Optical Single-Beam Non-Zero Magnetic Field Sensor for Biological Applications. *IEEE Magn. Lett.* **2021**, *12*, 8104605. [CrossRef]
27. Schultze, V.; Schillig, B.; IJsselsteijn, R.; Scholtes, T.; Woetzel, S.; Stolz, R. An Optically Pumped Magnetometer Working in the Light-Shift Dispersed Mz Mode. *Sensors* **2017**, *17*, 561. [CrossRef] [PubMed]
28. Bloom, A.L. Principles of Operation of the Rubidium Vapor Magnetometer. *Appl. Opt.* **1962**, *1*, 61–68. [CrossRef]
29. Groeger, S.; Bison, G.; Schenker, J.-L.; Wynands, R.; Weis, A. A High-Sensitivity Laser-Pumped Mx Magnetometer. *Eur. Phys. J.* **2006**, *38*, 239–247. [CrossRef]

30. Scholtes, T.; Schultze, V.; IJsselsteijn, R.; Woetzel, S.; Meyer, H.-G. Light-Narrowed Optically Pumped M_x Magnetometer with a Miniaturized Cs Cell. *Phys. Rev. A* **2011**, *84*, 043416. [CrossRef]
31. Popov, E.N.; Bobrikova, V.A.; Voskoboinikov, S.P.; Barantsev, K.A.; Ustinov, S.M.; Litvinov, A.N.; Vershovskii, A.K.; Dmitriev, S.P.; Kartoshkin, V.A.; Pazgalev, A.S.; et al. Features of the Formation of the Spin Polarization of an Alkali Metal at the Resolution of Hyperfine Sublevels in the 2S1/2 State. *JETP Lett.* **2018**, *108*, 513–518. [CrossRef]
32. Grujić, Z.D.; Weis, A. Atomic Magnetic Resonance Induced by Amplitude-, Frequency-, or Polarization-Modulated Light. *Phys. Rev. A* **2013**, *88*, 012508. [CrossRef]
33. Appelt, S.; Ben-Amar Baranga, A.; Young, A.R.; Happer, W. Light Narrowing of Rubidium Magnetic-Resonance Lines in High-Pressure Optical-Pumping Cells. *Phys. Rev. A* **1999**, *59*, 2078–2084. [CrossRef]
34. Vershovskii, A.K.; Pazgalev, A.S.; Petrenko, M.V. All-Optical Magnetometric Sensor for Magnetoencephalography and Ultralow Field Tomography. *Tech. Phys. Lett.* **2020**, *46*, 877–880. [CrossRef]
35. Budker, D.; Kimball, D.F.; Rochester, S.M.; Yashchuk, V.V.; Zolotorev, M. Sensitive Magnetometry Based on Nonlinear Magneto-Optical Rotation. *Phys. Rev. A* **2000**, *62*, 043403. [CrossRef]
36. Budker, D.; Gawlik, W.; Kimball, D.F.; Rochester, S.M.; Yashchuk, V.V.; Weis, A. Resonant Nonlinear Magneto-Optical Effects in Atoms. *Rev. Mod. Phys.* **2002**, *74*, 1153–1201. [CrossRef]
37. Wu, T.; Peng, X.; Chen, J.; Guo, H. Fiber-Coupled OPM in Purely Coil-Shielded Environment. In *Flexible High Performance Magnetic Field Sensors*; Springer: Berlin/Heidelberg, Germany, 2022; pp. 161–177.
38. Vershovskii, A.K.; Petrenko, M.V. Optical Magnetometric Sensor for Operation in Magnetoencephalographic Complexes. *Tech. Phys. Lett.* 2022, *to be printed*.
39. Kaminow, I. Polarization-Maintaining Fibers. *Appl. Sci. Res.* **1984**, *41*, 257–270. [CrossRef]
40. Vershovskii, A.K.; Petrenko, M.V. Methods of Parametric Resonance Excitation in the Scheme of an Optical Magnetometric Sensor. *Tech. Phys.* **2022**, *66*, 821–826. [CrossRef]
41. Savenkov, S.N. Jones and Mueller Matrices: Structure, Symmetry Relations and Information Content. In *Light Scattering Reviews 4: Single Light Scattering and Radiative Transfer*; Kokhanovsky, A.A., Ed.; Springer: Berlin/Heidelberg, Germany, 2009; pp. 71–119, ISBN 978-3-540-74276-0.
42. Born, M.; Wolf, E. *Principles of Optics: Electromagnetic Theory of Propagation, Interference and Diffraction of Light*; Elsevier: Amsterdam, The Netherlands, 2013.
43. Lee, S.-K.; Romalis, M. Calculation of Magnetic Field Noise from High-Permeability Magnetic Shields and Conducting Objects with Simple Geometry. *J. Appl. Phys.* **2008**, *103*, 084904. [CrossRef]

sensors MDPI

Article

Reliable Fast (20 Hz) Acquisition Rate by a TD fNIRS Device: Brain Resting-State Oscillation Studies

Rebecca Re [1,2,*], Ileana Pirovano [1,3], Davide Contini [1], Caterina Amendola [1], Letizia Contini [1], Lorenzo Frabasile [1], Pietro Levoni [1], Alessandro Torricelli [1,2] and Lorenzo Spinelli [2]

1 Dipartimento di Fisica, Politecnico di Milano, Piazza Leonardo da Vinci, 32, 20133 Milan, Italy
2 Istituto di Fotonica e Nanotecnologie, Consiglio Nazionale delle Ricerche, Piazza Leonardo da Vinci, 32, 20133 Milan, Italy
3 Istituto di Tecnologie Biomediche, Consiglio Nazionale delle Ricerche, via Fratelli Cervi 93, 20090 Segrate, Italy
* Correspondence: rebecca.re@polimi.it

Abstract: A high power setup for multichannel time-domain (TD) functional near infrared spectroscopy (fNIRS) measurements with high efficiency detection system was developed. It was fully characterized based on international performance assessment protocols for diffuse optics instruments, showing an improvement of the signal-to-noise ratio (SNR) with respect to previous analogue devices, and allowing acquisition of signals with sampling rate up to 20 Hz and source-detector distance up to 5 cm. A resting-state measurement on the motor cortex of a healthy volunteer was performed with an acquisition rate of 20 Hz at a 4 cm source-detector distance. The power spectrum for the cortical oxy- and deoxyhemoglobin is also provided.

Keywords: time domain; functional near infrared spectroscopy; diffuse optics; brain; hemodynamics; resting-state brain oscillation

Citation: Re, R.; Pirovano, I.; Contini, D.; Amendola, C.; Contini, L.; Frabasile, L.; Levoni, P.; Torricelli, A.; Spinelli, L. Reliable Fast (20 Hz) Acquisition Rate by a TD fNIRS Device: Brain Resting-State Oscillation Studies. *Sensors* **2023**, *23*, 196. https://doi.org/10.3390/s23010196

Academic Editor: Tad Brunye

Received: 16 November 2022
Revised: 21 December 2022
Accepted: 22 December 2022
Published: 24 December 2022

1. Introduction

By exploiting picosecond pulsed lasers and single photon detectors, the time-domain (TD) near infrared spectroscopy (NIRS) technique allows retrieval of the absolute values of biological tissues' optical properties, i.e., absorption (μ_a) and reduced scattering (μ_s') coefficients. The acquired photon distribution of time-of-flight (DTOF) can be time-gated in order to better discriminate between the contribution of late photons, which traveled to a greater depth, and early photons, which traveled mostly through the more superficial layer [1]. Due to the poor signal-to-noise ratio (SNR), most of the TD NIRS instruments operate at an acquisition rate < 2 Hz, which is typically enough for monitoring the task-related cortical hemodynamic response that usually occurs with time constants of a few seconds [2]. However, for some specific applications, such as the monitoring of brain connectivity or resting-state oscillations, that sampling rate is too low. Spontaneous ongoing global activity of the brain at rest is highly structured in spatio-temporal patterns called resting-state networks. These fluctuations of brain activity exist even in the absence of tasks or stimuli [3], and were originally characterized by indirect and slow measurements of neuronal activity by blood oxygen level-dependent (BOLD) functional MRI (fMRI) thanks to the neurovascular coupling mechanism [4]. A non-invasive estimate of brain oscillations can also be achieved with the functional NIRS (fNIRS) technique, that exploits the different absorption spectra of oxygenated hemoglobin (O_2Hb) and deoxygenated hemoglobin (HHb), as well as the penetration capability of NIR light in the human head [5]. By calculating the power spectral density of the signals related to the hemodynamic parameters in the frequency range <5 Hz, it is possible to study the presence of characteristic frequency peaks associated with physiological and/or pathological phenomena. Resting-state oscillation fNIRS studies were performed on patients with mild cognitive impairment [6], acute brain injuries [7] or autoregulation dysfunction [8]. By means of a multichannel setup, connectivity studies are also possible [9]. It is worth noting that all fNIRS studies were performed by a continuous wave (CW) or frequency domain (FD) approach, by which it was possible to reach a proper acquisition rate (e.g., 10 Hz) [10–12], differently from TD fNIRS.

During recent years, a huge development of the TD NIRS technique and instrumentation has been observed, with interesting advancements at the level of both research laboratories and companies; however, no sufficient SNR level was reached in order to increase the measurement acquisition rate [13]. It is worth noting that increasing the measurement SNR will not only allow faster acquisition rates, but also longer source-detector distance (ρ) measurements, as compared to the typical examples reported in the literature, i.e., <4 cm. Although the possibility to probe tissue in depth at null source-detector distance was also demonstrated by TD fNIRS, the use of larger ρ can be of help in overcoming non-idealities of the instrument response function (IRF) or in improving depth selectivity [14]. On the other hand, an increase in ρ corresponds to a decrease in signal at the detector (i.e., for an increase of 1 cm in the source-detector distance, we lose about one order of magnitude in the signal). The main technological bottlenecks from this point of view are: (i) the maximum average laser power is limited by safety regulations (<2 mW/mm^2 for $\lambda < 700$ nm and from 2 up to 4 mW/mm^2 for $700 < \lambda < 860$ nm [15]); (ii) the need for high stability in time at the sub-nanosecond level, to avoid cross-talk between time drift and the estimation of optical properties; (iii) a limited signal harvesting efficiency of the detection line, i.e., the responsivity of the system. Koga et al. [16] attempted to develop a high power TD NIRS system by modifying an existing device. This instrument was employed in the assessment of superficial and deep muscle deoxygenation kinetics during heavy intensity exercises. They reported that it was possible to detect differences in optical properties up to 3 cm depth in a phantom, performing measurements at different source-detector distances (from 3 to 7 cm, at 1 cm steps in separate trials) but only at a 0.5 Hz acquisition rate. During the in vivo measurements (ρ = 3 and 6 cm), it was possible to find differences in superficial and deep muscle deoxyhemoglobin kinetics following the onset of heavy intensity exercise. In a recent work, Jiang et al. [17] presented a TD fNIRS optical tomography system (NIROT Pioneer) based on supercontinuum laser sources and SPAD detectors reaching a sampling rate of 2.5 Hz, aimed to perform DOT acquisitions. In this case, a fiber optical switch alternatively selected 11 source positions. The probe geometry allowed them to obtain an FOV with 2.5 cm diameter (i.e., $\rho < 3$ cm). A commercial solution, named "Flow", was presented recently by Kernel (Kernel, Los Angeles, CA, USA, https://www.kernel.com/ (accessed on 22 December 2022)). They showed a multichannel wearable headset that measures brain activity [18], which potentially allows measurements up to a 100 Hz acquisition rate in the single channel, but now limited to 7.1 Hz to avoid cross-talk effects among channels. They also reported measurements performed at a 6 cm source-detector distance, but with a limited count rate (~10^4 counts/s).

For what concerns the possibility to perform brain resting-state oscillation studies with TD fNIRS, to the best of our knowledge, there are only two previous attempts reported in the literature. In 2004, there was a 12 Hz acquisition from Themelis et al. (three independent channels with ρ = 1, 2 and 3 cm) [19], where they reported the presence of the heartbeat in the 830 nm cortical signal, without presenting a spectrum. They also affirmed that the selection of a proper time delay in the detected signal could increase the sensitivity of the TD fNIRS system to contributions coming from the deeper brain regions. The second attempt was performed by Kacprzak et al. in 2019 [20]. They developed an instrument based on pulsed semiconductor lasers and time-correlated single photon counting electronics (TCSPC), step-index fibers with 400 μm diameter and an IRF with a full width at half maximum (FWHM) of 500 ps. They presented in vivo acquisitions of the brain resting-state oscillations for both healthy subjects and patients with severe neurovascular disorders, with a 10 Hz acquisition rate, ρ = 3 cm and two locations on the head. They evaluated light attenuation (which reflects the superficial variations) and variance (more related to the cerebral compartment) of the DTOFs performing an FFT analysis of their changes.

They showed interesting results for what concerns the presence of peaks in the frequency spectrum of the attenuation but, unfortunately, the measurement duration was set to only 10 min for the patients, and the frequency peaks in the variance spectrum were buried by the noise, indicating an insufficient SNR during the acquisition. In addition, they did not provide the same analysis for the hemodynamic parameters.

In this paper, we present a TD fNIRS setup where high power laser sources, hybrid photomultiplier tubes and custom detection bundles made of plastic optical fibers are employed. The instrument SNR is drastically improved with respect to previous analogue instrumentation, allowing acquisitions with a rate of 20 Hz and with source-detector distances up to 5 cm, both in phantom and in vivo applications. A first preliminary frequency domain analysis of the hemodynamic signals derived from TD fNIRS measurements is also provided.

2. Instrument Description

The TD fNIRS device is equipped with two high power pulsed diode lasers (LDH-P-C, Picoquant GmbH, Berlin, Germany) working respectively at 689.5 $\pm$ 0.5 nm (RED) and 828.5 $\pm$ 0.5 nm (IR). They are electronically driven at 80 MHz (PDL-828 Sepia II, Picoquant GmbH, Berlin, Germany) and emit pulses with a minimum pulse width of 72 ps (96 ps) for the RED (IR). The beam is coupled to step-index multimode glass optical fibers with a core/cladding diameter of 600/660 μm and NA = 0.22 (QMMJ-55-IRVIS-600/660-3-1.25, OZ Optics LTD., Ottawa, ON, Canada). It is possible to attenuate the beams by means of motorized and electronically driven continuous glass variable neutral density attenuators (NT43-770, Edmund Optics GmbH, Germany) inserted in a free beam region created by means of specific U-brackets (UB-12-11, OZ Optics LTD., Ottawa, ON, Canada). Before the sample, an optical beam combiner (FOBS-12P, OZ Optics LTD., Ottawa, ON, Canada) delays the IR wavelength and couples it with the RED one, in order to implement a time-multiplexing modality for the injection of light [21], i.e., both wavelengths interleaved in the same temporal window (12.5 ns) with a proper relative delay (6.4 ns). After the sample, diffused light is collected by means of four independent detection lines (D1–D4). Each detection line consists of: (i) a custom-made fiber optic bundle with 3 mm diameter and 1.25 m length, composed by 7 graded-index plastic optical fiber (POF) with NA = 0.3, core/cladding diameter of 900/1000 μm (FiberFin Inc., Yorkville, Illinois, USA) in hexagonal configuration; (ii) an attenuation stage provided by electronically driven continuous glass variable neutral density attenuators (NDC-50C-4-B, Thorlabs Inc., Newton, NJ, USA); (iii) a hybrid photomultiplier tube (PMA-50 Hybrid Series, Picoquant GmbH, Berlin, Germany). The DTOF acquisition is accomplished by a TCSPC unit (HydraHarp 400, Picoquant GmbH, Berlin, Germany) with short dead time (<80 ns), a maximum count rate per input channel of 12.5 $\times$ 10^6 cps and an overall sustained throughput of about 40 $\times$ 10^6 events/s, as summed over all channels. The whole system is controlled by a series of home-made units based on microcontrollers (DSPIC, Microchip Technology Inc., Chandler, AZ, USA), which also give an independent time basis and allow the synchronization with external hardware. In Figure 1, a scheme of the instrument is presented. The device was built with a modular structure, and it is equipped with a set of custom-made 3D printed probes made of a compatible material for diffuse optics applications [22]. The state-of-the-art instrument has a 1 $\times$ 4 configuration, i.e., one injection and four detection channels working in parallel.

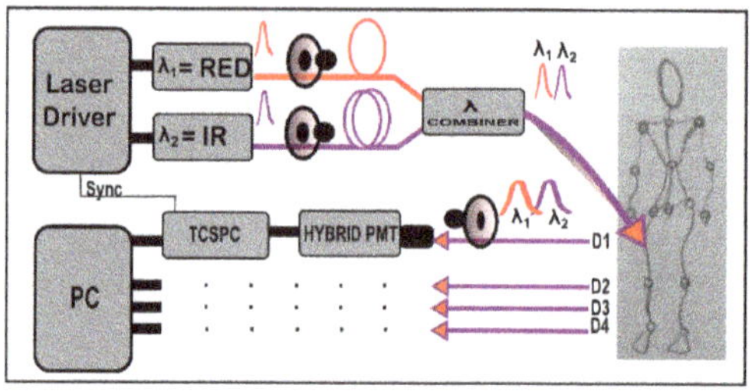

Figure 1. TD NIRS device scheme. λ = wavelength, PMT = Photomultiplier tube, TCSPC = Time-correlated single photon counting, D = Detection channel, Sync = Synchronization signal.

3. Characterization Protocols

In this section, different international standardized characterization protocols employed for assessing the performances of diffuse optics instruments are presented. We also present the assessment of the maximum allowed count rate and acquisition rate. In addition, an in vivo measurement on an arm muscle is presented to validate the use of the device on humans. In the following sections, the optical parameters are estimated by a non-linear fitting procedure based on the Levenberg-Marquardt algorithm that minimizes the error (chi-square) between the measured DTOF and a theoretical function obtained by the convolution between the IRF and the analytical solution of the diffusion equation in a semi-infinite homogeneous medium [23].

3.1. Basic Instrumental Performance (BIP)

In the BIP protocol, the basic characteristics of the instrument are explored [24]. The maximum power exiting from the injection fibers towards the tissue is 1.90 mW (7.9 mW) for the RED (IR). These power settings were chosen in order to obtain an IRF with a FWHM of 240 ± 11 ps and 236 ± 12 ps, respectively, for RED and IR, expressed as the average ± standard deviation among the four detection lines. The width at 1% of the peak was 920 ± 40 ps (1020 ± 60 ps) for the RED (IR). In the same way, we can express the average responsivity $S_{avg}(\lambda)$, obtaining: $S_{avg}(RED) = (2.8 \pm 1.4) \times 10^{-8}$ m^{2}sr and $S_{avg}(IR) = (1.5 \pm 0.7) \times 10^{-8}$ m^{2}sr. The average afterpulsing ratio $R_{ap}(\lambda)$ is: $R_{ap}(RED) = 1.2 \pm 0.8\%$ and $R_{ap}(IR) = 1.0 \pm 0.8\%$. The detector differential non-linearity is: $\varepsilon_{DNL} = 5.9 \pm 0.4\%$. The system requires a warm-up time of 110 (40) min in order to reach stability within ±1% (3%) of the final average values (counts, barycenter and FWHM of the IRF) calculated over the last 30 min of a 5 h acquisition. All the described parameters reflect those of previous TD fNIRS devices [25–27]. It is relevant that the setup characteristics, and in particular the choice of the proper optical fibers, allowed us to obtain IRFs with narrow FWHM, without undesired peaks due to internal reflections, as shown in Figure 2, providing the best conditions for a good fitting of the acquired data with the theoretical model [28].

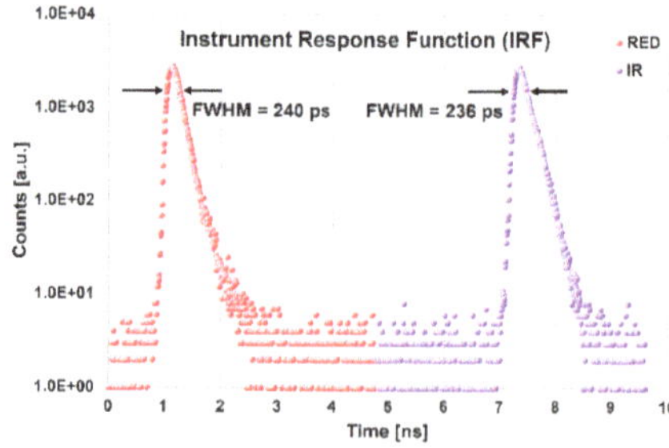

Figure 2. Typical acquisition window during an instrument response function (IRF) measurement. RED = 689 nm, IR = 828 nm, FWHM = Full width at half maximum, semi-logarithmic scale.

3.2. Assessment of the Maximum Count Rate and Acquisition Rate

The upper limit to the maximum count rate is set by the detection and acquisition chain (see Section 2). The employed hybrid photomultipliers have a recommended upper limit in terms of count rate of 10^7 counts/s (i.e., above this value, an electronic-controlled shutter automatically closes to prevent damage to the active area). The TCSPC system HydraHarp400, based on time-tagged time-resolved (TTTR) mode, guarantees a maximum count rate per input channel of 12.5×10^6 cps and an overall sustained throughput of about 4×10^7 events/second, as summed over all channels. On the other side, during the acquisition, the count rate is typically kept limited in order to operate in the single photon counting regime: it is necessary to guarantee that the count rate remains below 5% of the pulse rate, i.e., 4 MHz, since our lasers are working at 80 MHz, in order to avoid the "pile-up" effect [29]. Otherwise, the TCSPC system would register more than one photon per excitation cycle, causing a distortion of the DTOF and an error in the retrieval of the optical properties of the media under study. Recently, it was demonstrated, both with simulations and phantom acquisitions, that it is possible to work above the single photon statistics limit [30]. On these bases, we performed specific acquisitions in order to assess the maximum allowed count rate.

We then performed 10 repeated measurements, each with 1 s acquisition time and ρ = 3 cm, on a solid homogenous phantom (μ_a = 0.1 cm^{-1} and μ_s' = 10 cm^{-1} nominal optical properties) at different acquisition count rates: from 5×10^5 ph/s up to 1.1×10^7 ph/s on the board, with 5×10^5 steps. The retrieved μ_s', shown in Figure 3a, showed small variations with the increasing of the number of acquired photons with <3% error with respect to the average calculated among the acquisitions at lower counts (from 0.5 to 4×10^6) where the pile-up effect is negligible. For μ_a (Figure 3b), we observed an increasing trend with the increasing of the acquired counts, for both wavelengths. In order for the error not to exceed 3% with respect to its average value, calculated as for μ_s', it is necessary to set the injected photon count rate to a maximum of about 8×10^6 ph/s, on the board, or equivalently 4×10^6 ph/s for each wavelength with the relative DTOFs interleaved in the same temporal window.

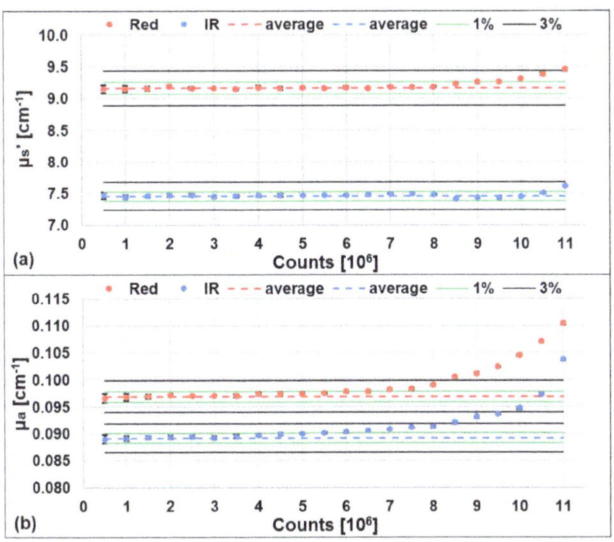

Figure 3. Averaged reduced scattering coefficient (μ_s', (**a**)) and absorption coefficient (μ_a, (**b**)) over 10 repetitions and relative error bars, for different count rates. Red: RED wavelength, blue: IR wavelength. The dashed lines are the average μ_s' and μ_a retrieved for counts from 0.5 to 4×10^6. The green and black lines represent the 1% and 3% error regions, respectively.

Thanks to the previous findings and to the availability of a high number of detectable photons, we also tested the possibility to increase the acquisition rate, while maintaining enough detected photons to guarantee an optimal retrieval of the optical properties. We performed 10 repeated acquisitions for each acquisition time, on the same phantom as before with $\rho = 3$ cm. The acquisition sampling times were set to: 1, 0.1, 0.05, 0.01, 0.005, 0.004, 0.003, 0.002 and 0.001 s. We repeated the measurements at two different count rates: 2×10^6 counts/s and 7×10^6 counts/s per board.

In Figure 4, the retrieved values for μ_a (first column) and μ_s' (second column) at the two wavelengths are shown as functions of the acquisition time, when the initial count rate is set to $2 \cdot 10^6$ counts/s (first row) or $7 \cdot 10^6$ counts/s (second row). The solid horizontal lines represent the average value over the acquisitions at 1, 0.1 and 0.05 s. It is evident that, when reducing the acquisition time, the optical properties are obtained with a larger deviation from the average values and a greater dispersion (i.e., standard deviation).

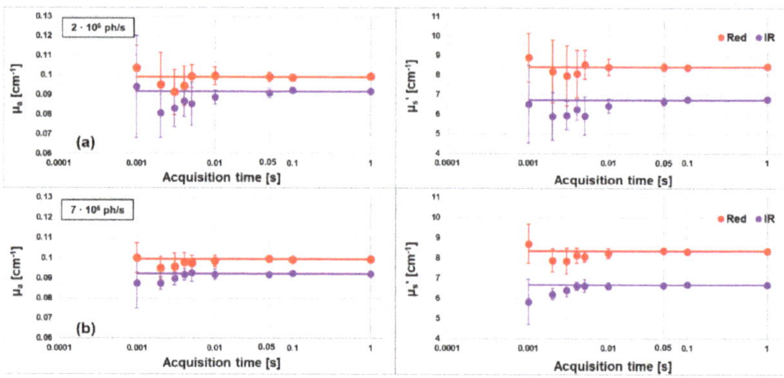

Figure 4. Absorption (μ_a) and reduced scattering (μ_s') coefficients for both wavelengths as function of the acquisition time. Initial count rate set at 2×10^6 counts/s (**a**) or 7×10^6 counts/s (**b**). The horizontal lines represent the average value over the acquisitions at 1, 0.1 and 0.05 s.

It is then possible to estimate the minimum number of photons in the acquired DTOFs which guarantees a sufficient SNR for a reliable estimation of the optical parameters. For this purpose, we calculated the percentage coefficient of variation (CV%), defined as the standard deviation of a quantity divided by its average value and multiplied by 100 [31]. To obtain a CV < 1% for both optical coefficients and both wavelengths, a count rate of around 1.6×10^5 counts/s for each acquisition, i.e., for each board, is necessary. That is equivalent, when wavelengths are interleaved, to 8.0×10^4 counts/s for each wavelength. To guarantee enough photons, as stated from the CV parameter, we can use a minimum acquisition time of 0.1 s (0.03 s) with a count rate of 2×10^6/s (7×10^6/s).

3.3. Further Characterizations

The reproducibility (i.e., the capability to reproduce consistent values for the optical properties of the same phantom among four different days) and the linearity (i.e., the capability to correctly estimate the linear change in the optical properties) of our instrument were tested according to the MEDPHOT protocol [31].

We found that μ_a and μ_s' values showed variations lower than 3% around their average values calculated among the different days, showing an excellent reproducibility.

The linearity was tested on a set of 32 solid homogenous phantoms labeled with numbers from 1 to 8 and letters from A to D, in order to represent the different μ_a and μ_s' values, respectively (nominal optical properties from 0.01 to 0.49 cm^{-1} in 0.07 cm^{-1} steps for the absorption coefficient, and from 5 to 20 cm^{-1} in 5 cm^{-1} steps for reduced scattering coefficient, at 660 nm). We performed 10 repeated measurements, each with 1 s acquisition time, in reflectance geometry with ρ = 3 cm and a number of counts in the DTOF sufficient to guarantee a CV < 1% (see Section 3.2). Linearity was tested for both coefficients and both wavelengths by a linear interpolation. The R^2 coefficients obtained were always >0.95, showing an excellent linearity, as shown in Figure S1 and Tables S1 and S2 of the Supplementary Materials.

Thanks to the increased SNR, we also investigated the possibility to perform acquisitions with different source-detector distances ρ (from 1 to 5 cm, at 1 cm steps) on the previous set of phantoms. During each measurement, we set the highest reachable count rate. In Figure 5, we show these count rates, for the RED, for all phantoms and source-detector distances (different colors). In this figure, we indicate the phantoms with their labels, and we set a horizontal line representing the value on the y-axis for the goal in terms of counts/s (8×10^4 counts/s) necessary to obtain a CV < 1% (see Section 3.2). We can notice that, for the less scattering phantoms (A) it is always possible to reach enough counts, except for the most absorbent (8) for ρ = 5 cm (black dot). For the less absorbent phantoms (2), it is always possible to reach the goal counts, increasing the scattering (A–D) or the ρ. Moving towards more scattering and absorbing media, the measurement at $\rho \geq 4$ cm is no longer achievable. Similar results were obtained for the IR wavelength, which shows in general a higher number of counts achievable, as shown during the BIP protocol. These data underline the improvement in terms of SNR of this TD fNIRS device over the previous ones published [21,27,32], with which it was typically not possible to measure phantoms D6 or D8 at ρ = 2 or 3 cm.

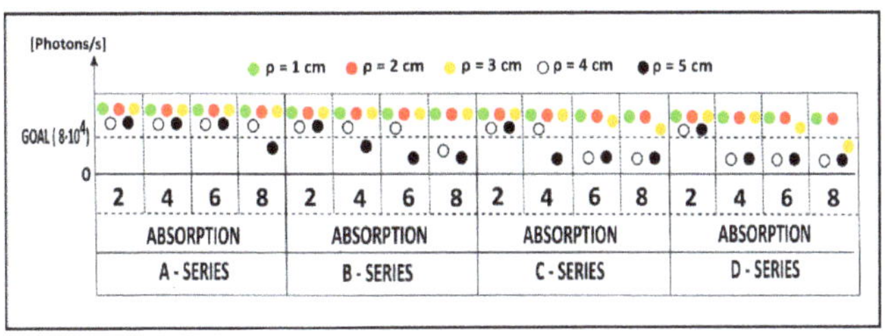

Figure 5. Number of photons/s for the RED wavelength, collected on solid phantoms for different values of absorption (labels 2, 4, 6 and 8: 0.07, 0. 21, 0.35, 0,49 cm^{-1}, respectively) and reduced scattering (series A, B, C and D: 5, 10, 15 and 20 cm^{-1}, respectively) with different source-detector distances ρ (different colors). In the figure, the count rate necessary to obtain a CV < 1% is shown as well.

3.4. In Vivo Characterization Protocol: Arm Muscle Arterial Occlusion

In this section, we present an in vivo protocol to understand the feasibility of measurements on human tissues with a high acquisition rate (20 Hz) and long source-detector distances (up to 5 cm).

An arterial cuff occlusion (250 mmHg) of the left arm of a healthy adult volunteer was performed. The probe was placed on the internal side of the forearm (Figure 6), along the muscle fibers. The acquisition rate was set to 20 Hz and the measurements were performed, simultaneously, at 4 source-detector distances: from 2 to 5 cm, at 1 cm steps. The protocol consisted of 120 s baseline, 180 s occlusion and 300 s recovery. We noticed that, according to the results obtained in Section 3.2, the signal was sufficient to perform reliable acquisitions at 20 Hz at all interfiber distances. The absolute values for μ_a and μ_s' were obtained as explained in Section 3 for each acquisition point. The Lambert-Beer law was applied to estimate the O_2Hb and HHb concentration at each time point during the experiment. A moving average of order 20 was applied to the retrieved hemodynamic parameters.

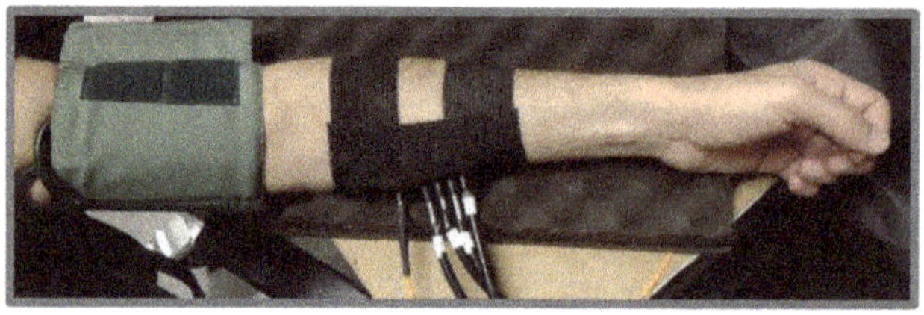

Figure 6. Probe placement during the in vivo occlusion on the arm muscle.

In Figure 7, the time courses of the relative variations obtained for O_2Hb and HHb are shown for all ρ-distances. The variations refer to the baseline values, calculated by averaging the concentration values found in the first 120 s of the experiment. As expected, during the occlusion the O_2Hb decreases, since both veins and arteries are occluded, and no other oxygenated blood can enter in the investigated region. Conversely, the HHb increases because the muscle oxidative metabolism continues during the occlusion period. After the release of the cuff, we can observe the typical hyperemic peak. The qualitative behavior of the time courses at all ρ-distances is the same, but for ρ = 5 cm the amplitude of the variations is smaller. This behavior is more pronounced for the HHb. We do not have a clear explanation for this phenomenon. At first glance, it cannot be due to possible measurement faults, such as a lack of photons, because the SNR was sufficient at all source-detector distances. A possible explanation may be the heterogeneity of the tissue sampled at different ρ. This hypothesis was partially confirmed by an ultrasound exam of that arm region, which showed that tissue composition was different above and below a depth of 2.3 cm.

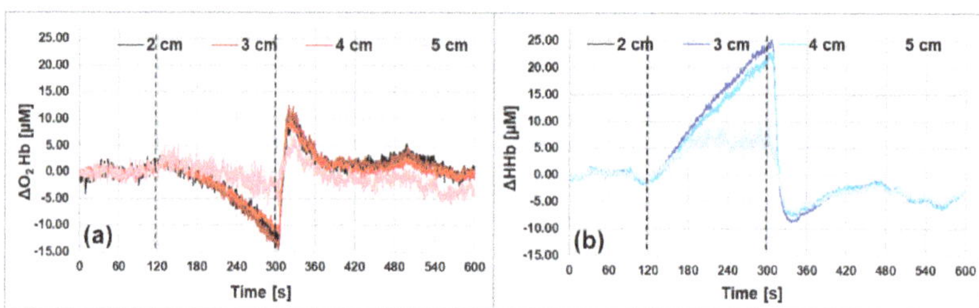

Figure 7. Hemodynamic parameters during an arterial arm occlusion. The dashed vertical lines indicate the start and the end of the occlusion period. The different lines represent the different source-detector distances (from 2 to 5 cm). (**a**) Oxyhemoglobin (O_2Hb). (**b**) Deoxyhemoglobin (HHb).

This preliminary measurement also demonstrates the feasibility of the application of the 20 Hz acquisition rate during in vivo measurements, with the possibility to follow big changes in absorption, such as the ones that occur during an arterial occlusion in the muscle. Changes around 15 µM for O_2Hb and 25 µM for HHb were, in fact, detectable.

4. Cortical Resting-State Oscillations: Results and Discussion

In this section, we show an in vivo measurement with 20 Hz acquisition rate on the brain motor cortex of a healthy volunteer during a resting-state period and the resulting power spectrum for cortical O_2Hb and HHb. This pilot study does not aim to explain the physiological origin of the peaks found in the frequency spectrum, but to demonstrate, for the first time, that it is possible to detect them by TD fNIRS.

We performed an acquisition on an adult healthy subject (male, 53 years old), in correspondence with the primary motor cortex area (C3 position according to the 10/20 EEG international system [33]). The subject relaxed in the supine position, with eyes closed, for 5 min. The acquisition rate was set to 20 Hz and the source-detector distance to 3 cm.

The previous custom probe was placed on the scalp with a black auto-adhesive bandage, guaranteeing a good adhesion and avoiding ambient light leakage. The count rate of the measurement allowed performance of the acquisitions at 20 Hz, according to the results obtained in Section 3.2.

In order to enhance the contribution of the photons coming from deeper regions (late photons) from those coming from the more superficial regions (early photons), we modeled the tissue as a two-layer medium (up layer, UP; down layer, DW) and we calculated the time-dependent mean photon pathlengths in the UP and DW layers as described in Zucchelli et al. [34]. These pathlengths were used to estimate the absolute values of the cortical O_2Hb and HHb hemoglobin concentrations, assuming a thickness of the upper layer of 1 cm (i.e., an equivalent thickness of the extra-cerebral tissue). For cortical O_2Hb and HHb, we found an average of 44.71 μM and 17.94 μM, respectively, calculated over the initial 5 s. In Figure 8, the time courses of the concentration of O_2Hb (red) and HHb (blue), after subtraction of the average values, are shown. A moving average of order 20 was applied to the retrieved hemodynamic parameters as well. As we can notice in Figure 8, a 1 s periodicity is clearly visible, superimposed on faster oscillations, for both hemoglobin species; this amplitude variability is higher for O_2Hb than for HHb [10].

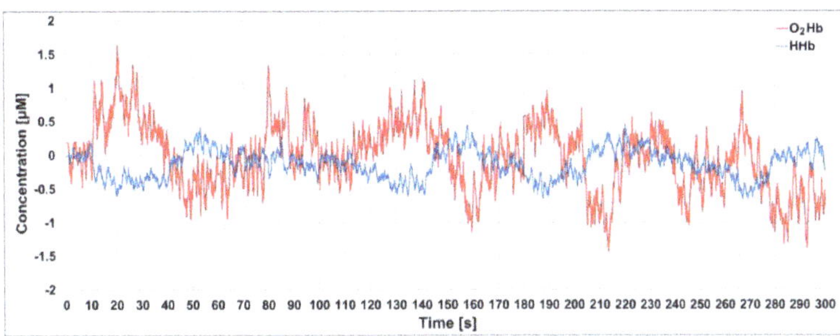

Figure 8. Time courses of the concentration of the cortical O_2Hb and HHb, after subtraction of the average over the initial 5 s.

We then calculated the power spectrum for the cortical O_2Hb and HHb with a custom-made code, based on the FFT algorithm (MATLAB 2021b, The MathWorks Inc., Natick, MA, USA), as shown in Figure 9. No filters were applied on the signal. At first, we can observe that the power spectrum amplitude is higher for O_2Hb with respect to HHb, as previously shown in the literature with CW fNIRS [10]. In both spectra, it is possible to see the typical peak of the cardiac activity (~1 Hz), more pronounced in O_2Hb as compared with HHb. Obrig et al. [10] have shown that the heartbeat produces changes in pressure, which are more visible in O_2Hb, stating that this parameter should be more sensitive to systemic variations. In the O_2Hb spectrum, a peak compatible with the respiration activity during rest (~0.2–0.3 Hz) can be recognized as well. In the HHb signal, a similar peak is present as well, but it is less evident. In addition, in previous CW-NIRS studies, the respiration peak could not always be visible [35].

In the figure inset, the power spectra for frequencies $\leq$ 0.5 Hz are shown. This frequency range is of particular interest since it includes the low frequency oscillations (LFOs, around 0.1 Hz) and the very low frequency oscillations (VLFOs, around 0.04 Hz) [10].

Firstly, we note that both in the O_2Hb and in the HHb spectra a peak around 0.1 Hz is present, related to the intrinsic myogenic activity of the vascular smooth cell. As stated by Yücel et al. [36], at this frequency two different effects may be superimposed: the Mayer waves and vasomotion-flowmotion waves. The former are defined as waves in the arterial blood pressure, which cause an oscillation more visible in the superficial O_2Hb [36]. Mayer waves should not be visible in the HHb [37]. On the contrary, vasomotion is defined as the oscillation in the blood vessels' tone, which causes the cross-section of the blood vessel to oscillate, giving rise to the flowmotion [38]. This oscillation should be visible both in the cortical O_2Hb and HHb, since in general the LFO amplitude should increase with the decreasing of the vessel diameter and the vessel diameter should decrease with the increasing of the depth (from the scalp to the cortex) [39]. The possibility to simultaneously quantify both hemoglobin species by TD fNIRS helps us in affirming that the peak at 0.1 Hz of the O_2Hb could consist in a superposition of the two effects, i.e., Mayer waves and vasomotion-flowmotion, while the one found for HHb should be due to the vasomotion-flowmotion effect only.

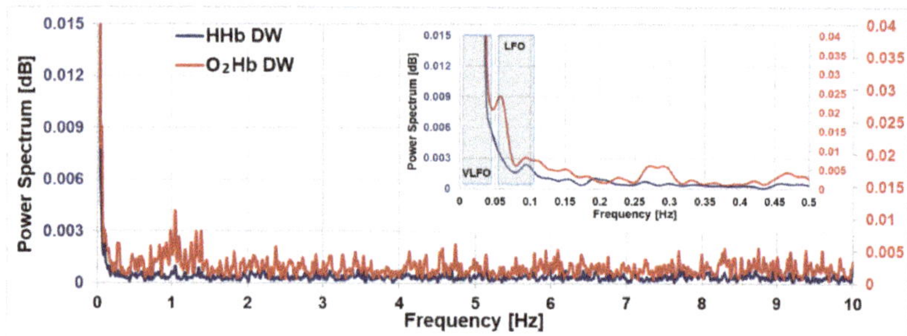

Figure 9. Power spectrum of the cortical O_2Hb (red) and HHb (blue) hemoglobin for 5 min's resting-state acquisition on the motor cortex. In the inset, a zoomed-in view of the frequencies ≤ 0.5 Hz is shown. LFO: Low frequency oscillation; VLFO: Very low frequency oscillation.

If we now consider frequencies <0.1 Hz, in the O_2Hb spectrum an oscillation around 0.06 Hz is clearly visible, possibly related to the neurogenic activity of the vessel walls. To better understand its origin, further experiments are required, where some physiological changes can be induced to observe the respective changes in the spectra. Of course, the concurrent acquisition of the main physiological parameters (such as heartbeat, respiratory rate, arterial blood pressure, blood volume pulses and others) can help in a better interpretation of the whole spectrum.

Finally, in Figure 9, it is possible to notice a strong frequency component at less than 0.04 Hz, for both hemoglobin species. This range covers the neurogenic activity of the vessel wall and the vascular endothelium function. In a future work, in order to remove the continuous component from the frequency spectrum and thus recover the characteristic peaks in this region, a detrending algorithm has to be applied. Furthermore, other methodologies should be used to obtain sharper peaks, more comparable with the previous literature findings, such as the power spectral density (PSD) estimate via Welch's method.

5. Conclusions

In this paper, we presented a TD fNIRS device reaching a higher SNR as compared with previous similar instruments, obtained by combining more powerful lasers and a more efficient detection system. As shown in Sections 3.2 and 3.3, it was possible to collect enough signal at the 20 Hz acquisition rate to reliably (CV < 1%) retrieve the optical properties of homogeneous phantoms with high absorption (0.35 cm^{-1}) and highly reduced scattering (20 cm^{-1}) coefficients at a 3 cm source-detector distance. Measurements with a source-detector distance up to 5 cm were also achievable on homogeneous phantoms mimicking the optical properties of a biological medium (μ_a = 0.1 cm^{-1} and $\mu_s{}'$ =10 cm^{-1}). In general, we demonstrated the possibility to perform measurements with an up to 5 cm interfiber distance in reflectance geometry, at a maximum acquisition rate of 20 Hz on diffusive samples with optical properties like those of biological tissues. This result, to the best of our knowledge, has never been reached by any other TD fNIRS instrument to date. Thanks to the four independent detection lines, it was possible to perform acquisitions in parallel in four different acquisition points. In Section 3.4, we showed the possibility to employ this device during in vivo measurements on the arm muscle, thus retrieving the absolute values of the hemodynamic parameters, employing source-detector distances up to 5 cm.

Furthermore, in Section 4, we showed the power spectra for the absolute values of both cortical O_2Hb and HHb obtained by a TD fNIRS acquisition. This preliminary acquisition, on a healthy subject, aimed to prove the feasibility of performing measurements on the cerebral cortex with high sampling rate (20 Hz) by TD fNIRS, rather than explain in depth each resulting spectral peak; to the best of our knowledge, this result has never been achieved to date, as already underlined in the Introduction. In particular, thanks to this acquisition we showed that by TD fNIRS: (1) it is possible to detect the intracranial heartbeat signal, in particular in the cortical O_2Hb signal; (2) it is possible to observe the intracranial respiration, at least in the O_2Hb signal, that was observed by Kacprazak et al. [20] in the superficial layer only; (3) we increased the SNR, obtaining a non-noisy spectrum by acquisitions of only 5 min. In the only previously published paper, they needed longer measurements (20 min) and affirmed that, for the patients, 10 min of acquisition at 10 Hz were not sufficient, and that some interesting frequencies were buried under the noise; (4) we were able to provide the spectra of the cerebral O_2Hb and HHb, starting from their absolute values, by one measurement at a single source-detector distance. Thanks to this opportunity, we were able to distinguish important spectral contributions in the frequency range below 0.5 Hz.

We think that this study opens up the possibility to perform TD fNIRS measurements at a high acquisition rate (up to 20 Hz), filling the gap with CW fNIRS instruments and other previous techniques such as fMRI. In particular, if only an exploration of the more superficial layer of the brain cortex with fNIRS is possible, there are a series of advantages in choosing this optical technique. It is possible to perform acquisitions at the bedside and to guarantee a continuous monitoring. fNIRS is less sensitive to motion artifacts and the signal does not present physiological noise due to respiratory and cardiac activities, which cause an unwanted modulation in fMRI signal [40,41]. In addition, if the fMRI signal carries information only about the BOLD, with fNIRS and in particular TD fNIRS, it is possible to decouple the contributions of the oxygenated and deoxygenated blood. In this way, the capability of this technique to provide a more accurate estimation of cortical hemodynamic parameters can also be fully exploited in cerebral resting-state oscillation studies and, in the future, by increasing the measurement points, in brain connectivity studies as well.

Of course, further work is necessary to understand the best analysis method for the extrapolation of the hemodynamics frequency spectra. Theoretical simulations will also be necessary to define constraints, if any, about the length of the experiment, the number of photons needed to distinguish two different peaks and other technical aspects. Furthermore, it will be necessary to employ additional physiological sensors, in order to acquire at least heartbeat and respiration rate, and to increase the number of subjects involved, for a better interpretation of the in vivo results.

Supplementary Materials: The following supporting information can be downloaded at: https://www.mdpi.com/article/10.3390/s23010196/s1, Figure S1: Linearity plots according to MEDPHOT protocol; Table S1: Linear interpolation goodness for absorption coefficient; Table S2: Linear interpolation goodness for scattering coefficient.

Author Contributions: R.R., D.C., A.T. and L.S. carried out conceptualization; R.R., I.P. and L.S. developed the instrument, R.R., C.A., L.C., L.F. and P.L. performed the experiments; R.R. and L.S. analyzed the data; L.S. contributed to the analysis tools; L.S. was responsible for the project; R.R., I.P., D.C., L.S. and A.T. contributed to writing the paper. All authors have read and agreed to the published version of the manuscript.

Funding: This research was partially funded by Regione Lombardia and Consiglio Nazionale delle Ricerche, III Accordo Quadro, grant FHfFC: Future Home for Future Communities.

Institutional Review Board Statement: The study was conducted according to the guidelines of the Declaration of Helsinki and approved by the Institutional Ethics Committee of Politecnico di Milano (protocol code 37/2020, 2/12/2020).

Informed Consent Statement: Informed consent was obtained from all subjects involved in the study.

Data Availability Statement: Data underlying the results presented in this paper are not publicly available at this time but may be obtained from the corresponding authors upon reasonable request.

Acknowledgments: This work was partially supported by Regione Lombardia project NEWMED (Grant No. POR FESR 2014–2020).

Conflicts of Interest: D.C. and A.T. are cofounders of pioNIRS S.r.l. (Italy). Other authors declare no conflicts of interest related to this article.

References

1. Torricelli, A.; Contini, D.; Pifferi, A.; Caffini, M.; Re, R.; Zucchelli, L.; Spinelli, L. Time domain functional NIRS imaging for human brain mapping. *Neuroimage* **2014**, *85*, 28–50. [CrossRef] [PubMed]
2. Cinciute, S. Translating the hemodynamic response: Why focused interdisciplinary integration should matter for the future of functional neuroimaging. *PeerJ* **2019**, *2019*, e6621. [CrossRef] [PubMed]
3. Niu, H.; He, Y. Resting-state functional brain connectivity: Lessons from functional near-infrared spectroscopy. *Neuroscientist* **2014**, *20*, 173–188. [CrossRef] [PubMed]
4. Deco, G.; Jirsa, V.K.; McIntosh, A.R. Emerging concepts for the dynamical organization of resting-state activity in the brain. *Nat. Rev. Neurosci.* **2010**, *12*, 43–56. [CrossRef]
5. Villringer, A.; Planck, J.; Hock, C.; Schleinkofer, L.; Dirnagl, U. Near infrared spectroscopy (NIRS): A new tool to study hemodynamic changes during activation of brain function in human adults. *Neurosci. Lett.* **1993**, *154*, 101–104. [CrossRef]
6. Yang, D.; Hong, K.S. Quantitative Assessment of Resting-State for Mild Cognitive Impairment Detection: A Functional Near-Infrared Spectroscopy and Deep Learning Approach. *J. Alzheimer's Dis.* **2021**, *80*, 647–663. [CrossRef]
7. Othman, M.H.; Bhattacharya, M.; Møller, K.; Kjeldsen, S.; Grand, J.; Kjaergaard, J.; Dutta, A.; Kondziella, D. Resting-State NIRS–EEG in Unresponsive Patients with Acute Brain Injury: A Proof-of-Concept Study. *Neurocrit. Care* **2021**, *34*, 31–44. [CrossRef]
8. Bindra, J.; Pham, P.; Aneman, A.; Chuan, A.; Jaeger, M. Non-invasive Monitoring of Dynamic Cerebrovascular Autoregulation Using Near Infrared Spectroscopy and the Finometer Photoplethysmograph. *Neurocrit. Care* **2016**, *24*, 442–447. [CrossRef]
9. Sasai, S.; Homae, F.; Watanabe, H.; Taga, G. Frequency-specific functional connectivity in the brain during resting state revealed by NIRS. *Neuroimage* **2011**, *56*, 252–257. [CrossRef]
10. Obrig, H.; Neufang, M.; Wenzel, R.; Kohl, M.; Steinbrink, J.; Einhäupl, K.; Villringer, A. Spontaneous Low Frequency Oscillations of Cerebral Hemodynamics and Metabolism in Human Adults. *Neuroimage* **2000**, *12*, 623–639. [CrossRef]
11. Tong, Y.; Deb Frederick, B. Time lag dependent multimodal processing of concurrent fMRI and near-infrared spectroscopy (NIRS) data suggests a global circulatory origin for low-frequency oscillation signals in human brain. *Neuroimage* **2010**, *53*, 553–564. [CrossRef] [PubMed]
12. Blaney, G.; Sassaroli, A.; Pham, T.; Krishnamurthy, N.; Fantini, S. Multi-Distance Frequency-Domain Optical Measurements of Coherent Cerebral Hemodynamics. *Photonics* **2019**, *6*, 83. [CrossRef] [PubMed]
13. Yamada, Y.; Suzuki, H.; Yamashita, Y. Time-Domain Near-Infrared Spectroscopy and Imaging: A Review. *Appl. Sci.* **2019**, *9*, 1127. [CrossRef]
14. Wabnitz, H.; Jelzow, A.; Mazurenka, M.; Steinkellner, O.; MacDonald, R.; Milej, D.; Zołek, N.; Kacprzak, M.; Sawosz, P.; Maniewski, R.; et al. Performance assessment of time-domain optical brain imagers, part 2: NEUROPt protocol. *J. Biomed. Opt.* **2014**, *19*, 086012. [CrossRef] [PubMed]
15. *IEC 60825-1:2007*; Safety or Laser Products—Part1: Equipment Classification and Requirements. International Electrotechnical Commission: Geneva, Switzerland, 2007.
16. Koga, S.; Barstow, T.J.; Okushima, D.; Rossiter, H.B.; Kondo, N.; Ohmae, E.; Poole, D.C. Validation of a high-power, time-resolved, near-infrared spectroscopy system for measurement of superficial and deep muscle deoxygenation during exercise. *J. Appl. Physiol.* **2015**, *118*, 1435–1442. [CrossRef] [PubMed]
17. Jiang, J.; Mata, A.D.C.; Lindner, S.; Lindner, S.; Lindner, S.; Charbon, E.; Wolf, M.; Kalyanov, A. 2.5 Hz sample rate time-domain near-infrared optical tomography based on SPAD-camera image tissue hemodynamics. *Biomed. Opt. Express* **2022**, *13*, 133–146. [CrossRef] [PubMed]
18. Ban, H.Y.; Barrett, G.M.; Borisevich, A.; Chaturvedi, A.; Dahle, J.L.; Dehghani, H.; Dubois, J.; Field, R.M.; Gopalakrishnan, V.; Gundran, A.; et al. Kernel Flow: A high channel count scalable time-domain functional near-infrared spectroscopy system. *J. Biomed. Opt.* **2022**, *27*, 074710. [CrossRef] [PubMed]
19. Themelis, G.; Selb, J.; Thaker, S.; Stott, J.J.; Custo, A.; Boas, D.; Franceschini, M.A. Depth of arterial oscillation resolved with NIRS time and frequency domain. In *Biomedical Topical Meeting*; Optica Publishing Group: Washington, DC, USA, 2004; p. WF2.

20. Kacprzak, M.; Sawosz, P.; Weigl, W.; Milej, D.; Gerega, A.; Liebert, A. Frequency analysis of oscillations in cerebral hemodynamics measured by time domain near infrared spectroscopy. *Biomed. Opt. Express* **2019**, *10*, 761–771. [CrossRef]
21. Re, R.; Contini, D.; Caffini, M.; Cubeddu, R.; Spinelli, L.; Torricelli, A. A compact time-resolved system for near infrared spectroscopy based on wavelength space multiplexing. *Rev. Sci. Instrum.* **2010**, *81*, 113101. [CrossRef]
22. Amendola, C.; Lacerenza, M.; Pirovano, I.; Contini, D.; Spinelli, L.; Cubeddu, R.; Torricelli, A.; Re, R. Optical characterization of 3D printed PLA and ABS filaments for diffuse optics applications. *PLoS ONE* **2021**, *16*, e0253181. [CrossRef]
23. Contini, D.; Martelli, F.; Zaccanti, G. Photon migration through a turbid slab described by a model based on diffusion approximation I Theory. *Appl. Opt.* **1997**, *36*, 4587. [CrossRef] [PubMed]
24. Wabnitz, H.; Taubert, D.R.; Mazurenka, M.; Steinkellner, O.; Jelzow, A.; Macdonald, R.; Milej, D.; Sawosz, P.; Kacprzak, M.; Liebert, A.; et al. Performance assessment of time-domain optical brain imagers, part 1: Basic instrumental performance protocol. *J. Biomed. Opt.* **2014**, *19*, 086010. [CrossRef]
25. Re, R.; Muthalib, M.; Zucchelli, L.; Perrey, S.; Contini, D.; Caffini, M.; Spinelli, L.; Kerr, G.; Torricelli, A. Multichannel time domain fNIRS mapping of cortical activation and superficial systemic responses during neuromuscular electrical stimulation. In *Proceedings of the Optics InfoBase Conference Papers*; Optica Publishing Group: Washington, DC, USA, 2013.
26. Re, R.; Martinenghi, E.; Mora, A.D.; Contini, D.; Pifferi, A.; Torricelli, A. Probe-hosted silicon photomultipliers for time-domain functional near-infrared spectroscopy: Phantom and in vivo tests. *Neurophotonics* **2016**, *3*, 045004. [CrossRef] [PubMed]
27. Re, R.; Pirovano, I.; Contini, D.; Spinelli, L.; Torricelli, A. Time domain near infrared spectroscopy device for monitoring muscle oxidative metabolism: Custom probe and in vivo applications. *Sensors* **2018**, *18*, 264. [CrossRef] [PubMed]
28. Spinelli, L.; Martelli, F.; Farina, A.; Pifferi, A.; Torricelli, A.; Cubeddu, R.; Zaccanti, G. Accuracy of the nonlinear fitting procedure for time-resolved measurements on diffusive phantoms at NIR wavelengths. In *Optical Tomography and Spectroscopy of Tissue VII*; SPIE: Bellingham, DC, USA, 2009; Volume 7174.
29. *Advanced Time-Correlated Single Photon Counting Applications*; Becker, W. (Ed.) Springer Series in Chemical Physics; Springer International Publishing: Cham, Switzerland, 2015; Volume 111, ISBN 978-3-319-14928-8.
30. Avanzi, E.; Behera, A.; Contini, D.; Spinelli, L.; Dalla Mora, A.; Di Sieno, L. Effects and correctability of pile-up distortion using established figures of merit in time-domain diffuse optics at extreme photon rates. *Sci. Rep.* **2022**, *12*, 5417. [CrossRef]
31. Pifferi, A.; Torricelli, A.; Bassi, A.; Taroni, P.; Cubeddu, R.; Wabnitz, H.; Grosenick, D.; Möller, M.; Macdonald, R.; Swartling, J.; et al. Performance assessment of photon migration instruments: The MEDPHOT protocol. *Appl. Opt.* **2005**, *44*, 2104–2114. [CrossRef]
32. Re, R.; Contini, D.; Turola, M.; Spinelli, L.; Zucchelli, L.; Caffini, M.; Cubeddu, R.; Torricelli, A. Multi-channel medical device for time domain functional near infrared spectroscopy based on wavelength space multiplexing. *Biomed. Opt. Express* **2013**, *4*, 2231–2246. [CrossRef]
33. Seeck, M.; Koessler, L.; Bast, T.; Leijten, F.; Michel, C.; Baumgartner, C.; He, B.; Beniczky, S. The standardized EEG electrode array of the IFCN. *Clin. Neurophysiol.* **2017**, *128*, 2070–2077. [CrossRef]
34. Zucchelli, L.; Contini, D.; Re, R.; Torricelli, A.; Spinelli, L. Method for the discrimination of superficial and deep absorption variations by time domain fNIRS. *Biomed. Opt. Express* **2013**, *4*, 2893. [CrossRef]
35. Tong, Y.; Hocke, L.M.; Licata, S.C.; Frederick, B.D. Low-frequency oscillations measured in the periphery with near-infrared spectroscopy are strongly correlated with blood oxygen level-dependent functional magnetic resonance imaging signals. *J. Biomed. Opt.* **2012**, *17*, 106004. [CrossRef]
36. Yücel, M.A.; Selb, J.; Aasted, C.M.; Lin, P.; Borsook, D.; Becerra, L.; Boas, D.A.; Kvernmo, H.D.; Stefanovska, A.; Bracic, M.; et al. Mayer waves reduce the accuracy of estimated hemodynamic response functions in functional near-infrared spectroscopy. *Biomed. Opt. Express* **2016**, *7*, 3078–3088. [CrossRef] [PubMed]
37. Schroeter, M.L.; Schmiedel, O.; Von Cramon, D.Y. Spontaneous low-frequency oscillations decline in the aging brain. *J. Cereb. Blood Flow Metab.* **2004**, *24*, 1183–1191. [CrossRef] [PubMed]
38. Sassaroli, A.; Pierro, M.; Bergethon, P.R.; Fantini, S. Low-Frequency Spontaneous Oscillations of Cerebral Hemodynamics Investigated With Near-Infrared Spectroscopy: A Review. *IEEE J. Sel. Top. Quantum Electron.* **2012**, *18*, 1478–1492. [CrossRef]
39. Intaglietta, M. Vasomotion and flowmotion: Physiological mechanisms and clinical evidence. *Vasc. Med. Rev.* **1990**, *vmr-1*, 101–112. [CrossRef]
40. Glover, G.H.; Li, T.-Q.; Ress, D. Image-Based Method for Retrospective Correction of Physiological Motion Effects in fMRI: Retroicor. *Magn. Reson. Med.* **2000**, *44*, 162–167. [CrossRef] [PubMed]
41. Mesquita, R.C.; Franceschini, M.A.; Boas, D.A.; Murphy, K.; Birn, R.M.; Handwerker, D.A.; Jones, T.B.; Bandettini, P.A.; Arieli, A.; Sterkin, A.; et al. Functional connectivity in single and multislice echoplanar imaging using resting-state fluctuations. *Philos. Trans. R. Soc. Lond. B Biol. Sci.* **2007**, *37*, 184–192.

sensors

MDPI

Article

Reliability of Mental Workload Index Assessed by EEG with Different Electrode Configurations and Signal Pre-Processing Pipelines

Alfonso Mastropietro [1,*], Ileana Pirovano [1], Alessio Marciano [2], Simone Porcelli [2] and Giovanna Rizzo [1]

1 Institute of Biomedical Technologies, National Research Council, Via Fratelli Cervi 93, 20054 Segrate, Italy
2 Department of Molecular Medicine, University of Pavia, Via Forlanini 6, 27100 Pavia, Italy
* Correspondence: alfonso.mastropietro@itb.cnr.it; Tel.: +39-02-26422220

Abstract: Background and Objective: Mental workload (MWL) is a relevant construct involved in all cognitively demanding activities, and its assessment is an important goal in many research fields. This paper aims at evaluating the reproducibility and sensitivity of MWL assessment from EEG signals considering the effects of different electrode configurations and pre-processing pipelines (PPPs). Methods: Thirteen young healthy adults were enrolled and were asked to perform 45 min of Simon's task to elicit a cognitive demand. EEG data were collected using a 32-channel system with different electrode configurations (fronto-parietal; Fz and Pz; Cz) and analyzed using different PPPs, from the simplest bandpass filtering to the combination of filtering, Artifact Subspace Reconstruction (ASR) and Independent Component Analysis (ICA). The reproducibility of MWL indexes estimation and the sensitivity of their changes were assessed using Intraclass Correlation Coefficient and statistical analysis. Results: MWL assessed with different PPPs showed reliability ranging from good to very good in most of the electrode configurations (average consistency > 0.87 and average absolute agreement > 0.92). Larger fronto-parietal electrode configurations, albeit being more affected by the choice of PPPs, provide better sensitivity in the detection of MWL changes if compared to a single-electrode configuration (18 vs. 10 statistically significant differences detected, respectively). Conclusions: The most complex PPPs have been proven to ensure good reliability (>0.90) and sensitivity in all experimental conditions. In conclusion, we propose to use at least a two-electrode configuration (Fz and Pz) and complex PPPs including at least the ICA algorithm (even better including ASR) to mitigate artifacts and obtain reliable and sensitive MWL assessment during cognitive tasks.

Keywords: mental workload; EEG; signal processing; reliability; cognitive performance; Simon task

Citation: Mastropietro, A.; Pirovano, I.; Marciano, A.; Porcelli, S.; Rizzo, G. Reliability of Mental Workload Index Assessed by EEG with Different Electrode Configurations and Signal Pre-Processing Pipelines. *Sensors* **2023**, *23*, 1367. https://doi.org/10.3390/s23031367

Academic Editor: Sung-Phil Kim

Received: 21 December 2022
Revised: 18 January 2023
Accepted: 21 January 2023
Published: 26 January 2023

1. Introduction

Mental workload (MWL) can be defined, as recently proposed by Longo et al. [1], as "the degree of activation of a finite pool of resources, limited in capacity, while cognitively processing a primary task over time, mediated by external stochastic environmental and situational factors, as well as affected by definite internal characteristics of a human operator, for coping with static task demands, by devoted effort and attention". Even if the latter seems, to date, the most comprehensive definition of MWL, more commonly, MWL is roughly defined as a multidimensional construct describing the relationship between the cognitive task demand, under specific conditions, and the actual resources that can be actively engaged by an individual during the execution of the task [2,3].

MWL is a relevant construct since it is involved in almost all human activities [4], from everyday life activities to the most complex cognitive tasks, when a certain degree of mental processing is required. Interestingly, MWL is correlated to task demand and performance, since it is usually considered that high, as well as low, levels of MWL may have a negative impact on task performance and increase the incidence of errors [5–7] during the execution of a task. Therefore, the assessment and quantification of MWL represent one of the main interests in ergonomics [8] with relevant potential impact in different fields such as aeronautics [9], automotive [10], education and training [11], clinical practice, and rehabilitation [12,13].

Among all the available assessment methods, physiological measurements have been proven to provide an objective and minimally invasive evaluation with high reliability. These techniques estimate MWL from changes in biological signals and their derived variables that are related to the cardiovascular and respiratory system, the ocular responses, and the electrodermal and brain activity [1,14].

In this context, electroencephalography (EEG) is a widely used technique for the estimation of MWL, since it allows obtaining a direct non-invasive measurement of brain activity in different conditions. The study of changes occurring within the characteristic EEG oscillation rhythms during the execution of specific tasks has revealed that an increase in MWL is associated with a decrease in alpha activity (8–13 Hz) in the parietal brain area and an increase in theta activity (4–8 Hz) in the frontal area [15–17]. In particular, a correlation between increased task complexity and the power spectra of EEG signals recorded at midline electrodes has been observed [18]. For this reason, a simple metric to quantify the MWL is the theta-to-alpha ratio, which is calculated by dividing the theta band power over the EEG midline frontal channel (Fz), and the alpha band power over the parietal channel (Pz) [18–20]. However, EEG signals have also been used to estimate the MWL, with different configurations and number of sensors, (e.g., CZ, pre-frontal and lateral fronto-parietal electrodes) according to the experimental setup [21]. In this scenario, to the best of our knowledge, a systematic evaluation of the influence of the employed electrodes on the quantitative estimation of MWL during a cognitive task is still lacking.

Another key factor that can influence the quantification of MWL from EEG power spectra is the pre-processing pipeline applied to remove the extracerebral components that affect the EEG recording [22]. Various pre-processing methods are currently proposed in the literature to extract MWL indicators from EEG signals, and a consensus is still missing among researchers. The bandpass filtering is typically used in most papers but with different cut-off frequencies [20,23,24]; the major artifacts are typically removed with Independent Component Analysis (ICA) [25,26], Artifact Subspace Reconstruction (ASR) algorithms [20,27] or other methods [28]; the signal is mainly re-referenced to the average of the electrodes [23,27] or the average of the mastoid electrodes [20]; the channel rejections are performed automatically [29] or manually [30]. Although some more general pipelines for EEG signal analysis exist, they are quite broad and not universally adopted [31]. Furthermore, they are not always suitable for real-time applications implying MWL estimation, since they are based on quite complex methods that are often time-consuming and do not allow automatic real-time analysis.

In recent years, there has been a relevant growth in EEG analysis methods. The high amount of available new tools leads to the need of developing guidelines to pursue research reproducibility and the robustness of results to increase consistency within the scientific literature. This issue is now being referred to as the "reproducibility crisis" [32]. Indeed, a lack of consistent EEG signal pre-processing techniques can affect the comparison of quantitative results from different studies, even if the same dataset is analyzed. The reliability of EEG biomarkers is particularly critical in the perspective of employing them in clinical practice for understanding human cognition [33,34]. As underlined in the Organization for Human Brain Mapping reports [35] in presenting best practices for specific neuroimaging methods, a single best analysis workflow does not exist, and the optimal solution has to be adapted for the specific application [34]. In the specific field of MWL estimation, the literature mainly focused on test–retest reliability in longitudinal studies and on the effects of EEG signal pre-processing on the performances of automatic MWL-level classification algorithms [36]. However, to the best of our knowledge, no evaluation has been systematically conducted on the MWL quantification by EEG biomarkers, i.e., the theta-to-alpha ratio values, disregarding the automatic load classification problem.

Considering what was previously introduced, the main aim of the paper is to evaluate the reliability of MWL assessment by EEG in terms of reproducibility and sensitivity to identify the best processing pipeline and electrode configuration for MWL quantification during cognitive tasks.

2. Related Works

The reliability of EEG analysis, and consequently the quantitative indexes derived, is a long-standing fundamental issue addressed by the scientific community. In the literature, test–retest studies have been conducted to assess the replicability of EEG-derived indexes over time. Ding and colleagues [37] tested the reproducibility of EEG spectral analysis at the electrode and source level during rest and imaginary tasks. Corsi-Cabrera et al. [38] conducted a longitudinal study on six women to assess within-subject reliability and inter-session stability of resting EEG over nine months in the estimation of the absolute power and inter- and intra-hemispheric coherent activity. However, these works did not take into consideration the effects of different pre-processing workflows on the results' replicability. In this context, a few works have tested the pre-processing influence on the longitudinal replicability of results. In 2017, Shirk et al. [39] tested the impact of subjective artifact removal on Event-Related Potential (ERP) results, estimating the inter-rater reliability of different subjective signal-cleaning approaches. The test–retest study by Suarez-Revelo and colleagues [40,41] compared different pre-processing of resting state EEG for the estimation of spectral power in six frequency bands. For specific MWL correlates estimation, a test–retest study was conducted in 2021 by Getzmann et al. [42] to assess the performance of the cEEGrids recordings, which are based on C-shaped electrode arrays positioned around the ear. However, no evaluation as regards the pre-processing technique was presented.

While the test–rest approach is valid to prove the stability of results, especially in longitudinal studies, it is not the most suitable test to assess the impact of pre-processing on quantitative estimation when repeated measurements are not provided. In this context, a series of papers have been recently published in which the performances of machine learning approaches to classify the MWL level after different signal pre-processing pipelines were compared [36,43–51]. These works are focused only on the automatic classification accuracy, considering several features extracted from all the EEG frequency bands and electrode signals, e.g., ERP, as input to the algorithm, whereas any direct evaluation of the EEG features extracted is provided.

To the best of our knowledge, in the published literature, no works are investigating how the pre-processing workflow choices affect the MWL quantitative correlates, i.e., the theta-to-alpha ratio tested in the present work.

3. Materials and Methods

3.1. Experimental Protocol

Thirteen young healthy adults (age: 27 ± 6; 9 males/4 females) were enrolled in the study. The study was conducted according to the principles expressed in the Declaration of Helsinki and was approved by the local ethics committee of the University of Pavia, Italy (2531CEMaugeri-27072021). The participants signed a written informed consent. Subjects were asked to avoid ingesting any caffeine-containing drink or nicotine and performing mentally demanding tasks for at least 3 h before the session started. Moreover, they were invited to sleep at least 7 h before the experiment. The volunteers were not allowed to take any medication before the experimental session, and they did not suffer from any type of neurological and psychiatric disease. The experiments were performed at controlled room temperature (18–20 °C) and air humidity (40–60%). The experimental session consisted of performing a 45 min cognitive-demanding task sitting in front of a computer screen, and it was composed of three consecutive blocks of 14 min and 30 s each (i.e., Task 1, Task 2, Task 3), interspersed with 30 s of rest. At the beginning of the sessions, a 3-min resting period with open eyes was proposed to the volunteers and used as a baseline signal. Simon's task was selected to elicit a cognitive demand in the volunteers. The Simon task is a behavioral measure of interference/conflict resolution [52,53]. The participants were asked to respond to visual stimuli by pressing a rightward keyboard button to the "right" stimulus and a leftward button to the "left" stimulus. The stimuli were randomly presented on the right side or the left side of the screen. Regardless of the spatial presentation of the stimuli, the subjects were asked to press the buttons corresponding to the letter shown by the visual stimulus. A schematic representation of the experimental protocol is displayed in Figure 1. Cognitive tasks were implemented and presented online using the PsyToolkit platform [54,55] (https://www.psytoolkit.org, accessed on 20 January 2023). To measure users' performance, Reaction Times (RT) and Error Rates (ERR%) were collected as behavioral data in the different blocks of tasks.

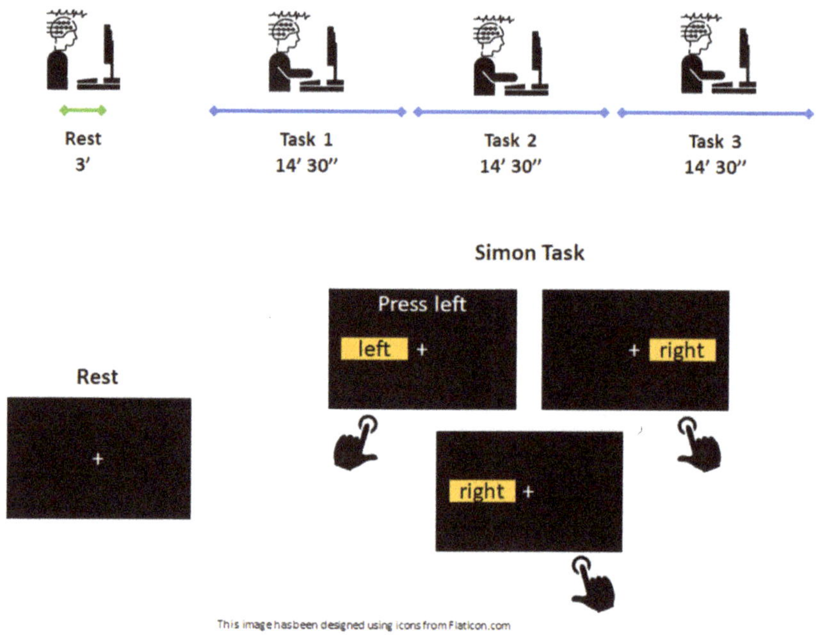

Figure 1. A schematic representation of the experimental protocol timing (**upper panel**) and an example depicting what was presented to the volunteers on the PC screen (**lower panel**).

3.2. EEG Acquisitions

Continuous EEG data were collected using a compact 32-channel system (eego™sports 32, ANT Neuro®, Enschede, The Netherlands). A gel-based electrode cap with sintered Ag/AgCl electrodes was used (Waveguard, ANT Neuro®, 10–20 system). The online reference was placed at the CPz electrode. Signal was acquired with eego sports acquisition software connected to a 24 bits amplifier at a sampling rate of 500 Hz. Impedances for all electrodes were kept below 20 kΩ. EEG signals were recorded across 30 channels: Fp1, Fpz, Fp2, F7, F3, Fz, F4, F8, FC5, FC1, FC2, FC6, T7, C3, Cz, C4, T8, CP5, CP1, CP2, CP6, P7, P3, Pz, P4, P8, POz, O1, Oz, and O2 excluding the mastoids electrodes (M1 and M2). The starting and ending points of each block composing the acquisition (Rest, Task 1, Task 2, Task 3) were manually labeled using the acquisition software.

3.3. EEG Pre-Processing

Four different processing pipelines were evaluated to assess their impact on the estimation of the MWL indicator. A schematic representation is displayed in Figure 2.

1. FILT—The first and simplest pipeline was characterized using band-pass filtering to mitigate the effects of the artifacts. In detail, EEG signals were band-pass filtered

in the range 1–40 Hz using a Hamming windowed sinc FIR filter. Bad channels were removed by evaluating the normed joint probability of the average log power across the channels [56]. Channels whose probability falls more than three standard deviations from the mean are removed as bad channels.

2. FILT + ASR—The second pipeline was implemented by adding the ASR algorithm to the FILT pipeline. ASR uses principal-component-like subspace decomposition to remove transient and high-amplitude artifacts, it provides a noiseless signal reconstruction using a reference signal fragment [57] and can be helpful for real-time artifact removal. ASR was used to interpolate artifact "bursts" with a variance higher than fifteen standard deviations different from the automatedly detected reference signal, as previously suggested [58].
3. FILT + ICA—The third pipeline was proposed by adding the ICA artifact rejection method to the first pipeline. ICA algorithms are typically used to detect and remove artifacts (such as eye movements and electrocardiographic signals) that usually overlay with brain activity in EEG recordings. The extended Infomax [59] ICA algorithm was used in this work. ICLabel [60] was used to automatically reject independent components having a probability to be plausible brain sources of less than 40%.
4. FILT + ASR + ICA—The last most complex pipeline included sequentially all the previous different approaches.

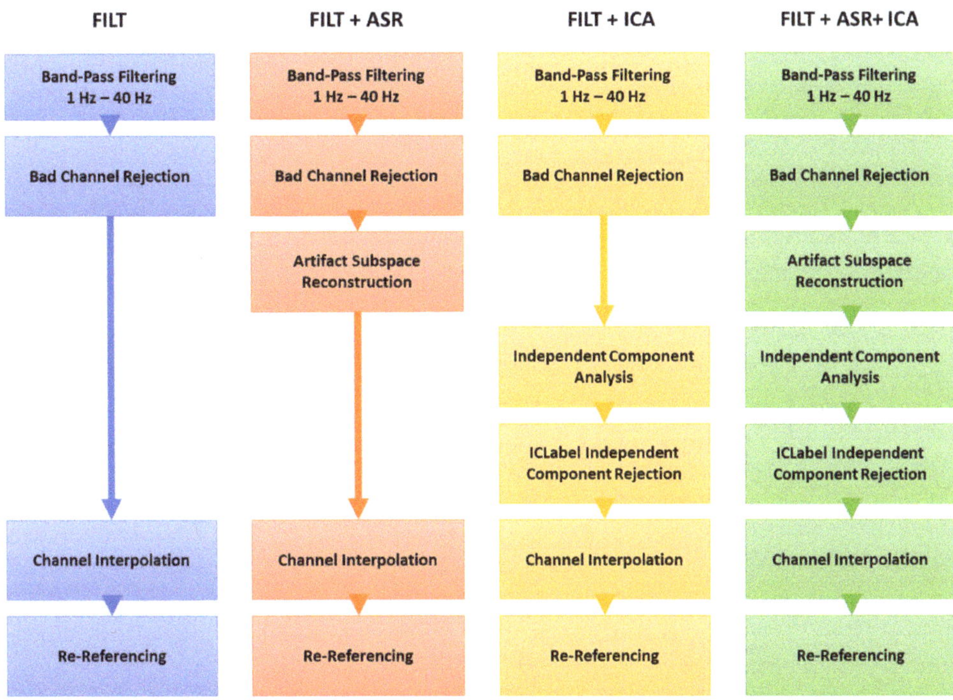

Figure 2. A schematic representation of the pre-processing pipelines that were evaluated in this study.

To complete all the previous pipelines, channels that were removed as "bad channels" were replaced by data interpolated from nearby "artifact-free" channels using a spherical function, and EEG signals were re-referenced to the average of the channels. Among all the analyzed EEG signals, on the whole, 3 channels were removed (specifically P7 in 1 subject and CP2 in 2 subjects).

All the pre-processing steps were implemented in MATLAB (R2021b, The MathWorks) using the EEGLAB toolbox [61].

3.4. MWL Assessment

The pre-processed EEG signals were analyzed in the frequency domain to extract the power spectra in the range 1–45 Hz using the Welch's power spectral density (PSD) estimate. The EEG signal was windowed using a Hamming window (1 s length, 500 samples, non-overlapping) and the periodogram was computed, for each segment, by using the discrete Fourier transform. The squared magnitude of the result was computed and the individual periodograms were averaged, separately for each of the three experimental blocks, to obtain the power spectra for each task. Subsequently, the integral of the power spectrum across frequencies in theta (4–8 Hz) and alpha (8–13 Hz) ranges was calculated to obtain the absolute band power for each channel. The MWL index of each block was then calculated by dividing the theta absolute power θ with the alpha absolute power α into three different electrode configurations.

1. Fz and Pz electrodes:

$$MWL_{Fz,Pz} = \frac{\theta_{Fz}}{\alpha_{Pz}} \tag{1}$$

2. Cz electrode:

$$MWL_{Cz} = \frac{\theta_{Cz}}{\alpha_{Cz}} \tag{2}$$

3. Frontal (F7, F3, Fz, F4, F8) and Parietal (P7, P3, Pz, P4, P8) electrodes:

$$MWL_{FP} = \frac{\theta_{Frontal}}{\alpha_{Parietal}} \tag{3}$$

where *θ_Frontal* and *α_Parietal* are the sum of the absolute powers in frontal and parietal electrodes.

The MWL index calculated during tasks was normalized to the value of the rest condition as follows:

$$MWL = \frac{(MWL_{Task} - MWL_{Rest})}{MWL_{Rest}} \tag{4}$$

3.5. Reproducibility Assessment

The reproducibility refers to the level of consistency and agreement in the estimation of MWL at different EEG electrode configurations and pre-processing pipelines with increasing levels of complexity. To assess the reproducibility, Intraclass Correlation Coefficient (ICC) was adopted as a descriptive statistical method. ICC reflects both the degree of correlation and the agreement between measurements [62]. The two-way mixed-effects model was selected to assess both consistency and agreement among different scenarios.

In particular, the two-way mixed effects, consistency, and single measurement ICC (3,1) index was defined as follows:

$$\frac{MS_R - MS_E}{MS_R + (k-1)MS_E} \tag{5}$$

whereas the two-way mixed effects, absolute agreement, and single measurement ICC (2,1) index was defined as follows:

$$\frac{MS_R - MS_E}{MS_R + (k-1)MS_E + \frac{k}{n}(MS_C - MS_E)} \tag{6}$$

where MSR = mean square for rows; MSE = mean square for error; MSC = mean square for columns; n = number of targets; k = number of ratings.

3.6. Statistical Analysis

To evaluate if there is a statistically significant interaction effect between the three within-subjects factors (pre-processing pipelines, electrode configurations, task blocks) in explaining differences in MWL metrics estimated in different conditions (e.g., electrode configurations, processing pipelines, task blocks), the repeated measure ANOVA test was adopted. Greenhouse–Geisser correction was applied to only within-subjects factors violating the sphericity assumption (with significant Mauchly's test p-value, $p \leq 0.05$).

To evaluate the sensitivity, which refers to the ability to discriminate changes in the MWL index at increasing cognitive loads and different experimental settings, multiple pairwise comparisons between groups were performed using the pairwise t-test, and the false discovery rate adjustment was applied to correct p-values. p-values ≤ 0.05 were considered significant. The statistical tests were performed in R (ver. 4.2.1) [63] embedded in RStudio (2022.07.1, Build 554).

4. Results

4.1. Reproducibility

Considering each specific electrode configuration individually, the consistency among MWL metrics, obtained using different pre-processing pipelines, exhibits values (averaged over tasks) higher than 0.81 in all the conditions. In particular, the mean consistency is 0.94 for FzPz, 0.94 for Cz and 0.88 for fronto-parietal configurations, respectively. The highest consistency can be observed between Filt + ASR, Filt + ICA and Filt + ASR + ICA (maximum consistency at 0.99), whereas the lowest values are those corresponding to the comparison of Filt with the other pre-processing pipelines (minimum consistency at 0.81). As to the consistency among electrode configurations, its mean values are 0.83 in the case of FzPz vs. Cz, 0.85 in the case of FzPz vs. fronto-parietal and 0.74 in the case of Cz vs. fronto-parietal configurations, respectively.

Regarding the absolute agreement, the tendency is similar to that described above for consistency but with lower values. In particular, as regards the absolute agreement among pre-processing pipelines in each electrode's configuration, the mean absolute agreement is 0.92 for FzPz, 0.91 for Cz and 0.78 for fronto-parietal configurations, respectively. The highest absolute agreement can be observed between Filt + ASR, Filt + ICA and Filt + ASR + ICA (maximum absolute agreement at 0.99), whereas the lowest values are those corresponding to the comparison of Filt with the other pre-processing pipelines (minimum consistency at 0.58). As to the absolute agreement among electrode configurations, its mean values are 0.73 in the case of FzPz vs. Cz, 0.77 in the case of FzPz vs. fronto-parietal and 0.49 in the case of Cz vs. fronto-parietal configurations, respectively. A concise representation of the results is shown in Figure 3.

4.2. Impact of Experimental Factors on MWL

To investigate the impact of the within-subjects' factors (i.e., electrode configurations, pre-processing pipelines and tasks) in discriminating differences among MWL indexes, we explored the results of the three-way repeated measures ANOVA test (as summarized in Table 1). Considering the single factors individually (i.e., pipeline, configuration, task), significant differences within them were observed ($p < 0.05$). As to the interaction of two factors (i.e., pipeline and configuration, pipeline and task, configuration and task), statistically significant differences were shown when pipelines and configurations along with tasks, respectively, ($p < 0.05$) were considered, whereas a significant difference was not observed when the combined effect of pipelines and configuration was considered. Moreover, as shown in Table 1, there is a statistically significant three-way interaction between pipelines, configurations and tasks, $F(18, 216) = 2.225$, $p = 0.004$.

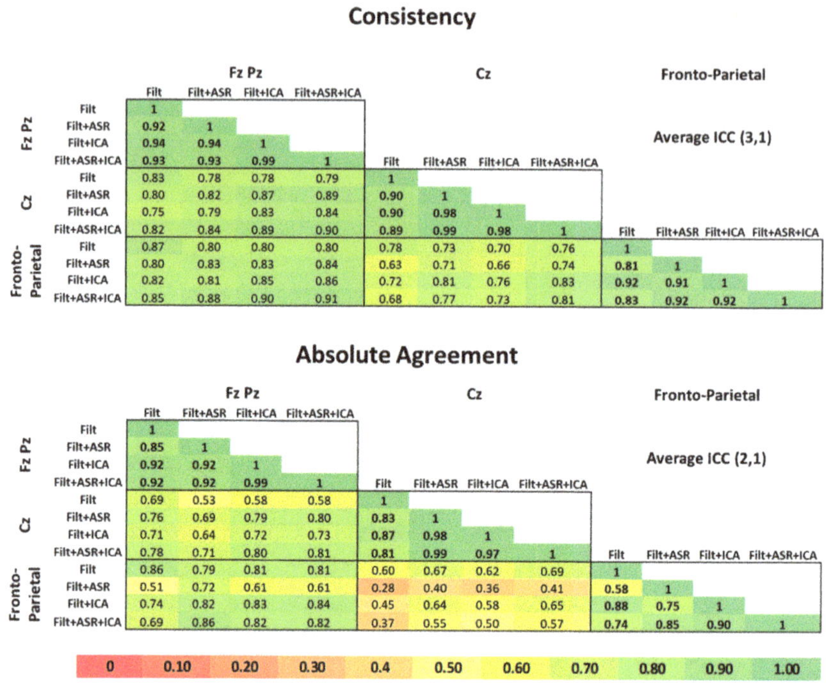

Figure 3. ICC values are shown for both consistency (**upper panel**) and absolute agreement (**lower panel**). Colors range from red (no consistency/absolute agreement) to green (highest consistency/absolute agreement) as shown in the color bar at the bottom of the figure.

Table 1. Summary of results of the ANOVA three-way repeated measures test. Under the "*Effect*" column are listed all the factors included in the study; *DFn* is the acronym of "degrees of freedom in the numerator"; *DFd* is the acronym of "degrees of freedom in the denominator"; F is the test statistic for ANOVA; *p* is the *p*-value; under the "$p < 0.05$" column, there is an asterisk when the *p*-value is less than 0.05; *ges* is the "generalized eta squared".

Effect	DFn	DFd	F	p	$p < 0.05$	Ges
Pipelines	1.27	15.27	4.253	0.049	*	0.016
Configurations	2	24	3.91	0.034	*	0.038
Tasks	1.49	17.9	10.47	0.002	*	0.166
Pipelines × Configurations	2.59	31.12	2.635	0.075	n.s.	0.003
Pipelines × Tasks	1.51	18.12	3.876	0.05	*	0.006
Configurations × Tasks	2.24	26.9	3.485	0.04	*	0.014
Pipelines × Configurations × Tasks	18	216	2.225	0.004	*	0.000987

4.3. Sensitivity to MWL Changes during Prolonged Simon Task

In Figure 4, the population's average MWL indexes calculated during the three consecutive experimental blocks, considering the three different electrode configurations and the four pre-processing pipelines, are represented. Regardless of the method/electrodes evaluated, we observe a common trend of the MWL index during the execution of the Simon task over time. Specifically, in all cases, we found an initial relevant increase in MWL compared to the rest condition in the first 15-min block of task execution. Afterward, in the second and third blocks, a decrease in MWL is observed even though it still remained higher than MWL calculated at baseline. Considering the users' performances during the Simon task, the average RT decreases over time and blocks, ranging from 541 ± 33 ms to 515 ± 36 ms whereas, conversely, the ERR%s increases, ranging from 3.1% ± 2.1% to 4.0% ± 2.4% as shown in Figure S1.

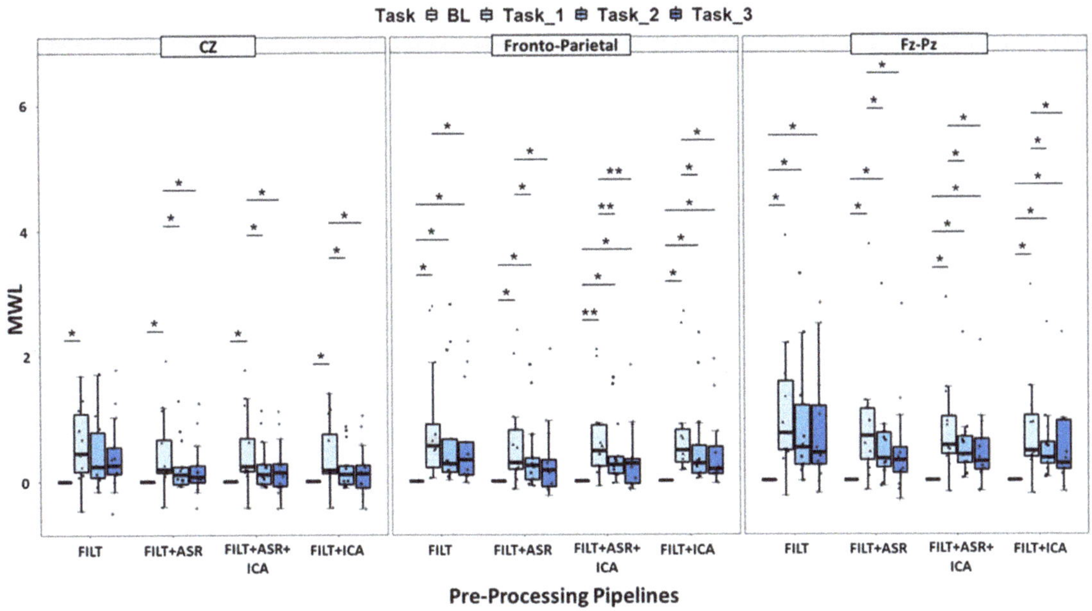

Figure 4. Representation of MWL in different experimental conditions (electrode configurations and pre-processing pipelines) and tasks. Asterisks refer to statistically significant differences ($p < 0.05$ *; $p < 0.01$ **).

In detail, considering the multiple pairwise comparisons results shown in Figure 4, statistically significant differences were observed in most of the conditions. In particular, exploring the differences among tasks and rest, significant MWL differences between Task 1 and Rest were found in all the electrode configurations and pre-processing approaches, whereas significant MWL differences between Task 2 and Rest were observed just in fronto-parietal and FzPz configurations considering all pre-processing pipelines. Finally, significant MWL differences between Task 3 and Rest were found in fronto-parietal and FzPz configurations in the case of FILT, FILT + ICA and FILT + ASR + ICA pipelines.

As to the differences among tasks, significant differences between Task 1 and Task 2, as well as between Task 1 and Task 3, were observed in all configurations in the case of FILT + ASR, FILT + ICA and FILT + ASR pipelines. No significant differences were found between Task 2 and Task 3.

Globally, the conditions in which the maximum number of differences (five out of six) were found are those where the fronto-parietal and FzPz electrodes are considered and the FILT + ICA and FILT + ASR + ICA pipelines were used to process the EEG signals.

A summary of descriptive statistical features and the list of p-values and effect sizes related to the between-groups pairwise comparisons are reported in Table S1 and Table S2, respectively.

5. Discussion

This paper evaluated the reproducibility of MWL estimation from EEG signals considering different processing pipelines and electrode configurations as well as the sensitivity of the MWL metric to discriminate among different cognitive loads during a prolonged cognitive task. Furthermore, this work aimed also at providing guidelines for the quantitative estimation of the MWL changes taking into consideration a few aspects that are usually overlooked in the literature and, when results are available, they lack consistency.

To assess the reliability of EEG-based MWL estimation, we requested the volunteers to perform a cognitive task, i.e., the Simon task, eliciting MWL changes related to mental processes, such as working memory and attentional control, associated with the execution of the task goal during the congruent/incongruent stimuli presentation [64]. Even though this work neglected the investigation of the neurophysiological mechanisms underlying task-related mental constructs, our results show that the Simon task was able to elicit an increase in MWL if compared to the rest condition. Furthermore, a temporal effect influences the response; in fact, the initial increase in MWL, during the first block of tasks, is followed by a reduction in the following tasks, which is probably due to the onset of mental fatigue related to the prolonged mental demand. Therefore, the MWL index appears to be sensitive to the Simon effect and its elicited changes in mental effort.

Although MWL variations were well observed in most conditions, we found a dependence of the quantification and statistical identification of changes on both acquisitions, i.e., electrode position, and pre-processing approaches. In the literature, the investigation of different electrode configurations and pre-processing pipelines focuses on the influence of these factors on MWL classification accuracy through automatic algorithms based on machine learning and deep learning [50,51]. To our knowledge, no works assessed the reliability directly in MWL indexes derived from EEG signals. This paper wants to put an accent on this quantitative aspect and provide suggestions to choose the methodological aspects that will guarantee the most reliable outcome.

Considering each single electrode configuration independently, the reproducibility expressed in terms of consistency was good or very good across all the processing pipelines used to pre-process the EEG signals in every condition. As to the absolute agreement, it exhibited lower values and moderate to very good reliability, especially in the fronto-parietal configuration. This is most likely due to the wider extension of the fronto-parietal configuration being that more prone to be corrupted by artifacts if compared to the electrodes that are placed in the midline [65]. For that reason, the MWL estimation in the fronto-parietal configuration is more susceptible to the choice of the pre-processing pipeline whereas the FzPz and Cz configurations, which exhibited the best consistency and absolute agreement among pre-processing pipelines, are less susceptible to that factor. As to the pre-processing pipelines, the most complex algorithms (e.g., FILT + ASR + ICA, FILT-ICA, and FILT-ASR) were those showing the highest values of reproducibility.

Considering the reproducibility evaluated across different electrode configurations, the lowest values of consistency and absolute agreement were found when comparing Cz with fronto-parietal configurations. Conversely, the best reliability was obtained between FzPz and fronto-parietal configuration. In general, the single electrode configuration (Cz) is that with the lowest reliability when compared to the others. Finally, even in this case, the most complex algorithms are those showing the highest consistency and agreement.

Regarding the factors that can affect the assessment of the MWL index, pre-processing pipelines and electrode configurations can be chosen independently of each other, since there is no statistically significant interaction between them. On the contrary, there is a significant interaction between tasks and electrode configurations or between tasks and pre-processing pipelines; indeed, the choice of electrode configurations and pre-processing pipelines independently affects the sensitivity of MWL to discriminate different cognitive loads during tasks.

In particular, the best electrode configurations in terms of sensitivity to MWL changes are those with the highest number of electrodes (e.g., fronto-parietal and FzPz), probing both frontal and parietal lobes. The use of Cz, even though proposed in recent work for its ease of use [20] and its potential application with single-electrode systems in real-time MWL monitoring, is not the best choice in terms of sensitivity and has the lowest reliability if compared to the other electrode configurations.

The MWL index appears to be more reliable if information is taken from both frontal and parietal electrodes rather than from a single channel probing. Indeed, the results shown in this paper support the use of at least Fz and Pz electrodes, as previously performed in other works investigating changes in MWL [18,19,66], as the minimum set of sensors suitable for obtaining reliable and sensitive estimation.

As for the pre-processing pipelines, the approaches allowing the best discrimination among tasks are those including the ICA method (with or without ASR). Our results agree with those obtained by Kingphai and Moshfeghi [50], who evaluated the accuracy of MWL classification after different signal pre-processing procedures. In fact, they found that the most complete pipelines including the ICA technique provide the best classification accuracy. However, they did not evaluate the introduction of ASR as a prior step, despite being used in other classification works [27].

A limitation to the generalization of our results could be represented by the fact that we analyzed signals obtained in a controlled experimental protocol, where subjects were requested to avoid relevant movements while performing the cognitive tasks. The influence of the pre-processing pipelines could be more significant in free-moving conditions, and the results could slightly differ from those presented in this paper. However, we assume that the midline electrode signals could provide repeatable results even in the more complex experimental setup, since movement artifacts usually less affect these electrodes. Another limitation of the present work can be represented by the low number of subjects involved but, considering that the statistical analysis pointed out significant differences even applying the correction for multiple comparisons, we are confident that the results presented in this paper could be generalized.

As for the analysis pipeline, we propose here a set of four different approaches that try to include all the pre-processing steps that are most frequently employed in the EEG literature. Anyway, variations in the choice of filters and algorithms parameters could induce different outcomes.

In the future, the evaluation of MWL reliability should be assessed also during physical exercises or free-moving experiments.

6. Conclusions

This work showed how the assessment of MWL using EEG signals depends on both the pre-processing pipelines and the electrode configurations. Therefore, each experimental protocol definition must be well pondered, since it can affect both the reproducibility and the sensitivity. Furthermore, comparisons of quantitative results between works implementing different methods should be carefully dealt with.

This paper suggests that using both frontal and parietal electrodes provides more robust performances in the detection of MWL changes during a cognitive task if compared to a single-electrode configuration. However, larger electrode configurations could be more prone to artifacts, be time-consuming, and be challenging in some experimental conditions (those involving non-collaborative subjects or those which involve the execution of tasks during movement).

Most complex pre-processing pipelines have been proven to be more suitable to ensure good inter-rater reliability and sensitivity in all experimental conditions.

In conclusion, our work provides a practical analysis framework for quantitative EEG-based MWL evaluation studies. We propose to use at least a two-electrode configuration (Fz and Pz) and complex pre-processing pipelines including at least the ICA algorithm (even better if ASR is included) to mitigate artifacts and obtain reliable and sensitive MWL assessment during cognitive tasks.

Supplementary Materials: The following supporting information can be downloaded at: https://www.mdpi.com/article/10.3390/s23031367/s1.

Author Contributions: Conceptualization, A.M. (Alfonso Mastropietro), S.P., G.R.; methodology, A.M. (Alfonso Mastropietro), I.P., A.M. (Alessio Marciano), S.P.; software, A.M. (Alfonso Mastropietro), I.P.; formal analysis, A.M. (Alfonso Mastropietro), I.P.; resources, G.R.; data curation, A.M. (Alfonso Mastropietro); writing—original draft preparation, A.M. (Alfonso Mastropietro); writing—review and editing, A.M. (Alfonso Mastropietro), I.P., A.M. (Alessio Marciano), S.P., G.R.; funding acquisition, G.R. All authors have read and agreed to the published version of the manuscript.

Funding: This work was supported by INAIL (Istituto Nazionale per l'Assicurazione contro gli Infortuni sul Lavoro) within the project PR19-SV-P1 "Rip@rto"—Simulatore di guida per assistere operatori nella valutazione delle capacità di guida dell'utente e nella scelta degli ausili di cui dotare l'automobile.

Institutional Review Board Statement: The study was conducted according to the principles expressed in the Declaration of Helsinki and was approved by the local ethics committee of the University of Pavia, Italy (2531CEMaugeri-27072021).

Informed Consent Statement: Informed consent was obtained from all subjects involved in the study.

Data Availability Statement: Data are available upon reasonable request to the corresponding author.

Conflicts of Interest: The authors declare no conflict of interest.

References

1. Longo, L.; Wickens, C.D.; Hancock, P.A.; Hancock, G.M. Human Mental Workload: A Survey and a Novel Inclusive Definition. *Front. Psychol.* **2022**, *13*, 2962. [CrossRef] [PubMed]
2. Cain, B. A Review of the Mental Workload Literature. *Defence Research and Development Toronto (Canada)*. 2007. Available online: https://apps.dtic.mil/sti/citations/ADA474193 (accessed on 20 January 2023).
3. Babiloni, F. Mental Workload Monitoring: New Perspectives from Neuroscience. *Commun. Comput. Inf. Sci.* **2019**, *1107*, 3–19. [CrossRef]
4. Mitchell, D.K. Mental Workload and ARL Workload Modeling Tools. *Army Research Lab Aberdeen Proving Ground MD*. 2000. Available online: https://apps.dtic.mil/sti/citations/ADA377300 (accessed on 20 January 2023).
5. Fan, J.; Smith, A.P. The Impact of Workload and Fatigue on Performance. *Commun. Comput. Inf. Sci.* **2017**, *726*, 90–105. [CrossRef]
6. Lysaght, R.J.; Hill, S.G.; Dick, A.O.; Plamondon, B.D.; Linton, P.M. Operator Workload: Comprehensive Review and Evaluation of Operator Workload Methodologies. *Analytics Inc Willow Grove PA*. 1989. Available online: https://apps.dtic.mil/sti/citations/ADA212879 (accessed on 20 January 2023).
7. Kantowitz, B.H. Attention and Mental Workload. *Proc. Hum. Factors Ergon. Soc. Annu. Meet.* **2016**, *44*, 456–459. [CrossRef]
8. Young, M.S.; Brookhuis, K.A.; Wickens, C.D.; Hancock, P.A. State of Science: Mental Workload in Ergonomics. *Ergonomics* **2015**, *58*, 1–17. [CrossRef]
9. Wilson, G.F. An Analysis of Mental Workload in Pilots During Flight Using Multiple Psychophysiological Measures. *Int. J. Aviat. Psychol.* **2009**, *12*, 3–18. [CrossRef]
10. Chihara, T.; Kobayashi, F.; Sakamoto, J. Evaluation of Mental Workload during Automobile Driving Using One-Class Support Vector Machine with Eye Movement Data. *Appl. Ergon.* **2020**, *89*, 103201. [CrossRef]
11. Gerjets, P.; Walter, C.; Rosenstiel, W.; Bogdan, M.; Zander, T.O. Cognitive State Monitoring and the Design of Adaptive Instruction in Digital Environments: Lessons Learned from Cognitive Workload Assessment Using a Passive Brain-Computer Interface Approach. *Front. Neurosci.* **2014**, *8*, 385. [CrossRef]
12. Marchand, C.; de Graaf, J.B.; Jarrassé, N. Measuring Mental Workload in Assistive Wearable Devices: A Review. *J Neuroeng. Rehabil.* **2021**, *18*, 1–15. [CrossRef]
13. Longo, L. Mental Workload in Medicine: Foundations, Applications, Open Problems, Challenges and Future Perspectives. In Proceedings of the IEEE 29th International Symposium on Computer-Based Medical Systems (CBMS), Belfast and Dublin, Ireland, 20–24 June 2016; pp. 106–111. [CrossRef]
14. Charles, R.L.; Nixon, J. Measuring Mental Workload Using Physiological Measures: A Systematic Review. *Appl. Ergon.* **2019**, *74*, 221–232. [CrossRef]
15. Gevins, A.; Smith, M.E.; McEvoy, L.; Yu, D. High-Resolution EEG Mapping of Cortical Activation Related to Working Memory: Effects of Task Difficulty, Type of Processing, and Practice. *Cereb. Cortex* **1997**, *7*, 374–385. [CrossRef] [PubMed]
16. Missonnier, P.; Deiber, M.P.; Gold, G.; Millet, P.; Gex-Fabry Pun, M.; Fazio-Costa, L.; Giannakopoulos, P.; Ibáñez, V. Frontal Theta Event-Related Synchronization: Comparison of Directed Attention and Working Memory Load Effects. *J. Neural Transm.* **2006**, *113*, 1477–1486. [CrossRef] [PubMed]
17. Stipacek, A.; Grabner, R.H.; Neuper, C.; Fink, A.; Neubauer, A.C. Sensitivity of Human EEG Alpha Band Desynchronization to Different Working Memory Components and Increasing Levels of Memory Load. *Neurosci. Lett.* **2003**, *353*, 193–196. [CrossRef] [PubMed]

18. Holm, A.; Lukander, K.; Korpela, J.; Sallinen, M.; Müller, K.M.I. Estimating Brain Load from the EEG. *ScientificWorldJournal* **2009**, *9*, 639–651. [CrossRef]
19. Gevins, A.; Smith, M.E. Neurophysiological Measures of Cognitive Workload during Human-Computer Interaction. *Theor. Issues Ergon. Sci.* **2003**, *4*, 113–131. [CrossRef]
20. Kartali, A.; Janković, M.M.; Gligorijević, I.; Mijović, P.; Mijović, B.; Leva, M.C. Real-Time Mental Workload Estimation Using EEG. *Commun. Comput. Inf. Sci.* **2019**, *1107*, 20–34. [CrossRef]
21. Borghini, G.; Astolfi, L.; Vecchiato, G.; Mattia, D.; Babiloni, F. Measuring Neurophysiological Signals in Aircraft Pilots and Car Drivers for the Assessment of Mental Workload, Fatigue and Drowsiness. *Neurosci. Biobehav. Rev.* **2014**, *44*, 58–75. [CrossRef]
22. Bigdely-Shamlo, N.; Mullen, T.; Kothe, C.; Su, K.-M.; Robbins, K.A. The PREP pipeline: Standardized preprocessing for large-scale EEG analysis. *Front. Neuroinform.* **2015**, *9*, 16. [CrossRef]
23. di Flumeri, G.; Borghini, G.; Aricò, P.; Sciaraffa, N.; Lanzi, P.; Pozzi, S.; Vignali, V.; Lantieri, C.; Bichicchi, A.; Simone, A.; et al. EEG-Based Mental Workload Neurometric to Evaluate the Impact of Different Traffic and Road Conditions in Real Driving Settings. *Front. Hum. Neurosci.* **2018**, *12*, 509. [CrossRef]
24. das Chakladar, D.; Dey, S.; Roy, P.P.; Dogra, D.P. EEG-Based Mental Workload Estimation Using Deep BLSTM-LSTM Network and Evolutionary Algorithm. *Biomed. Signal Process. Control* **2020**, *60*, 101989. [CrossRef]
25. Dasari, D.; Shou, G.; Ding, L. ICA-Derived EEG Correlates to Mental Fatigue, Effort, and Workload in a Realistically Simulated Air Traffic Control Task. *Front. Neurosci.* **2017**, *11*, 297. [CrossRef] [PubMed]
26. Qu, H.; Shan, Y.; Liu, Y.; Pang, L.; Fan, Z.; Zhang, J.; Wanyan, X. Mental Workload Classification Method Based on EEG Independent Component Features. *Appl. Sci.* **2020**, *10*, 3036. [CrossRef]
27. Lim, W.L.; Sourina, O.; Wang, L.P. STEW: Simultaneous Task EEG Workload Data Set. *IEEE Trans. Neural Syst. Rehabil. Eng.* **2018**, *26*, 2106–2114. [CrossRef] [PubMed]
28. Rosanne, O.; Albuquerque, I.; Gagnon, J.F.; Tremblay, S.; Falk, T.H. Performance Comparison of Automated EEG Enhancement Algorithms for Mental Workload Assessment of Ambulant Users. In Proceedings of the 2019 9th International IEEE/EMBS Conference on Neural Engineering (NER), San Francisco, CA, USA, 20–23 March 2019; pp. 61–64. [CrossRef]
29. di Flumeri, G.; Aricò, P.; Borghini, G.; Sciaraffa, N.; Ronca, V.; Vozzi, A.; Storti, S.F.; Menegaz, G.; Fiorini, P.; Babiloni, F. EEG-Based Workload Index as a Taxonomic Tool to Evaluate the Similarity of Different Robot-Assisted Surgery Systems. *Commun. Comput. Inf. Sci.* **2019**, *1107*, 105–117. [CrossRef]
30. Bilalpur, M.; Kankanhalli, M.; Winkler, S.; Subramanian, R. EEG-Based Evaluation of Cognitive Workload Induced by Acoustic Parameters for Data Sonification. In Proceedings of the ICMI '18: 20th ACM International Conference on Multimodal Interaction, Boulder, CO, USA, 16–20 October 2018; Association for Computing Machinery (ACM): New York, NY, USA; pp. 315–323. [CrossRef]
31. Makoto, M. Makoto's Preprocessing Pipeline. *SCCN*. Available online: https://sccn.ucsd.edu/wiki/Makoto%27s_preprocessing_pipeline (accessed on 26 October 2022).
32. Höller, Y.; Uhl, A.; Bathke, A.; Thomschewski, A.; Butz, K.; Nardone, R.; Fell, J.; Trinka, E. Reliability of EEG measures of interaction: A paradigm shift is needed to fight the reproducibility crisis. *Front. Hum. Neurosci.* **2017**, *11*, 441. [CrossRef] [PubMed]
33. Botvinik-Nezer, R.; Wager, T.D. Reproducibility in neuroimaging analysis: Challenges and solutions. *Biol. Psychiatry Cogn. Neurosci. Neuroimaging* **2023**, *in press*. [CrossRef]
34. Pernet, C.; Garrido, M.I.; Gramfort, A.; Maurits, N.; Michel, C.M.; Pang, E.; Salmelin, R.; Schoffelen, J.M.; Valdes-Sosa, P.A.; Puce, A. Issues and recommendations from the OHBM COBIDAS MEEG committee for reproducible EEG and MEG research. *Nat. Neurosci.* **2020**, *23*, 1473–1483. [CrossRef]
35. Pernet, C.; Garrido, M.; Gramfort, A.; Maurits, N.; Michel, C.M.; Pang, E.; Salmelin, R.; Schoffelen, J.M.; Valdes-Sosa, P.A.; Puce, A. Best Practices in Data Analysis and Sharing in Neuroimaging Using MEEG. *Preprint at OSF*. Available online: https://osf.io/a8dhx2018 (accessed on 20 January 2023).
36. Zhou, Y.; Huang, S.; Xu, Z.; Wang, P.; Wu, X.; Zhang, D. Cognitive workload recognition using EEG signals and machine learning: A review. *IEEE Trans. Cogn. Dev. Syst.* **2022**, *14*, 799–818. [CrossRef]
37. Ding, L.; Duan, W.; Wang, Y.; Lei, X. Test-retest reproducibility comparison in resting and the mental task states: A sensor and source-level EEG spectral analysis. *Int. J. Psychophysiol.* **2022**, *173*, 20–28. [CrossRef]
38. Corsi-Cabrera, M.; Galindo-Vilchis, L.; Del-Río-Portilla, Y.; Arce, C.; Ramos-Loyo, J. Within-subject reliability and inter-session stability of EEG power and coherent activity in women evaluated monthly over nine months. *Clin. Neurophysiol.* **2007**, *118*, 9–21. [CrossRef]
39. Shirk, S.D.; McLaren, D.G.; Bloomfield, J.S.; Powers, A.; Duffy, A.; Mitchell, M.B.; Ezzati, A.; Ally, B.A.; Atri, A. Inter-rater reliability of preprocessing EEG data: Impact of subjective artifact removal on associative memory task ERP results. *Front. Neurosci.* **2017**, *11*, 322. [CrossRef] [PubMed]
40. Suarez-Revelo, J.; Ochoa-Gomez, J.; Duque-Grajales, J. Improving test-retest reliability of quantitative electroencephalography using different preprocessing approaches. In Proceedings of the 2016 38th Annual International Conference of the IEEE Engineering in Medicine and Biology Society (EMBC), Orlando, FL, USA, 16–20 August 2016; pp. 961–964. [CrossRef]
41. Suárez-Revelo, J.X.; Ochoa-Gómez, J.F.; Tobón-Quintero, C.A. Validation of EEG pre-processing pipeline by test-retest reliability. In *Workshop on Engineering Applications*; Springer: Cham, Switzerland, 2018; pp. 290–299.

42. Getzmann, S.; Reiser, J.E.; Karthaus, M.; Rudinger, G.; Wascher, E. Measuring correlates of mental workload during simulated driving using ceegrid electrodes: A test–retest reliability analysis. *Front. Neuroergonomics* **2021**, *2*, 729197. [CrossRef]
43. Arico, P.; Borghini, G.; Di Flumeri, G.; Colosimo, A.; Graziani, I.; Imbert, J.P.; Granger, G.; Benhacene, R.; Terenzi, M.; Pozzi, S.; et al. Reliability over time of EEG-based mental workload evaluation during Air Traffic Management (ATM) tasks. In Proceedings of the 2015 37th Annual International Conference of the IEEE Engineering in Medicine and Biology Society (EMBC), Milan, Italy, 25–29 August 2015; pp. 7242–7245. [CrossRef]
44. Dehais, F.; Duprès, A.; Blum, S.; Drougard, N.; Scannella, S.; Roy, R.N.; Lotte, F. Monitoring pilot's mental workload using ERPs and spectral power with a six-dry-electrode EEG system in real flight conditions. *Sensors* **2019**, *19*, 1324. [CrossRef] [PubMed]
45. Longo, L. Modeling cognitive load as a self-supervised brain rate with electroencephalography and deep learning. *Brain Sci.* **2022**, *12*, 1416. [CrossRef] [PubMed]
46. Kutafina, E.; Heiligers, A.; Popovic, R.; Brenner, A.; Hankammer, B.; Jonas, S.M.; Mathiak, K.; Zweerings, J. Tracking of Mental Workload with a Mobile EEG Sensor. *Sensors* **2021**, *21*, 5205. [CrossRef] [PubMed]
47. Xiong, R.; Kong, F.; Yang, X.; Liu, G.; Wen, W. Pattern recognition of cognitive load using eeg and ecg signals. *Sensors* **2020**, *20*, 5122. [CrossRef] [PubMed]
48. Zhu, X.; Korivand, S.; Hamill, K.; Jalili, N.; Gong, J. A comprehensive decoding of cognitive load. *Smart Health* **2022**, *26*, 100336. [CrossRef]
49. Sarailoo, R.; Latifzadeh, K.; Amiri, S.H.; Bossaghzadeh, A.; Ebrahimpour, R. Assessment of Instantaneous Cognitive Load Imposed by Educational Multimedia using EEG Signals. *Front. Neurosci.* **2022**, *16*, 1275. [CrossRef]
50. Kingphai, K.; Moshfeghi, Y. On EEG Preprocessing Role in Deep Learning Effectiveness for Mental Workload Classification. *Commun. Comput. Inf. Sci.* **2021**, *1493*, 81–98. [CrossRef]
51. Zhang, J.; Li, S. A Deep Learning Scheme for Mental Workload Classification Based on Restricted Boltzmann Machines. *Cogn. Technol. Work.* **2017**, *19*, 607–631. [CrossRef]
52. Simon, J.; Wolf, J.D. Choice reaction time as a function of angular stimulus-response correspondence and age. *Ergonomics* **1963**, *6*, 99–105. [CrossRef]
53. Hommel, B. Inverting the Simon Effect by Intention. *Psychol. Res.* **1993**, *55*, 270–279. [CrossRef]
54. Stoet, G. PsyToolkit: A Novel Web-Based Method for Running Online Questionnaires and Reaction-Time Experiments. *Teach. Psychol.* **2017**, *44*, 24–31. [CrossRef]
55. Stoet, G. PsyToolkit: A Software Package for Programming Psychological Experiments Using Linux. *Behav. Res. Methods* **2010**, *42*, 1096–1104. [CrossRef] [PubMed]
56. Gabard-Durnam, L.J.; Leal, A.S.M.; Wilkinson, C.L.; Levin, A.R. The Harvard Automated Processing Pipeline for Electroencephalography (HAPPE): Standardized Processing Software for Developmental and High-Artifact Data. *Front. Neurosci.* **2018**, *12*, 97. [CrossRef]
57. Mullen, T.; Kothe, C.; Chi, Y.M.; Ojeda, A.; Kerth, T.; Makeig, S.; Cauwenberghs, G.; Jung, T.P. Real-Time Modeling and 3D Visualization of Source Dynamics and Connectivity Using Wearable EEG. In Proceedings of the Annual International Conference of the IEEE Engineering in Medicine and Biology Society (EMBC), Osaka, Japan, 3–7 July 2013; pp. 2184–2187. [CrossRef]
58. Chang, C.Y.; Hsu, S.H.; Pion-Tonachini, L.; Jung, T.P. Evaluation of Artifact Subspace Reconstruction for Automatic EEG Artifact Removal. In Proceedings of the 40th Annual International Conference of the IEEE Engineering in Medicine and Biology Society (EMBC), Honolulu, HI, USA, 18–21 July 2018; pp. 1242–1245. [CrossRef]
59. Bell, A.J.; Sejnowski, T.J. An Information-Maximization Approach to Blind Separation and Blind Deconvolution. *Neural Comput.* **1995**, *7*, 1129–1159. [CrossRef]
60. Pion-Tonachini, L.; Kreutz-Delgado, K.; Makeig, S. ICLabel: An Automated Electroencephalographic Independent Component Classifier, Dataset, and Website. *Neuroimage* **2019**, *198*, 181–197. [CrossRef]
61. Delorme, A.; Makeig, S. EEGLAB: An Open Source Toolbox for Analysis of Single-Trial EEG Dynamics Including Independent Component Analysis. *J. Neurosci. Methods* **2004**, *134*, 9–21. [CrossRef]
62. Koo, T.K.; Li, M.Y. A Guideline of Selecting and Reporting Intraclass Correlation Coefficients for Reliability Research. *J. Chiropr. Med.* **2016**, *15*, 155–163. [CrossRef]
63. R Core Team. *R: A Language and Environment for Statistical Computing*; R Foundation for Statistical Computing: Vienna, Austria; Available online: https://www.R-project.org/ (accessed on 26 October 2022).
64. Borgmann, K.W.U.; Risko, E.F.; Stolz, J.A.; Besner, D. Simon Says: Reliability and the Role of Working Memory and Attentional Control in the Simon Task. *Psychon. Bull. Rev.* **2007**, *14*, 313–319. [CrossRef]
65. McEvoy, L.K.; Smith, M.E.; Gevins, A. Test–Retest Reliability of Cognitive EEG. *Clin. Neurophysiol.* **2000**, *111*, 457–463. [CrossRef] [PubMed]
66. Radüntz, T.; Fürstenau, N.; Mühlhausen, T.; Meffert, B. Indexing Mental Workload During Simulated Air Traffic Control Tasks by Means of Dual Frequency Head Maps. *Front. Physiol.* **2020**, *11*, 300. [CrossRef] [PubMed]

sensors

MDPI

Article

Predicting Emotion with Biosignals: A Comparison of Classification and Regression Models for Estimating Valence and Arousal Level Using Wearable Sensors

Pekka Siirtola *, Satu Tamminen, Gunjan Chandra, Anusha Ihalapathirana and Juha Röning

Biomimetics and Intelligent Systems Group, University of Oulu, P.O. Box 4500, FI-90014 Oulu, Finland
* Correspondence: pekka.siirtola@oulu.fi

Abstract: This study aims to predict emotions using biosignals collected via wrist-worn sensor and evaluate the performance of different prediction models. Two dimensions of emotions were considered: valence and arousal. The data collected by the sensor were used in conjunction with target values obtained from questionnaires. A variety of classification and regression models were compared, including Long Short-Term Memory (LSTM) models. Additionally, the effects of different normalization methods and the impact of using different sensors were studied, and the way in which the results differed between the study subjects was analyzed. The results revealed that regression models generally performed better than classification models, with LSTM regression models achieving the best results. The normalization method called baseline reduction was found to be the most effective, and when used with an LSTM-based regression model it achieved high accuracy in detecting valence (mean square error = 0.43 and R^2-score = 0.71) and arousal (mean square error = 0.59 and R^2-score = 0.81). Moreover, it was found that even if all biosignals were not used in the training phase, reliable models could be obtained; in fact, for certain study subjects the best results were obtained using only a few of the sensors.

Keywords: emotion detection; valence; arousal; wearable sensors; regression; classification; machine learning

Citation: Siirtola, P.; Tamminen, S.; Chandra, G.; Ihalapathirana, A.; Röning, J. Predicting Emotion with Biosignals: A Comparison of Classification and Regression Models for Estimating Valence and Arousal Level Using Wearable Sensors. *Sensors* **2023**, *23*, 1598. https://doi.org/10.3390/s23031598

Academic Editor: Yuk-Ming Tang

Received: 13 December 2022
Revised: 30 January 2023
Accepted: 30 January 2023
Published: 1 February 2023

1. Introduction

Wearable wrist-worn sensors are commonly used to monitor human motion based on inertial sensors such as accelerometers, gyroscopes, and and magnetometers. In addition, wearables can include sensors for measuring biosignals. Nowadays, wrist-worn wearable devices can house a wide range of biosensors, including photoplethysmography (BVP) to measure the blood volume pulse, heart rate (HR), and heart rate variability (HRV), thermometers (ST) to measure skin temperature, and electrodermal activity (EDA) sensors to measure galvanic skin responses. Based on these, it is possible to monitor human motion along with monitor other aspects of human behavior and events occurring inside the human body.

Articles have shown good results in detecting stress and affect states based on the data provided by wearable sensors. For instance, in [1], eight affect states (excited, happy, calm, tired, bored, sad, stressed, and angry) were detected based on acceleration, electrocardiogram, blood volume pulse, and body temperature signals. The results were promising, especially when personal models were used in the recognition process. Similarly, in [2], heart rate, blood volume pulse, and skin conductance were used to detect seven affective states (fun, challenge, boredom, frustration, excitement, anxiety, and relaxation); using artificial neural networks, most of these could be detected with accuracy over 80%. In [3], a classifier able to detect high and low stress as well as non-stressful situations in laboratory conditions was developed based on wearable wrist-work sensors. The results showed that the two stress classes could be detected with an accuracy of 83%. In [4], a binary classifier was trained to detect stress and non-stressed state, and it was noted that stress could be detected using the sensors in commercial smartwatches. There have been several other studies showing that stress detection based on classification models can be performed with high accuracy using user-independent models; see for instance [5,6].

In the past, automatic emotion recognition from biosensor data has focused on detecting discrete classes of emotion. However, humans have hundreds of emotions, and id discrete classes can be recognized it is possible to recognize a limited number of them. In addition, it is important both to recognize the affected state and to recognize the level of the affected state, such as a person being slightly happy, extremely happy, or anything between these two. In fact, psychological studies have suggested that the full spectrum of human emotion can be characterized by just a few dimensions. One common strategy to express human emotion in discrete classes is to divide emotions into valence and arousal ([7,8]). Valence is the horizontal extent, ranging from displeasure to pleasure, and arousal is the vertical extent, ranging from deactivation to activation. By combining valence and arousal, every human emotion can be expressed; often, these are visualized using Russell's circumplex model of emotions (see Figure 1).

In this article, valence and arousal levels are predicted based on biosignal data collected using wearable wrist-worn sensors. The novel contributions of our paper are as follows:

1. It is shown that fine-grained levels of valence and arousal can be detected based on the data of wrist-worn sensors with high reliability when the data and the target values provided by the study subjects are normalized in the right way.
2. We compare the suitability of different classification and regression models for detecting the levels of valence and arousal.
3. We explore the finding that the same features and sensors may not be optimal for every individual.

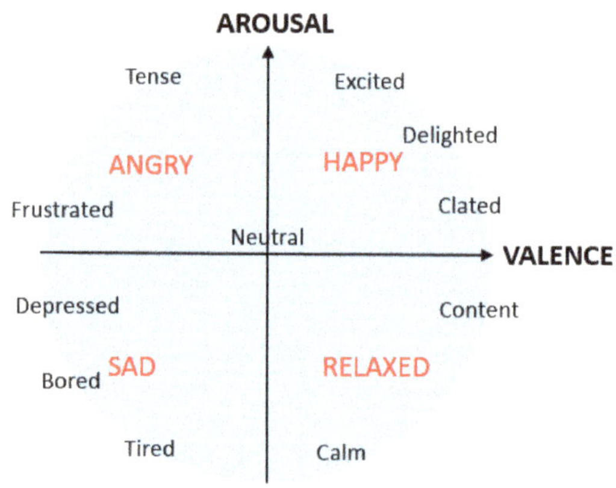

Figure 1. Russell's Circumplex Model of Emotions.

The rest of this article is organized as follows. Related works are introduced in Section 2, and the data used in the experiments are explained in Section 3. Section 4 introduces the methods used in the study, and Section 5 explains the experimental setup, the applied methods, and the obtained results. Finally, a discussion is presented in Section 6 followed by our conclusions and prospects for future work in Section 7.

2. Related Work

In [9], valence and arousal detection using regression models were studied based on audio data. There have been several audio- and video data-based studies, for instance, [9,10], on detecting valance and arousal based on continuous response values and regression models. However, when it comes to wearable sensors, emotion recognition has in the past focused on identifying discrete classes of emotion. There have been a number of studies in which valence and arousal levels were detected; however, in these, the detection was based on dividing valance and arousal values into discrete classes, meaning that the prediction models were based on classification methods. Often, valance and arousal values are only divided into two discrete classes, high or low arousal, for instance in [11–14]; however, there have been studies in which more fine-grained classes were studied as well.

In [15], valence and arousal were divided into three classes: low, neutral, and high arousal/valance. However, in the final classification only the low and high arousal/valence observations were used. A dataset was collected from 21 study subjects playing games with increasing difficulty and self-reporting their valence and arousal levels while playing. The participants were wearing OpenBCI headsets and JINS MEME eyewear, and these were used to collect electroencephalography (EEG), electrooculography (EOG), and kinematic motion data (acceleration from the head and glasses, and gyroscope data from the glasses). Classification methods such as ensemble learning and random forest were used as classifiers. In the study, ten-fold cross-validation was used instead of leave-one-subject-out cross-validation. The best accuracies were obtained using an ensemble learner, and in the binary case, these were 73% for arousal and 80% for valence. In [16], two datasets were studied: the publicly open CASE dataset, containing electrocardiogram (ECG), BVP, EDA, respiratory rate, ST, and EMG (electromyography) signals, and another dataset called MERCA containing data from the autonomic nervous system (HR, HRV, ST, and EDA) and oculomotor nerve system (pupil dilation, saccadic amplitude, and saccadic velocity) collected using an Empatica E4 wrist-worn sensor and wearable eye tracker. The aim of the study was to detect the levels of valence and arousal. For this purpose, three scenarios were studied: a binary case in which valence and arousal were divided into two classes (high and low), a three-class case (low, neutral, and high arousal/valance), and a four-class case (high valence + high arousal, high valence + low arousal, low valence + high arousal, and low valence + low arousal). Different machine learning and deep learning models were used in the experiments; when leave-one-subject-out cross-validation was used, the recognition rate for the binary case was around 70%, while it was lower for the cases with three and four classes.

However, as the level of valence and arousal can be high, low, or anything between these, dividing the level of valence and arousal into discrete classes is not the best option, and there is evidence that wearable sensors can be used to predict continuous affect state values as well. In [17], a significant correlation was found between valence levels and cortisol levels, which have been accepted as a reliable physical measure of emotions. In addition, it was found that EEG (electroencephalography) signals collected using a wearable device correlated with valence levels, showing that the signals of wearable devices correlates with cortisol levels. The study did not rely on machine learning methods; however, what was noticeable was that valence levels were not divided into discrete classes, meaning that the study shows that wearables, in this case EEG sensors, can be used to detect continuous valence level values.

When it comes to detecting continuous target values such as the level of valence and arousal using machine learning and artificial intelligence methods, prediction needs to be built by relying on regression models instead of classification models. However, it seems that there are not many studies where valance and arousal levels are estimated using regression models. In [18], arousal level was estimated based on ECG, respiratory rate, EDA, and ST signals using only simple a linear regression model, which is not an up-to-date regression model. Moreover, the authors did not study valence at all. On the other hand, regression has been applied to other related problems. For instance, in [19] the authors used statistical methods such as regression to predict anxiety based on wearable data (in this case BVP, ST, EDA, and microphone data), and in other studies it has been shown that continuous stress levels can be estimated based on regression and statistical methods (for instance [20–23]).

These related studies are presented in Table 1. What can be noted from the table is what is not studied, how well the data of wrist-worn wearable sensors can be used to estimate the level of both valence and arousal, and how well modern regression models can predict valence and arousal levels compared to classification models when the valence and arousal levels are divided into fine-grained discrete classes.

Table 1. Previous research utilizing wearable sensor data to identify valence, arousal, and other related affect states.

Article	Signals	What Is Recognized	Continuous/Discrete Targets	Methods
Hsu et al. (2017) [11]	ECG	valence and arousal	discrete (high/low)	least squares SVM
Romeo et al. (2019) [12]	2 settings: 1. EEG + 13 other biosignal, 2. GSR, IBI and ST	valence and arousal	discrete (high/low)	Multiple Instance Learning, SVM, Random Forest, Naive Bayes
Li et al. (2020) [13]	EEG, EOG, EMG, EDA, BVP, ST, and respiration	valence and arousal	discrete (high/low)	DNN
Zhao et al. [14]	BVP, EDA, ST	valence and arousal	discrete (high/low)	SVM
Diaz-Romero et al. (2021) [15]	EEG, EOG, motion	valence and arousal	discrete (2 and 3 class cases)	SVM, logistic regression, Random Forest, and ensemble classifier
Zhang et al. (2020) [16]	2 settings: 1. ECG, BVP, EDA, respiratory rate, ST, and EMG, 2. HR, HRV, ST, and EDA + data from wearable eyetracker	valence and arousal	discrete (2, 3 and 4 class cases)	correlation-based emotion recognition algorithm (CorrNet)
Hwang et al. (2018) [17]	EEG	valence	continuous	statistical methods, correlation
Nirjhar et al. (2020) [19]	BVP, ST, EDA, audio	anxiety	continuous	statistical methods such as least square regression
Grundlehner et al. (2009) [18]	ECG, respiration rate, EDA and ST	arousal	continuous	linear regression
Siirtola & Röning (2020) [20]	BVP, EDA, ST	stress	continuous	Bagged tree based ensemble for regression
Healey & Picard (2005) [21]	ECG, EMG, EDA, and respiration rate	stress	continuous	statistical methods such as correlation
El Haouij (2018) [22]	EDA from both wrists, HR, respiratory rate, posture	stress	continuous	Random Forest for regression
Vos et al. (2022) [23]	BVP, HR, EDA, ST	stress	continuous	(XGBoost) XGB, (artificial neural networks) ANN, XGB+ANN

3. Experimental Dataset

The experiments in this study were carried out based on the open WESAD dataset [24], which was gathered using an Empatica E4 [25] wrist device and chest-worn RespiBAN device. In this study, only the Empatica E4 data was used. This device includes sensors to measure acceleration (ACC), skin temperature (ST), electrodermal activity (EDA), blood volume pulse (BVP), heart rate (HR), and heart rate variability (HRV).

WESAD contains data from 15 participants; from each study subject, it contains baseline data, data from a stressful situation, data from a state of amusement, and data from two meditation states. The purpose of the meditation sessions was to relax the study subject after each task, and the meditation data were not used in this study. The baselines were collected at the beginning of the data-gathering session while the participants were sitting/standing at a table and reading neutral magazines. During the gathering of stress data, participants had two tasks: (1) they had to provide a public presentation, and (2) they had to solve arithmetic tasks. Data from the state of amusement was collected while the participants were watching funny videos. The length of the stressful situation was approximately 10 min, the amused situation was 6.5 min, and the relaxed situation (baseline) was 20 min. After each task, the subjects were asked to fill in a self-report consisting of three types of questionnaires: PANAS [26], shortened STAI [27], and SAM [28]. PANAS asks whether a person had certain positive and negative moods during tasks, STAI concentrates on questioning how strong a person's feelings of anxiety are, and SAM is used to ask about a person's level of valance and arousal on a scale of 1–9. Therefore, when the models to predict valence and arousal levels were trained, the labels used in the training process were based on these subjective estimations. In this study, answers to the SAM questionnaire were scaled to $[-4, 4]$ and used to define the correct target variables. This means that when classification methods are used, instances are classified into nine classes.

In the pre-processing stage, as suggested in [29], not all of the baseline data were used in the experiments. In fact, the first half of the baseline data were removed from the dataset, as it is possible that immediately after starting to gather data for the baseline the study subject's body may not be in a relaxed state. Moreover, pre-processing of the EDA signal was carried out following the guidelines in [24]. A 5 Hz low-pass filter was applied to the raw EDA signal, then it was divided into phasic and tonic parts [30] using `cvxEDA` (https://github.com/lciti/cvxEDA). BVP and ST signals were used as they were.

For the model training, signals were divided into windows and features were extracted from these. A window size of 60 s was used in the experiment, which is the same as used in [4,24] for stress detection. The slide between two adjacent windows was 30 s. Different statistical features (min, max, mean, std, percentiles) per signal (BVP, ST, EDA, phasic and tonic parts of EDA), and physiological features were extracted from the HRV and PPG signal using `HeartPy` (https://python-heart-rate-analysis-toolkit.readthedocs.io/en/latest/). In addition, the slope was calculated from the ST. Features were not extracted from the accelerometer signal, as the accelerometer measures movement. Due to this, different activities performed during different tasks may be visible in acceleration data, leading to situations where accidental activities are detected instead of emotions. Therefore, only features extracted from biosignals were used to train the models. The full list of extracted features is shown in Table 2. These features and libraries are commonly used and recommended for extracting features and pre-processing biosignal data [31]. After the features were extracted, rows with NaN- and Inf-values were removed from the dataset.

The feature matrix was scaled to level 0–1, and is visualized in Figure 2 using t-SNE, a dimensionality reduction technique that can be used to project high-dimensional data into a low-dimensional space [32]. The figure shows that t-SNE effectively clusters data from different valance and arousal estimates reported by study subjects into own clusters, as shown in Figure 2a,b. However, the high number of clusters and wide spread of samples within the same class but different clusters makes analyzing this dataset challenging. This indicates a high level of variation within classes, and could be due to differences among study subjects. In addition, the high number of clusters indicates that the dataset has a complex underlying structure. Therefore, it may require advanced techniques in order to achieve good results. Moreover, the figures show that the dataset is highly imbalanced. Most of the samples are quite neutral, as their label is close to zero, and the dataset does not contain many extreme values. This makes the data analysis process even more challenging.

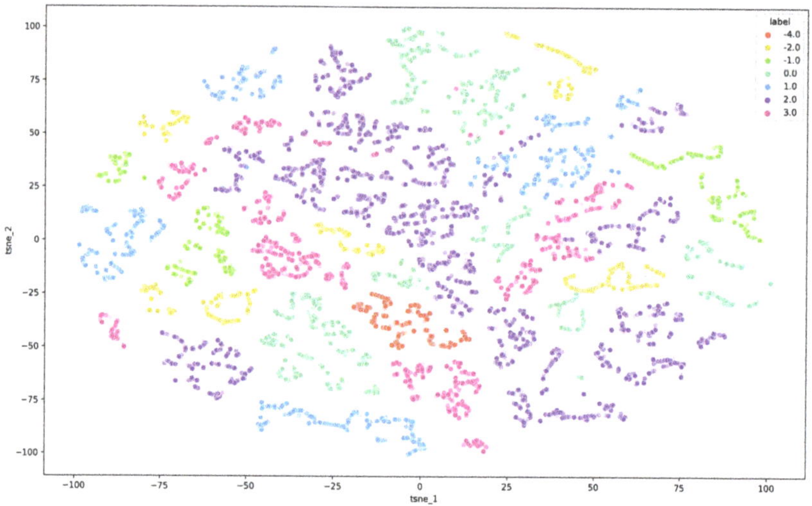

(**a**)

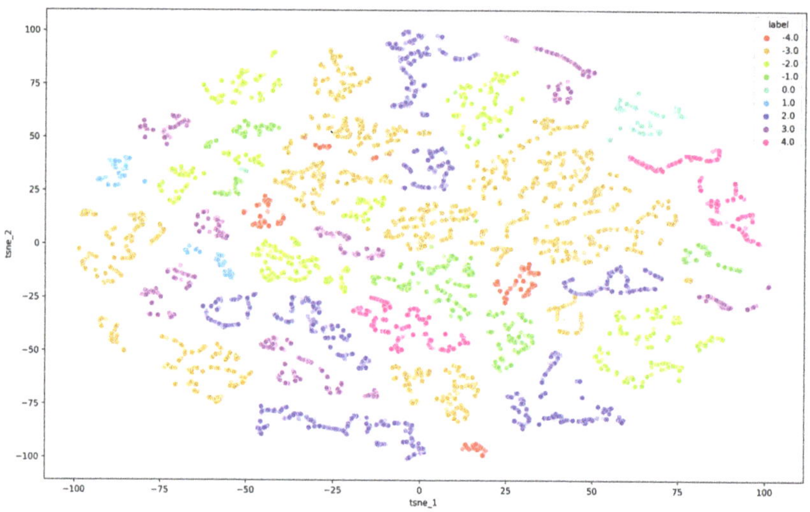

(**b**)

Figure 2. WESAD dataset illustrated using t-SNE: (**a**) data visualized using t-SNE and valence levels as targets; (**b**) data visualized using t-SNE and arousal levels as targets.

Table 2. List of extracted features.

Signal	Features Extracted from the Pre-Processed Biosignals
Electrodermal activity (EDA)	mean, std, min, max, median, range, percentile_95, percentile_5, percentile_75, percentile_25, phasic_mean, phasic_std, phasic_min, phasic_max, phasic_median, phasic_range, phasic_percentile_95, phasic_percentile_5, phasic_percentile_75, phasic_percentile_25, tonic_mean, tonic_std, tonic_min, tonic_max, tonic_median, tonic_range, tonic_percentile_95, tonic_percentile_5, tonic_percentile_75, tonic_percentile_25,
Skin temperature (ST)	mean, std, min, max, percentile_95, percentile_5, percentile_75, percentile_25, slope
Blood volume pulse (BVP)	mean, std, min, max, percentile_95, percentile_5, percentile_75, percentile_25 HR_mean, NNi_mean, NNi_std, NNi_length, NNi50, pNN50, TINN, rms, LF/HF_ratio, VLF_relative_power, LF_relative_power, HF_relative_power, VHF_relative_power, total_power, UHF_energy, LF_energy, HF_energy, breathing_rate

4. Methods

This section introduces the normalization methods compared in this study as well as the classification and regression models and the performance metrics used to compare them.

4.1. Normalization

People are different, and due to this, biosignals collected from the study subjects differ from individual to individual. In addition, biosignals are affected by daily changes, for instance those caused by sleep quality and chronic stress. Therefore, the difference in biosignals between the subjects can be considerable, and this may pose challenges for prediction models. Due to this, the prediction power of recognition models trained using raw data can vary a great deal between individuals.

Data normalization is often found to be an effective method to remove participant-specific effects on the data, such as daily changes and different natural ranges, and is a good way to make trained models more generalized to any study subject [33]. Due to this, normalization is a powerful method to adapt models to the current status of the study subject's body as well as to the calibration status of the sensor itself. Moreover, by regularly calculating the required parameters for normalization, for instance, every morning, normalization can be used as a tool to adapt models to the changes happening inside the human body, which affects the sensor readings as well. In this study, four datasets were created in order to experiment with the effects of different normalization methods. Figure 3 explains how these were created.

Person-specific *z*-score normalization has been found to be the most effective way to normalize biosignals [33,34] when classifying affect and stress stages. Due to this, it was used in this study. The *z*-score normalized value z_i for observation x_i can be calculated using the equation

$$z_i = \frac{x_i - \mu}{\sigma}, \tag{1}$$

where the μ is mean and the σ is standard deviation calculated from the whole signal $X, x_i \in X$ collected from an individual. This normalization is performed separately for each collected biosignal (EDA, BVP, HRV, and ST).

Baseline reduction is commonly used as a normalization method when dealing with data on emotions. Individual valence and arousal value estimations for the baseline data were reduced from all the target values for valence and arousal; thus, the target value for each individual for valence and arousal at the baseline was zero, as the baseline is considered a neutral stage. Because of the normalization, subject-wise differences from the target values could be removed.

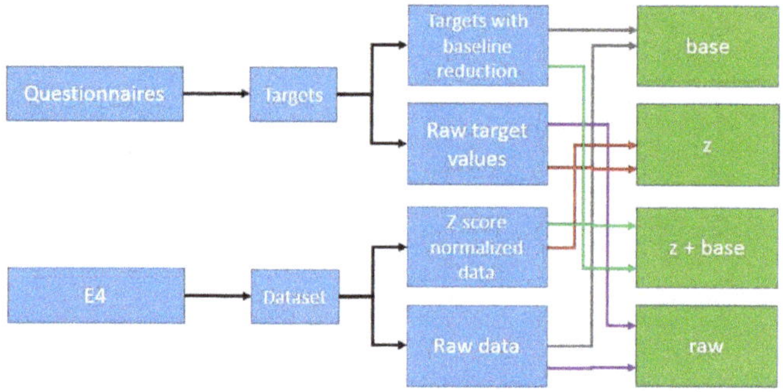

Figure 3. Four datasets (base, z, z+base, and raw) were created to experiment with different normalization methods.

The third normalization method tested in this study was a combination of the first two approaches; z-score normalization was used to remove individual differences from the signals, then baseline reduction was used to remove individual differences from the labels.

To determine the benefits of normalization, these three normalization methods were compared to situations where signals and labels were not normalized and the models were instead trained based on raw data which was not normalized at all.

It should be noted that baseline reduction has an effect on the number of target variables. When raw questionnaire data were used in the modeling process, the number of classes for valence level prediction was seven (none of the study subjects reported valence levels −3 and 4), and for arousal level prediction it was nine. However, after baseline reduction the number of classes for valence level prediction was eight and for arousal level prediction it was six.

4.2. Prediction Methods

As one of the purposes of our study was to compare classification and regression methods, several different classification and regression methods were assessed. Long Short-Term Memory (LSTM) is a variant of recurrent neural networks; it is highly suitable for time-series prediction, as it is capable of learning long-term dependencies from the data [35]. In this study, one hidden layer was used, as it has been shown in [36] that an LSTM with one layer provides better results than one with two layers when studying wearable sensor data. The LSTM layer of the model used in this study had 64 units, and the model had around 17,000 parameters to train. AdaBoost [37], Random Forest [38], XGBoost (eXtreme Gradient Boosting) [39], and Histogram-GBM (inspired by LightGBM [40]) are ensemble methods that train a group of weak learners, usually decision trees, and make a final prediction that is a combination of these. In comparison to decision trees, linear regression, and LDA (linear discriminant analysis), the latter are simpler methods. In this article, LSTM, AdaBoost, Random Forest, and XGBoost were used for both classification and regression, LDA and decision tree were used for classification only, and linear regression and Histogram-GBM were used for regression only.

4.3. Performance Metrics

All regression models were evaluated using two different types of evaluation parameters commonly used with regression models, namely, R^2 and the mean squared error (MSE) [41]. In addition, certain models were evaluated in greater detail using the root mean squared error (RMSE) and mean average error (MAE) [41]. Normally, the results of classification models are evaluated using a confusion matrix and performance metrics calculated from it; because in this case the classification classes were fine-grained and ordinal, the performance of the classification methods was evaluated using R^2 and MSE as well. When analyzing model performance using these metrics, it is important to note that a value of zero is optimal for MSE, RMSE, and MAE, while a value of one is optimal for R^2.

Traditionally, the performance of classification methods is analyzed using performance metrics such as accuracy, sensitivity, specificity, etc. However, as in this article the idea is to compare classification and regression methods and regression models cannot be analyzed using these metrics, both classification and regression models were analyzed using MSE, RMSE, R^2, and MAE. In fact, as valence and arousal are continuous phenomena and the targets for them are ordinal, it is natural that all the models be analyzed using these metrics. Moreover, evaluating both classification and regression models using the same performance metrics makes their comparison easier.

5. Experimental Setup and Results

The results of the experiments are presented in this section. All the results were calculated using the leave-one-subject-out method, meaning that one study subject's data is used for testing and all the other data is used for training, with the process then repeated in turn (Figure 4). Due to this, the trained models are user-independent. When the results for all the study subjects were obtained, they were combined as one sequence and MSE and R^2 values were calculated from these combined sequences. As most of the models used in this article contain random elements, the models were trained five times. All of the results presented in this section are averages from these runs, with the standard deviation between the runs shown in parenthesis. The scale of the target variables was $[-4, 4]$; if the estimated value was outside of this scale, it was replaced with -4 or 4.

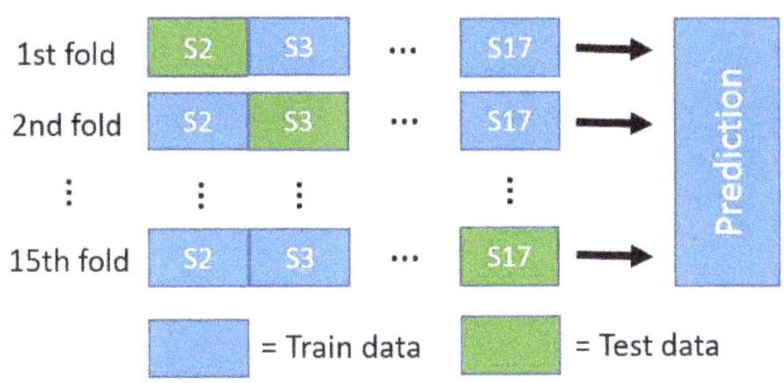

Figure 4. Leave-one-subject-out method used in the experiments. In turns, the data of one study subject are is used for testing while all other data are used for [42].

5.1. Comparison of Prediction Models and Normalization Methods

The results presented in Table 3 show how well different classification and regression models can predict valence and arousal levels based on raw sensor data and how the normalization of signals and target values affects the recognition rates. From these results, it can be noted that it is possible to reliably estimate valence and arousal levels based on data from wrist-worn wearable sensors and up-to-date prediction models. Moreover, this estimation is especially reliable when the prediction is made based on the LSTM model. It should be noted that the LSTM outperforms other classification and regression algorithms. The best results were obtained using the LSTM regression model with a baseline reduction as the normalization method, in this case, for valence level estimation MSE = 0.43 and R^2 = 0.71 and for arousal level estimation MSE = 0.59 and R^2 = 0.81.

Table 3. Comparison of classification (C) and regression (R) models to predict valence and arousal level based on MSE and R^2 scores (standard deviation in parentheses). Rows show results using different prediction models and columns show results using different normalization methods (raw = no normalization, z = z-score normalization, base = baseline reduction, z + base = z-score normalization and baseline reduction).

User-Independent Prediction Rates for Valence Level Estimation.					
Method	**Type**	**Raw: MSE/R^2**	**z: MSE/R^2**	**Base: MSE/R^2**	**z + Base: MSE/R^2**
LSTM	C	1.75 (0.53)/0.42 (0.18)	3.33 (0.16)/0.13 (0.07)	0.57 (0.11)/0.55 (0.10)	1.94 (0.18)/−0.04 (0.07)
LDA	C	4.88 (0.00)/0.39 (0.00)	2.96 (0.04)/−0.20 (0.02)	4.16 (0.0)/−0.20 (0.0)	2.63 (0.00)/−0.68 (0.00)
Decision Tree	C	4.04 (0.41)/−0.26 (0.09)	4.46 (0.12)/−0.32 (0.03)	2.54 (0.13)/−0.45 (0.04)	3.23 (0.10)/−1.07 (0.10)
AdaBoost	C	3.78 (0.04)/−0.54 (0.05)	3.83 (0.06)/−0.52 (0.04)	1.63 (0.02)/−0.37 (0.09)	3.21 (0.09)/−1.30 (0.05)
Random Forest	C	4.48 (0.14)/−0.75 (0.06)	3.41 (0.04)/−0.46 (0.07)	1.85 (0.04)/−0.63 (0.11)	2.59 (0.12)/−1.34 (0.07)
XGBoost	C	3.79 (0.10)/−1.15 (0.06)	3.02 (0.03)/−0.44 (0.02)	1.79 (0.05)/−0.61 (0.04)	2.02 (0.10)/−0.58 (0.01)
LSTM	R	0.74 (0.20)/ 0.73 (0.08)	0.70 (0.09)/**0.75 (0.04)**	**0.43 (0.05)**/0.71 (0.04)	0.57 (0.06)/0.61 (0.05)
AdaBoost	R	4.51 (0.19)/−0.79 (0.14)	3.41 (0.08)/−0.40 (0.00)	1.54 (0.04)/−1.07 (0.08)	2.05 (0.06)/−0.80 (0.05)
Histogram-based GBM	R	2.86 (0.05)/−0.91 (0.08)	2.79 (0.14)/−0.36 (0.04)	1.74 (0.03)/−0.55 (0.02)	2.03 (0.03)/−0.59 (0.01)
Linear model	R	3.28 (0.00)/0.10 (0.00)	2.20 (0.00)/−0.26 (0.00)	4.60 (0.0)/−0.27 (0.0)	1.47 (0.0)/−0.37 (0.0)
Random Forest	R	3.43 (0.06)/−1.14 (0.05)	3.06 (0.04)/−0.38 (0.03)	1.85 (0.04)/−0.62 (0.03)	2.08 (0.02)/−0.59 (0.01)
XGBoost	R	3.57 (0.11)/−1.65 (0.08)	2.86 (0.06)/−0.51 (0.03)	1.70 (0.02)/−0.87 (0.03)	2.01 (0.04)/−0.68 (0.06)
User-Independent Prediction Rates for Arousal Level Estimation.					
Method	**Type**	**Raw: MSE/R^2**	**z: MSE/R^2**	**Base: MSE/R^2**	**z + Base: MSE/R^2**
LSTM	C	2.51 (0.79)/0.64 (0.11)	8.41 (0.44)/−0.17 (0.07)	0.81 (0.10)/0.75 (0.03)	1.75 (0.19)/0.50 (0.05)
LDA	C	9.48 (0.00)/−0.56 (0.00)	4.86 (0.00)/0.24 (0.00)	3.08 (0.0)/0.00 (0.0)	2.37 (0.0)/0.29 (0.0)
Decision Tree	C	6.72 (0.54)/−0.27 (0.04)	7.08 (0.04)/−0.21 (0.01)	3.16 (0.16)/−0.20 (0.05)	2.81 (0.08)/0.02 (0.02)
Random Forest	C	6.22 (0.21)/−0.25 (0.07)	7.38 (0.04)/−0.21 (0.01)	3.08 (0.09)/−0.125 (0.02)	3.02 (0.07)/0.01 (0.02)
AdaBoost	C	6.05 (0.20)/−0.19 (0.03)	6.57 (0.22)/−0.06 (0.02)	2.81 (0.10)/0.11 (0.05)	2.56 (0.26)/0.23 (0.07)
XGBoost	C	6.31 (0.23)/−0.18 (0.06)	6.79 (0.07)/−0.11 (0.01)	3.37 (0.12)/−0.10 (0.025)	2.75 (1.56)/0.36 (0.20)
LSTM	R	1.57 (0.14)/0.73 (0.03)	2.56 (0.36)/0.58 (0.07)	**0.59 (0.06)/0.81 (0.02)**	1.24 (0.24)/0.59 (0.07)
AdaBoost	R	6.01 (0.25)/−0.16 (0.03)	6.32 (0.34)/−0.04 (0.05)	2.63 (0.16)/0.18 (0.05)	2.78 (0.05)/0.16 (0.03)
Histogram-based GBM	R	5.98 (0.16)/−0.18 (0.06)	4.96 (0.15)/−0.14 (0.03)	2.48 (0.09)/−0.06 (0.02)	2.55 (0.07)/−0.08 (0.02)
Linear model	R	7.18 (0.00)/0.03 (0.00)	3.53 (0.00)/0.19 (0.00)	2.37 (0.0)/0.09 (0.0)	1.71 (0.0)/0.20 (0.0)
Random Forest	R	5.90 (0.09)/−0.48 (0.02)	4.79 (0.10)/−0.18 (0.02)	2.18 (0.06)/0.02 (0.03)	2.51 (0.01)/−0.17 (0.01)
XGBoost	R	5.43 (0.25)/−0.26 (0.05)	4.75 (0.06)/−0.16 (0.02)	2.23 (0.07)/−0.09 (0.04)	2.58 (0.02)/0.19 (0.01)

A comparison of the results from the classification and regression models shows that, in general, the regression models performed better than the classification models, and only in very few cases did classification models perform better than regression models. This is not surprising, as valence and arousal are continuous phenomena and are not discrete, meaning that they should be analyzed using regression methods, not classification methods. However, in certain cases classification using LSTM worked very well. For instance, when the valence level is recognized, the LSTM-based classification model with baseline reduction normalization (mean MSE = 0.57 and R^2 = 0.55) performs nearly as well as the LSTM-based regression model with baseline reduction normalization, which has the overall best MSE score (0.43). In addition, when the arousal level is predicted using the LSTM-based classification model with baseline reduction normalization, the performance of the model is nearly as good as when using the LSTM-based regression model with baseline reduction normalization (MSE = 0.81 and R^2 = 0.75 compared to MSE = 0.59 and R^2 = 0.81). Therefore, it is not possible to conclude based on MSE and R^2 that LSTM-based regression models are better than LSTM-based classification models. To study the performance of the LSTM-based models in more detail and compare their classification and regression versions, Table 4 presents a comparison using MSE and R^2 along with RMSE and MAE. According to these results, baseline reduction is the best normalization method, supporting the findings based on the results of Table 3. Moreover, according to Table 4, in the case of valence recognition the difference between LSTM-based classification and regression models with baseline reduction is small when MSE, R^2, RMSE, and MAE values are compared. Nonetheless, when all four performance metrics are compared, the LSTM regression model with baseline reduction is better than the most similar classification model according to three metrics out of four. In the case of arousal recognition, the difference is clear, and again the LSTM based regression model with baseline reduction is the best model according to three metrics out of four.

Table 4. Detailed analysis of LSTM classification (C) and regression (R) models with different normalization methods.

User-Independent Prediction Rates for Valence Level Estimation.					
Method	**Type**	**MSE**	R^2	**RMSE**	**MAE**
LSTM raw	R	0.74 (0.20)	0.73 (0.08)	0.82 (0.12)	0.42 (0.08)
LSTM base	R	**0.43 (0.05)**	0.71 (0.04)	**0.65 (0.04)**	0.34 (0.05)
LSTM z	R	0.70 (0.09)	**0.75 (0.04)**	0.83 (0.06)	0.41 (0.06)
LSTM z+base	R	0.57 (0.06)	0.61 (0.05)	0.76 (0.04)	0.40 (0.04)
LSTM raw	C	1.75 (0.53)	0.42 (0.18)	1.42 (0.16)	0.63 (0.12)
LSTM base	C	0.57 (0.11)	0.55 (0.10)	0.75 (0.08)	**0.30 (0.04)**
LSTM z	C	3.33 (0.16)	0.13 (0.07)	1.90 (0.07)	1.27 (0.40)
LSTM z+base	C	1.94 (0.18)	−0.04 (0.07)	1.34 (0.07)	0.55 (0.05)
User-Independent Prediction Rates for Arousal Level Estimation.					
Method	**Type**	**MSE**	R^2	**RMSE**	**MAE**
LSTM raw	R	1.57 (0.14)	0.73 (0.03)	1.25 (0.06)	0.60 (0.08)
LSTM base	R	**0.59 (0.06)**	**0.81 (0.02)**	**0.77 (0.04)**	0.30 (0.03)
LSTM z	R	2.56 (0.36)	0.58 (0.07)	1.60 (0.11)	0.95 (0.10)
LSTM z+base	R	1.24 (0.24)	0.59 (0.07)	1.11 (0.11)	0.51 (0.06)
LSTM raw	C	2.51 (0.79)	0.64 (0.11)	1.48 (0.11)	0.55 (0.07)
LSTM base	C	0.81 (0.10)	0.75 (0.03)	0.90 (0.05)	**0.28 (0.03)**
LSTM z	C	8.41 (0.44)	−0.17 (0.07)	2.90 (0.08)	1.90 (0.12)
LSTM z+base	C	1.75 (0.19)	0.50 (0.05)	1.32 (0.07)	0.55 (0.06)

According to Table 4, the two best models are LSTM regression and classification models with baseline reduction. To obtain more insight into these models, Figures 5 and 6 illustrate how the predicted valance and arousal estimates follow the user-reported target variables when these models are used in prediction. The figures are drawn based on the results of the best runs; in the case of the regression model, MSE and R^2 for valence estimation were 0.38 and 0.74, respectively, while for arousal estimation they were 0.51 and 0.84, respectively. For the classification model, the MSE and R^2 for valence estimation were 0.42 and 0.69, respectively, while for arousal estimation they were 0.68 and 0.80, respectively. In the figures, predictions using an LSTM-based regression model are shown with a blue line, those using a classification model are shown using a green line, and the true arousal level is shown in orange. Due to subjective differences, the estimation is not as good for all subjects; however, these figures show that in general prediction is highly accurate with both models. In fact, for a number of study subjects the prediction is almost perfect. However, while the difference between LSTM regression and classification models according to the MSE and R^2 is minimal, Figures 5 and 6 reveal differences. It can be noted that the WESAD data does not contain very many samples from cases in which the level of valence is very high or very low, and it contains very few negative arousal cases. In fact, Figure 5 shows that the models have difficulty detecting high valence values; in particular, the classification models seem to suffer due to this lack of training data for high valence values. According to Figure 5, the classification model performs badly for samples in which valence is above zero, while the regression model has fewer such problems. Similarly, Figure 6 shows that the classification model has problems detecting high arousal values; here, the problems are not as severe as in the case of valence recognition, as the training data contain more cases with high values for arousal than for valence. In addition, according to Figure 6, neither model detects negative arousal samples.

Earlier results have already shown that baseline reduction is the most effective normalization method. However, when different normalization methods are compared, it is especially interesting to see the effects different normalization methods have on LSTM models, as these outperform other models. This is visualized in Figure 7. The results of this figure are taken from Table 4 by calculating the average performance of each normalization method when LSTM classification and regression models are used to detect valence and arousal levels. The figure clearly shows that there are large differences between the normalization methods; no matter which performance metric is used, baseline reduction always provides the best results. For MSE, RMSE, and MSE the error is the lowest and for R^2 the value is highest when using baseline reduction. In fact, according to Table 4, for the cases of both valance, and arousal the best results are obtained when using baseline reduction as the normalization method. Both classification and regression models benefit from this, showing that normalization should be used instead of analyzing raw data. The low performance of z-score normalization is surprising; it provides good results only rarely, and in this study, the only good results using z-score normalization were obtained when the valence level was detected using an LSTM-based regression model (MSE = 0.70 and R^2 = 0.75, see Table 4). While z-score normalization does not perform well compared to baseline reduction, it is a much better option than analyzing data without any normalization. In fact, Figure 7 shows that, on average, the worst results were obtained from raw data, with the performance of non-normalized data being especially bad according to the RMSE value.

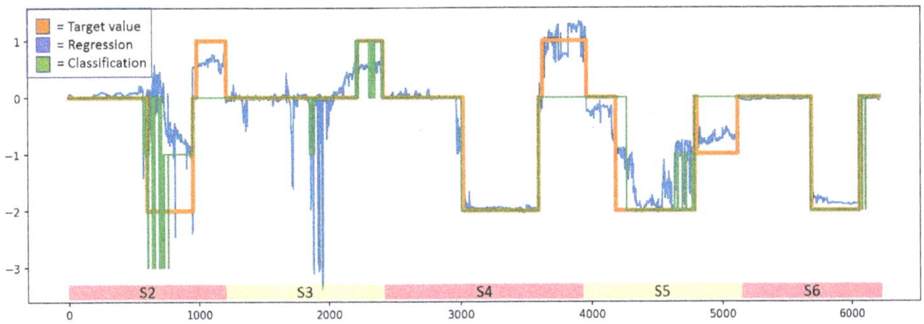

(a)

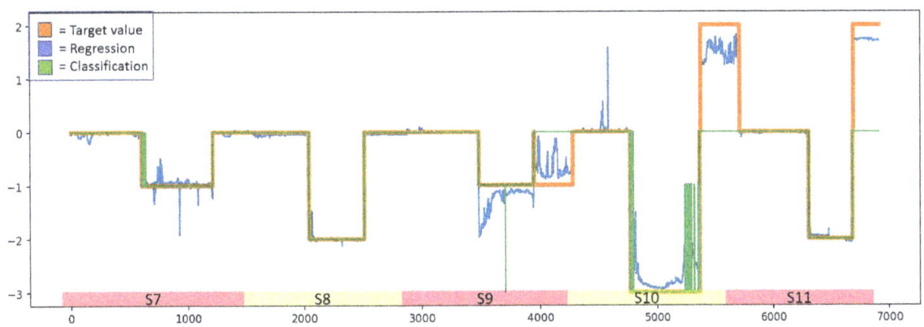

(b)

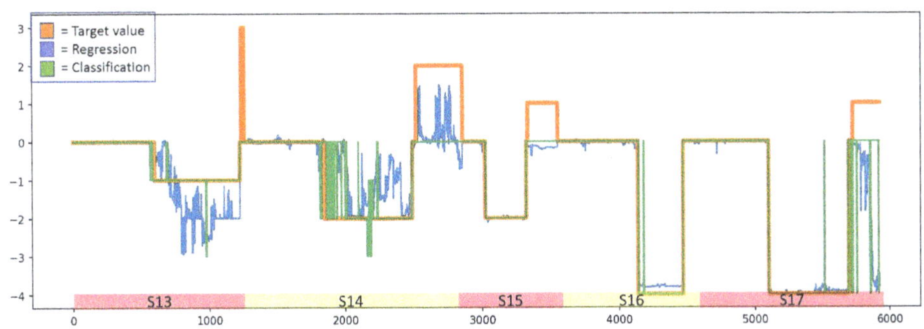

(c)

Figure 5. True and predicted valence levels for the WESAD dataset using LSTM-based regression (blue line) and classification (green line) models with baseline normalization. The true valence level is shown in orange. (**a**) Valence for study subjects 2, 3, 4, 5, and 6. (**b**) Valence for study subjects 7, 8, 9, 10, and 11. (**c**) Valence for study subjects 13, 14, 15, 16, and 17.

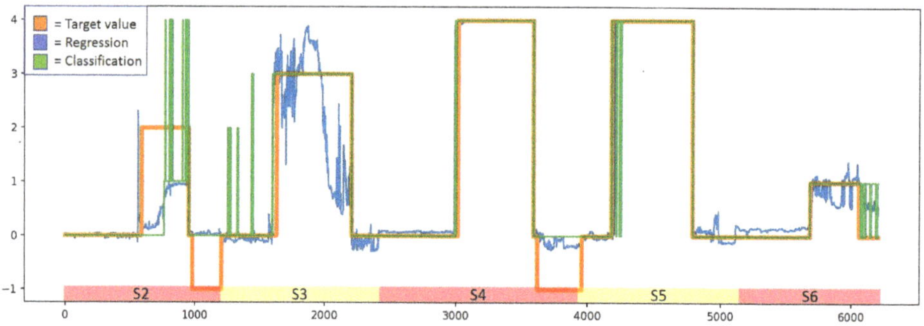

(**a**)

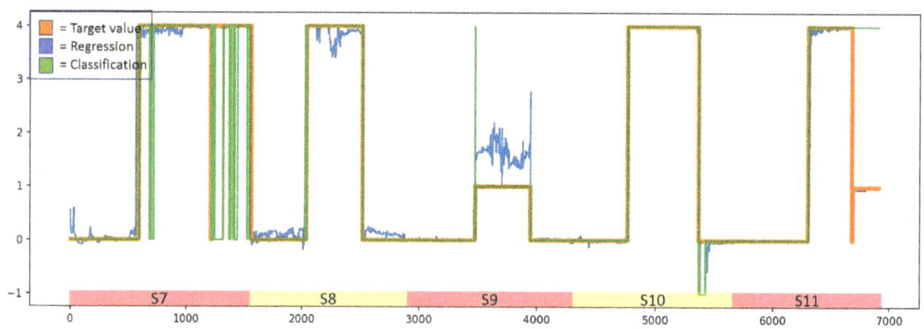

(**b**)

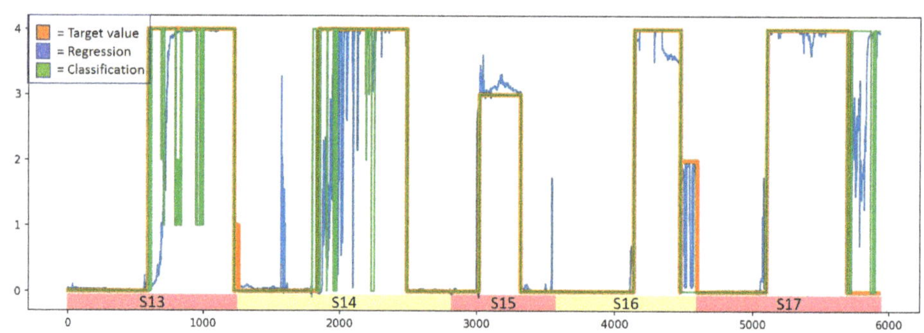

(**c**)

Figure 6. True and predicted arousal levels for the WESAD dataset using LSTM-based regression (blue line) and classification (green line) models with baseline normalization. The true arousal level is shown in orange color. (**a**) Arousal for study subjects 2, 3, 4, 5, and 6. (**b**) Arousal for study subjects 7, 8, 9, 10, and 11. (**c**) Arousal for study subjects 13, 14, 15, 16, and 17.

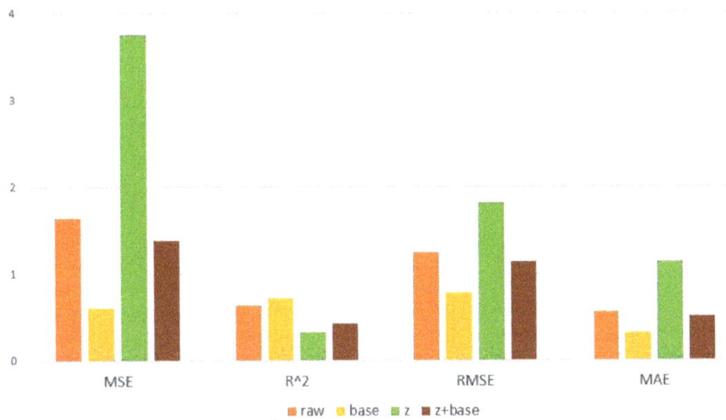

Figure 7. Effect of different normalization methods on the performance of the LSTM models.

Figures 5 and 6 show that the valence and arousal levels can be estimated with high reliability when studied separately, and an LSTM-based regression model with baseline reduction is the best method to do it. However, the most important thing is to understand how well emotions can be estimated when valence and arousal estimations are combined and visualized using Russell's circumplex model of emotions (see Figure 1). Figure 8 shows this visualization for different emotion classes; these estimations are from the run that provided the best results when the target values were normalized using baseline reduction. Therefore, they are the same ones shown in Figures 5 and 6 for the LSTM-based regression model. In Figure 8, the estimated values are shown in blue and the target values provided by the study subjects are visualized using red dots. As baseline reduction is used, in the case of the baseline class the target value for valence and arousal is zero. Figure 8a shows that the baseline emotion can be estimated with high accuracy, as almost all the estimations are close to the origin. In this case, the average estimated valence is -0.01 and the average estimated arousal is 0.04. According to Figure 1, strong negative emotions are located at the top left quarter of Russell's circumplex model of emotions, which is exactly where estimations of stress-class observations are located based on the models presented in this article (see Figure 8b). Moreover, the target values obtained from the study subjects are located in the same place. In fact, the predicted values and target values are very close to each other. Observations from the amusement class are estimated to be located close to the origin (Figure 8c) or to the right bottom quarter of Russell's circumplex model of emotions, where relaxed emotions are located. While the model estimates only slightly relaxed emotions during the amusement class, and the detected emotions are not as strong as those recognized from the stress class, this does not mean that the model performs badly in this case. Indeed, when predicted values are compared to the target values, it can again be noted that they are distributed in the same area on the valence–arousal graph. Therefore, prediction models based on the LSTM regression model and baseline reduction can estimate the valence and arousal levels for each emotion class with high accuracy, making it emotion-independent based on this analysis.

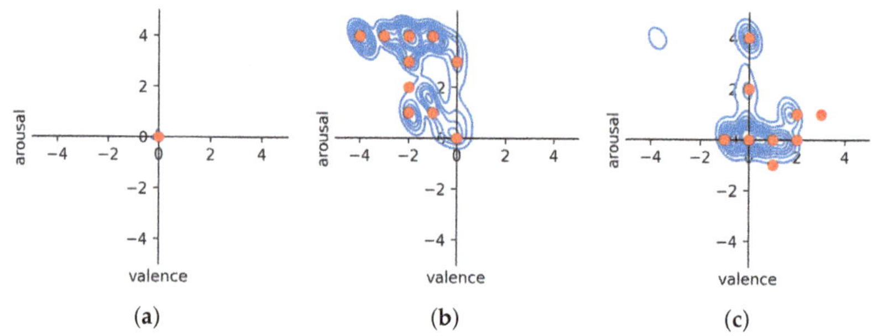

Figure 8. Results from different emotion classes. The results are presented on a valence–arousal graph, with valence on the x-axis and arousal on the y-axis. Study subjects' reported valence and arousal levels are shown as red dots, while the blue graphs indicate the distribution of estimated values across the graph: (**a**) baseline; (**b**) stress; (**c**) amusement.

5.2. Subject-Wise Results

Subject-wise valence and arousal level estimation results from the best-performing regression models are presented in Table 5, where LSTM models without any normalization and with baseline reduction are compared to AdaBoost and Random Forest models with baseline reduction. It should be noted that, according to Table 3, the AdaBoost and Random Forest models perform much worse on average than the LSTM models. The results in Table 5 show that for most of the study subjects the levels of valence and arousal can be predicted by all of these models, as well as with the AdaBoost and Random Forest models. There are even cases in which AdaBoost and Random Forest perform better than LSTM. However, the largest difference between AdaBoost, Random Forest, and LSTM is that in certain cases AdaBoost and Random Forest perform very badly, while the variance between the prediction rates for different study subjects is much smaller using LSTM. For instance, when the valence of subject 11 was predicted using the AdaBoost regression model, the R^2-score was -131.95, and for Random Forest the R^2-score was -127.81. These naturally have a huge effect on the average values presented in Table 3.

The results in Table 5 show that certain study subjects have data that are more difficult to predict. For instance, each model has difficulty predicting the valence of study subjects 14 and 17 and the arousal of study subjects 2, 14, and 17. There may be problems with the data of study subjects 14 and 17, or their bodies may react differently to stimuli compared to other study subjects. If the differences are caused by different stimuli, this suggests that it would be possible to obtain better results via model personalization. In addition, there are model-specific differences. For instance, the valence level of study subject 4 is not predicted well by LSTM when the model uses raw data; however, when the same person's data is predicted with the LSTM model trained using baseline reduction normalized data, the prediction is highly accurate. This shows the importance of normalization. Moreover, while LSTM performs well in most cases, for certain subjects the R^2-score is negative.

For this experiment, all of the models were trained five times; the results presented in this section are averages from these runs, with the standard deviation from these runs for each individual presented in parentheses in Table 5. When the standard deviations are studied in detail, it can be noted that for certain study subjects the results differ a great deal between different runs, especially when it comes to valence level detection. For instance, for study subjects 2, 5, and 14 the standard deviation of the R^2 score is greater than 1 when valence is detected using LSTM and baseline reduction.

Table 5. Subject-wise prediction rates (standard deviation in parentheses) for valence and arousal detection using different prediction models and normalization methods.

Subject-Wise Prediction Rates (Standard Deviation in Parentheses) for Valence Detection.				
Subject	**LSTM Raw: MSE/R^2**	**LSTM Base: MSE/R^2**	**AdaBoost Base: MSE/R^2**	**Random Forest Base: MSE/R^2**
2	1.50 (0.20)/−2.92 (2.26)	0.83 (0.25)/−1.31 (1.10)	0.71 (0.01)/−0.37 (0.05)	0.51 (0.20)/0.07 (0.45)
3	0.51 (0.25)/−0.02 (0.16)	0.50 (0.15)/−0.26 (0.18)	1.48 (0.12)/−0.65 (0.09)	1.62 (0.26)/−0.37 (0.20)
4	0.80 (0.19)/0.57 (0.12)	0.22 (0.27)/0.79 (0.28)	0.42 (0.08)/0.75 (0.04)	0.78 (0.28)/0.74 (0.03)
5	0.06 (0.02)/0.88 (0.02)	0.41 (0.17)/−1.52 (2.27)	0.94 (0.07)/−2.07 (0.76)	0.96 (0.08)/−1.50 (0.91)
6	0.05 (0.03)/0.92 (0.05)	0.02 (0.01)/0.97 (0.01)	0.18 (0.09)/0.59 (0.25)	0.10 (0.09)/0.78 (0.26)
7	0.10 (0.10)/0.83 (0.11)	0.03 (0.03)/0.88 (0.08)	0.35 (0.09)/0.56 (0.06)	0.29 (0.11)/0.57 (0.04)
8	0.00 (0.00/1.00 (0.00)	0.01 (0.00)/0.99 (0.00)	0.31 (0.19)/0.21 (0.79)	0.35 (0.17)/0.08 (0.71)
9	0.04 (0.04)/0.82 (0.17)	0.19 (0.09)/0.60 (0.11)	0.62 (0.04)/0.32 (0.02)	0.79 (0.17)/0.31 (0.02)
10	0.52 (0.53)/0.90 (0.10)	0.20 (0.08)/0.93 (0.04)	2.77 (0.17)/−5.19 (2.12)	2.91 (0.19)/−8.41 (2.75)
11	0.22 (0.14)/0.85 (0.10)	0.04 (0.01)/0.98 (0.01)	1.85 (0.08)/−131.95 (201.36)	2.10 (0.28)/−127.81 (203.78)
13	0.48 (0.13)/0.60 (0.05)	0.78 (0.13)/0.31 (0.05)	0.67 (0.03)/0.24 (0.04)	0.49 (0.18)/0.23 (0.05)
14	2.87 (1.48)/−1.90 (1.89)	1.28 (0.19)/−1.56 (1.12)	2.27 (0.14)/−26.71 (20.60)	2.06 (0.28)/−13.33 (9.17)
15	0.05 (0.01)/0.97 (0.01)	0.50 (0.09)/0.53 (0.02)	0.80 (0.32)/0.47 (0.20)	3.19 (2.30)/−0.29 (0.87)
16	0.06 (0.05)/0.98 (0.02)	0.06 (0.02)/0.98 (0.01)	3.28 (0.56)/−11.16 (8.54)	3.13 (0.69)/−10.55 (8.97)
17	2.18 (0.96)/0.71 (0.14)	1.24 (0.33)/0.65 (0.07)	6.09 (0.56)/−20.94 (9.45)	6.58 (0.76)/−17.29 (7.96)
Subject-Wise Prediction Rates (Standard Deviation in Parentheses) for Arousal Detection.				
Subject	**LSTM Raw: MSE/R^2**	**LSTM Base: MSE/R^2**	**AdaBoost Base: MSE/R^2**	**Random Forest Base: MSE/R^2**
2	1.48 (0.32)/−2.04 (1.19)	0.97 (0.11)/−9.65 (8.49)	0.89 (0.09)/−0.13 (0.25)	1.21 (0.38)/0.25 (0.30)
3	2.21 (0.71)/0.61 (0.09)	1.15 (0.41)/0.31 (0.21)	1.19 (0.10)/0.62 (0.03)	0.97 (0.11)/0.67 (0.01)
4	6.64 (0.92)/−1.73 (0.64)	0.45 (0.09)/0.88 (0.03)	0.67 (0.04)/0.87 (0.01)	0.63 (0.02)/0.86 (0.02)
5	0.35 (0.16)/0.95 (0.02)	0.67 (0.33)/0.80 (0.10)	0.69 (0.09)/0.81 (0.02)	30.45 (0.28)/0.87 (0.07)
6	0.10 (0.06)/−0.75 (0.13)	0.07 (0.03)/0.57 (0.20)	3.65 (0.04)/−0.08 (0.01)	3.09 (0.59)/−0.04 (0.04)
7	0.02 (0.02)/1.00 (0.03)	0.30 (0.08)/0.91 (0.03)	3.74 (0.07)/0.02 (0.01)	4.47 (0.76)/−0.78 (0.81)
8	0.18 (0.17)/0.94 (0.06)	0.04 (0.01)/0.99 (0.00)	1.82 (0.12)/−1.20 (0.28)	2.00 (0.29)/−1.92 (1.01)
9	0.18 (0.13)/0.71 (0.15)	0.19 (0.10)/0.73 (0.09)	2.29 (0.13)/0.20 (0.01)	1.39 (0.97)/0.39 (0.20)
10	0.14 (0.13)/0.99 (0.01)	0.08 (0.06)/0.98 (0.01)	1.71 (0.20)/0.41 (0.08)	2.48 (0.60)/−0.19 (0.54)
11	0.38 (0.64)/0.77 (0.39)	0.06 (0.03)/0.98 (0.01)	2.28 (0.30)/−0.79 (0.25)	1.94 (0.31)/−0.91 (0.34)
13	2.57 (2.11)/0.43 (0.58)	0.97 (0.26)/0.73 (0.07)	2.54 (0.38)/−0.53 (0.41)	1.39 (1.12)/0.24 (0.75)
14	1.62 (0.84)/0.69 (0.21)	1.75 (0.53)/0.44 (0.18)	5.21 (0.75)/−205.86 (339.33)	4.47 (1.25)/−208.87 (337.77)
15	0.22 (0.14)/0.87 (0.09)	0.28 (0.26)/0.87 (0.12)	1.24 (0.29)/0.53 (0.13)	2.77 (1.37)/−2.14 (2.60)
16	0.24 (0.14)/0.97 (0.02)	0.23 (0.15)/0.93 (0.04)	1.03 (0.0)/0.49 (0.0)	0.93 (0.10)/0.57 (0.08)
17	5.24 (1.57)/0.46 (0.23)	1.29 (0.64)/0.63 (0.19)	7.16 (0.31)/−22.30 (6.73)	5.70 (1.39)/−18.67 (8.54)

5.3. Experimenting with Sensor Combinations

Different sensor combinations were compared to study the effects of different sensors on the recognition results. The results calculated using the LSTM regression model with baseline reduction are shown in Table 6. Table 6 shows that not all of the Empatica E4's sensors are needed to estimate valence levels reliably, and arousal can be estimated at a high rate without using all the sensors as well. In fact, when using just the BVP and EDA sensors the valence levels can be estimated with the same detection rate as when using all the sensors. When these results are studied subject-wise, it can be noted that the variance between the study subjects is smaller when using only BVP and EDA sensors instead of all the sensors (see Table 7). When the LSTM regression model with baseline reduction ws used with all the sensors to recognize valence level, the R^2 score was negative for four study subjects (see Table 5). However, according to Table 7, the R^2 score is negative only for one study subject when using only the BVP and EDA sensors. In addition, the variance within the study subjects is smaller when using just the BVP and EDA sensors instead of all the sensors; for instance, the variance of the R^2-score varies from 0 to 0.14 depending on the study subject when using only the BVP and EDA sensors, while when using all the sensors it varies from 0.01 to 1.10.

Table 6. Average recognition rates (standard deviation in parentheses) using LSTM regression model with baseline reduction with different sensor combinations.

Sensors	Valence: MSE/R^2	Arousal: MSE/R^2
EDA + ST + BVP	0.43 (0.05)/0.71 (0.04)	0.59 (0.06)/0.81 (0.02)
EDA + BVP	0.42 (0.05)/0.72 (0.05)	1.13 (0.10)/0.61 (0.03)
BVP + ST	0.72 (0.12)/0.52 (0.09)	0.92 (0.15)/0.71 (0.05)
EDA + ST	0.82 (0.24)/0.44 (0.18)	0.87 (0.11)/0.70 (0.04)
EDA	0.91 (0.08)/0.39 (0.05)	1.17 (0.22)/0.58 (0.08)
BVP	0.72 (0.07)/0.48 (0.05)	1.07 (0.08)/0.64 (0.03)
ST	1.76 (0.09)/−0.62 (0.08)	3.60 (0.27)/−0.70 (0.15)

Table 7. Results from different emotion classes when using features extracted from only some of the sensors (EDA and BVP for valence and EDA and ST for arousal) and LSTM regression model with baseline reduction.

Subject	Valence (EDA + BVP): MSE/R^2	Arousal (EDA + ST): MSE/R^2
2	0.61 (0.04)/0.00 (0.00)	0.69 (0.08)/−1.31 (2.56)
3	2.03 (0.41)/−0.55 (0.14)	0.69 (0.46)/0.62 (0.27)
4	0.16 (0.03)/0.85 (0.02)	0.36 (0.04)/0.91 (0.01)
5	0.15 (0.04)/0.79 (0.01)	0.19 (0.07)/0.94 (0.02)
6	0.04 (0.01)/0.95 (0.01)	0.14 (0.15)/0.79 (0.13)
7	0.04 (0.00)/0.83 (0.02)	0.44 (0.13)/0.86 (0.04)
8	0.02 (0.00)/0.97 (0.00)	0.08 (0.00)/0.97 (0.00)
9	0.28 (0.00)/0.39 (0.07)	0.04 (0.03)/0.86 (0.04)
10	0.11 (0.12)/0.96 (0.03)	0.05 (0.02)/0.98 (0.01)
11	0.09 (0.06)/0.94 (0.05)	0.18 (0.06)/0.94 (0.02)
13	0.81 (0.02)/0.36 (0.06)	2.26 (0.59)/−0.48 (1.05)
14	0.77 (0.04)/0.43 (0.08)	3.81 (0.96)/−1.14 (0.70)
15	0.38 (0.01)/0.59 (0.00)	0.07 (0.01)/0.96 (0.00)
16	0.10 (0.03)/0.96 (0.01)	0.56 (0.01)/0.82 (0.00)
17	0.23 (0.00)/0.93 (0.00)	1.12 (0.32)/0.68 (0.10)

The recognition rate of the arousal level is slightly lower when using just EDA and ST instead of all the sensors (Table 6). In this case, the selection of only certain sensors does not have a similar positive effect on standard deviation as does for valence; again, however, the results in this case are better for certain individuals.

6. Discussion

The results presented in Section 5 show that fine-grained valence and arousal levels can be estimated reliably based on wrist-worn wearable sensor data and machine learning methods. According to the results shown in Table 3, LSTM models are superior to other methods. This is because LSTM is the only prediction model in our experiments capable of learning long-term dependencies from the data. Moreover, LSTM is the most advanced among these prediction models, which is why its superior performance on the WESAD dataset is not surprising. As shown in Figure 2, the WESAD dataset is complex, and it requires powerful methods for analysis. Especially in the case of estimating the arousal level, the difference between LSTM and other prediction methods is very large.

Table 3 shows that regression models perform better for the task in general than classification models when the performance of the models is measured using MSE and R^2 values. This is expected, as valence and arousal are continuous phenomena and are not discrete. Moreover, as classification methods treat valence and arousal levels as distinct categories, they are unable to take into account the ordinal nature of these levels and use it during the model training process. This means that if the training data do not contain samples from all the possible levels of valence and arousal, as in the case of our data, classification models cannot detect these from unseen datasets as well. Regression models do not have this limitation. Despite their limitations, in certain cases classification methods performed well in our experiments. In fact, according to MSE and R^2 values, the LSTM-based classification model with baseline reduction performs equally as well as the regression-based LSTM model, though the RMSE and MAE measures (Table 4) and visualization of the results (Figures 5 and 6) show that there are in fact differences between these models and that the LSTM regression model is more reliable than the LSTM classification model. The biggest difference is that the LSTM-based classification methods seem to have problems detecting positive valence values. In fact, the dataset does not contain many such samples, showing that the LSTM classification model is more vulnerable to limited and imbalanced datasets than the LSTM regression model. Moreover, when the predicted values were visualized using Russell's circumplex together with target values provided by the study subjects (see Figure 8), it can be noted that the results are almost identical. This shows that the models presented in this article are highly capable of recognizing the level of valence and arousal and that the valence and arousal level estimates provided by the study subjects contain information that can be used as target values for the recognition models when data are pre-processed and normalized correctly. However, it seems that the targets reported by the study subjects are not always reliable. For instance, subject 3 reported a valence label of 7 after a stress condition, which was the same as this person reported for baseline valence, claiming that he was looking forward to the next condition and was therefore cheerful. The models did not manage to predict this correctly; the results using LSTM, AdaBoost, and Random Forest are as shown in Figure 9. However, as the target value defined by the study subject was not reliable, most likely the estimations made by the models are closer to the truth than the target variables. Moreover, it is possible that people might not always know how they feel [43,44], and for this reason it seems that in certain cases the models are better at describing feelings than the study subjects themselves.

In addition, different methods for data normalization were used: z-score, baseline reduction, and z-score with baseline reduction. These were compared to the case where features were extracted from raw (non-normalized) data. The results were surprising, in that the best results were obtained using baseline reduction normalization. It was expected that z-score normalization would provide the best results, as it has been shown to improve the detection rates when stress or other affective states are recognized from wearable sensor data [45]. However, in this study no similar effect was noted. The reason for this could be that the previous studies concentrated on detecting discrete human emotions and not on estimating continuous valence and arousal values, as this study does. A baseline reduction-based z-score normalization has other advantages over z-score normalization as well; in order to normalize signals using *z*-score normalization, the participant-specific mean and standard deviation need to be calculated from each study subject's full data signal, meaning that normalization can only be carried out after data gathering [46]. Due to this, it is not suitable for real-time application. However, baseline reduction does not have a similar limitation, as only baseline data need to be collected, and the study subject can report his/her valence and arousal levels at the same time. Moreover, as baseline reduction improves recognition rates and *z*-score normalization does not, this means that the differences between the individuals are more related to differences in subjective target variable estimations than to differences in the signals themselves.

Our in-depth subject-wise study (see Table 5) shows that for most of the study subjects the AdaBoost and Random Forest models perform almost as well as LSTM. However, the biggest difference between these models is that for certain subjects the AdaBoost and Random Forest models perform very badly, while the LSTM-based model is more evenly reliable for each individual. Therefore, LSTM does not suffer from much variance between the study subjects, resulting in much better average prediction rates than all the other experimented models. These results show that machine learning is a powerful tool for detecting valence and arousal levels, and thereby for recognizing emotions. While the results using the LSTM regression model are good on average, even in the case of the LSTM regression model there is variation between the study subjects and different runs. This means that there is a need to personalize models and experiment with larger datasets that have more variation. Moreover, the results presented in Table 7 show that when using only certain sensors it is possible to obtain estimations that are just as good as when using all the sensors. For certain individuals, the results are even better, especially when it comes to detecting the valence level, as in this case the variance between the study subjects can be reduced when using only the EDA and BVP sensors instead of all the sensors. Therefore, prediction models could be personalized by selecting a unique sensor combination for each individual, which could improve the results and reduce the variance between study subjects.

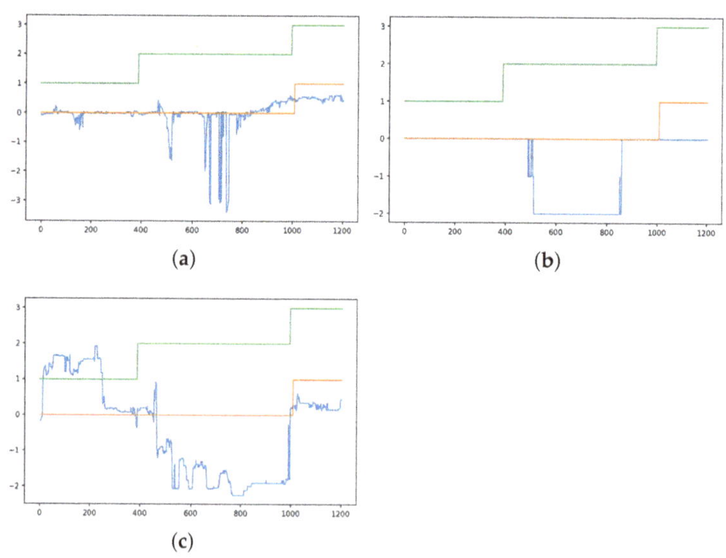

Figure 9. Subject 3 might have provided the wrong answer regarding the level of valence during a stressful stage. The blue line shows the predicted valence level, the orange line is the ground truth for the valence level, and the green line indicates the emotion class: 1 = baseline, 2 = stress, and 3: amusement. (**a**) LSTM, (**b**) AdaBoost, (**c**) Random Forest.

7. Conclusions

This study aimed to predict emotions using biosignals collected via wrist-worn sensor and to evaluate the accuracy of different prediction models. Two dimensions of emotions were considered, namely, valence and arousal. In this study, valence and arousal levels were estimated using machine learning methods based on an open-access WESAD dataset containing biosensor data from wrist-worn sensors. These data included skin temperature, electrodermal activity, blood volume pulse, heart rate, and heart rate variability collected from 15 study subjects. Study subjects were exposed to different stimuli (baseline, stress, and amusement); after each stimulus, they reported their valence and arousal levels. These estimates were used in this study as target variables. In fact, in the study it was shown that the level of valence and arousal can be predicted with high reliability with the help of these user-reported valence and arousal levels by using the LSTM regression model and normalizing target values through baseline reduction. However, while on average the results are very good, for certain individuals the results are much weaker. Moreover, it was found that reliable models could be obtained even if all biosignals were not used in the training phase; in fact, for certain study subjects the best results were obtained using only a few of the sensors.

To date, the field of emotion detection has mainly focused on identifying a limited number of discrete emotions or treating valence and arousal as coarse-grained discrete variables. However, this study demonstrates the ability to reliably detect fine-grained valence and arousal levels by analyzing data using advanced machine learning models. Additionally, this research suggests that regression models are more effective for this task than classification models. By recognizing emotions through the dimensions of valence and arousal rather than discrete emotions, this study takes a step towards a more sophisticated and nuanced understanding of emotion detection using biosignals and machine learning. This shift towards analyzing continuous variables rather than discrete emotions is expected to be the focus of future research in this field.

However, this study has weaknesses as well. When subject-wise results were studied, it was noted that there was variance between the recognition rates for the different individuals; therefore, the recognition rates were not equally good for each study subject. In certain cases, this variance between individuals could be reduced by personalizing the models by selecting a unique sensor combination for each individual. In fact, one future tasks is to study model personalization in more detail; for instance, incremental learning could be an effective method to personalize models based on streaming data [47]. Moreover, feature selection needs to be studied in order to provide reliable estimates for each individual. For instance, a sequential backward floating search has been found to be an effective feature selection method for biosignals [34].

Another weakness of this study is that the experiments were based on only one dataset; due to this, future work needs to include experimenting with other datasets. In particular, it should be studied how well negative arousal levels can be estimated, as the WESAD dataset used in this study contained very few negative arousal values. Moreover, the LSTM model used in this article was quite simple; it had one hidden layer, with 64 units and around 17,000 trainable parameters. However, if the aim is to build a model that can be run in real-time in a wrist-work device, a complexity analysis should be carried out in order to better understand how much calculation capacity this type of model requires. In addition, the parameters of the prediction models should be tuned in order to optimize the results, as in this study the prediction models were trained without parameter tuning.

Author Contributions: Conceptualization, P.S.; Funding acquisition, J.R.; Methodology, P.S.; Software, P.S.; Supervision, J.R.; Validation, P.S. and S.T.; Writing—original draft, P.S.; Writing—review and editing, S.T., G.C. and A.I. All authors have read and agreed to the published version of the manuscript.

Funding: This research received no external funding.

Acknowledgments: The authors are grateful for Infotech Oulu.

Conflicts of Interest: The authors declare no conflict of interest.

References

1. Zenonos, A.; Khan, A.; Kalogridis, G.; Vatsikas, S.; Lewis, T.; Sooriyabandara, M. HealthyOffice: Mood recognition at work using smartphones and wearable sensors. In Proceedings of the 2016 IEEE International Conference on Pervasive Computing and Communication Workshops (PerCom Workshops), Sydney, Australia, 14–18 March 2016; pp. 1–6.
2. Yannakakis, G.N.; Martínez, H.P.; Jhala, A. Towards affective camera control in games. *User Model. User-Adapt. Interact.* **2010**, *20*, 313–340. [CrossRef]
3. Gjoreski, M.; Gjoreski, H.; Luštrek, M.; Gams, M. Continuous Stress Detection Using a Wrist Device: In Laboratory and Real Life. In Proceedings of the 2016 ACM International Joint Conference on Pervasive and Ubiquitous Computing: Adjunct (UbiComp '16), Heidelberg, Germany, 12–16 September 2016; ACM: New York, NY, USA, 2016; pp. 1185–1193. [CrossRef]
4. Siirtola, P. Continuous stress detection using the sensors of commercial smartwatch. In Proceedings of the 2019 ACM International Joint Conference on Pervasive and Ubiquitous Computing and 2019 ACM International Symposium on Wearable Computers, London, UK, 11–13 September 2019; pp. 1198–1201.
5. Schmidt, P.; Reiss, A.; Dürichen, R.; Laerhoven, K.V. Wearable-Based Affect Recognition—A Review. *Sensors* **2019**, *19*, 4079. [CrossRef]
6. Can, Y.S.; Arnrich, B.; Ersoy, C. Stress detection in daily life scenarios using smart phones and wearable sensors: A survey. *J. Biomed. Inform.* **2019**, *92*, 103139. [CrossRef] [PubMed]
7. Russell, J.A. A circumplex model of affect. *J. Personal. Soc. Psychol.* **1980**, *39*, 1161. [CrossRef]
8. Russell, J.A. Core affect and the psychological construction of emotion. *Psychol. Rev.* **2003**, *110*, 145. [CrossRef]
9. Wöllmer, M.; Eyben, F.; Reiter, S.; Schuller, B.; Cox, C.; Douglas-Cowie, E.; Cowie, R. Abandoning emotion classes-towards continuous emotion recognition with modelling of long-range dependencies. In Proceedings of the 9th Interspeech 2008 Incorp. 12th Australasian International Conference on Speech Science and Technology SST 2008, Brisbane, Australia, 22–26 September 2008; pp. 597–600.
10. Mitenkova, A.; Kossaifi, J.; Panagakis, Y.; Pantic, M. Valence and arousal estimation in-the-wild with tensor methods. In Proceedings of the 2019 14th IEEE International Conference on Automatic Face & Gesture Recognition (FG 2019), Lille, France, 14–18 May 2019; pp. 1–7.
11. Hsu, Y.L.; Wang, J.S.; Chiang, W.C.; Hung, C.H. Automatic ECG-Based Emotion Recognition in Music Listening. *IEEE Trans. Affect. Comput.* **2017**, *11*, 85–99. [CrossRef]
12. Romeo, L.; Cavallo, A.; Pepa, L.; Berthouze, N.; Pontil, M. Multiple instance learning for emotion recognition using physiological signals. *IEEE Trans. Affect. Comput.* **2019**, *13*, 389–407. [CrossRef]
13. Li, F.; Shirahama, K.; Nisar, M.A.; Huang, X.; Grzegorzek, M. Deep Transfer Learning for Time Series Data Based on Sensor Modality Classification. *Sensors* **2020**, *20*, 4271. [CrossRef]
14. Zhao, B.; Wang, Z.; Yu, Z.; Guo, B. EmotionSense: Emotion recognition based on wearable wristband. In Proceedings of the 2018 IEEE SmartWorld, Ubiquitous Intelligence & Computing, Advanced & Trusted Computing, Scalable Computing & Communications, Cloud & Big Data Computing, Internet of People and Smart City Innovation (SmartWorld/SCALCOM/UIC/ATC/CBDCom/IOP/SCI), Guangzhou, China, 8–12 October 2018; pp. 346–355.
15. Diaz-Romero, D.J.; Rincón, A.M.R.; Miguel-Cruz, A.; Yee, N.; Stroulia, E. Recognizing emotional states with wearables while playing a serious game. *IEEE Trans. Instrum. Meas.* **2021**, *70*, 1–12. [CrossRef]
16. Zhang, T.; El Ali, A.; Wang, C.; Hanjalic, A.; Cesar, P. Corrnet: Fine-grained emotion recognition for video watching using wearable physiological sensors. *Sensors* **2020**, *21*, 52. [CrossRef]
17. Hwang, S.; Jebelli, H.; Choi, B.; Choi, M.; Lee, S. Measuring workers' emotional state during construction tasks using wearable EEG. *J. Constr. Eng. Manag.* **2018**, *144*, 04018050. [CrossRef]
18. Grundlehner, B.; Brown, L.; Penders, J.; Gyselinckx, B. The Design and Analysis of a Real-Time, Continuous Arousal Monitor. In Proceedings of the 2009 Sixth International Workshop on Wearable and Implantable Body Sensor Networks, Berkeley, CA, USA, 3–5 June 2009; pp. 156–161. [CrossRef]
19. Nirjhar, E.H.; Behzadan, A.; Chaspari, T. Exploring Bio-Behavioral Signal Trajectories of State Anxiety During Public Speaking. In Proceedings of the ICASSP 2020—2020 IEEE International Conference on Acoustics, Speech and Signal Processing (ICASSP), Barcelona, Spain, 4–8 May 2020; pp. 1294–1298. [CrossRef]
20. Siirtola, P.; Röning, J. Comparison of Regression and Classification Models for User-Independent and Personal Stress Detection. *Sensors* **2020**, *20*, 4402. [CrossRef] [PubMed]
21. Healey, J.A.; Picard, R.W. Detecting stress during real-world driving tasks using physiological sensors. *IEEE Trans. Intell. Transp. Syst.* **2005**, *6*, 156–166. [CrossRef]
22. El Haouij, N. Biosignals for Driver's Stress Level Assessment: Functional Variable Selection and Fractal Characterization. Ph.D. Thesis, Université Paris-Saclay (ComUE), Gif-sur-Yvette, France, 2018.
23. Vos, G.; Trinh, K.; Sarnyai, Z.; Azghadi, M.R. Ensemble Machine Learning Model Trained on a New Synthesized Dataset Generalizes Well for Stress Prediction Using Wearable Devices. *arXiv* **2022**, arXiv:2209.15146.

24. Schmidt, P.; Reiss, A.; Duerichen, R.; Marberger, C.; Van Laerhoven, K. Introducing WESAD, a Multimodal Dataset for Wearable Stress and Affect Detection. In Proceedings of the ACM 2018 on International Conference on Multimodal Interaction, Boulder, CO, USA, 16–20 October 2018; pp. 400–408.
25. Empatica E4. 2019. Available online: https://www.empatica.com/e4-wristband (accessed on 7 January 2021).
26. Watson, D.; Clark, L.A.; Tellegen, A. Development and validation of brief measures of positive and negative affect: The PANAS scales. *J. Personal. Soc. Psychol.* **1988**, *54*, 1063. [CrossRef] [PubMed]
27. Marteau, T.M.; Bekker, H. The development of a six-item short-form of the state scale of the Spielberger State—Trait Anxiety Inventory (STAI). *Br. J. Clin. Psychol.* **1992**, *31*, 301–306. [CrossRef]
28. Bradley, M.M.; Lang, P.J. Measuring emotion: The self-assessment manikin and the semantic differential. *J. Behav. Ther. Exp. Psychiatry* **1994**, *25*, 49–59. [CrossRef]
29. Mishra, V.; Pope, G.; Lord, S.; Lewia, S.; Lowens, B.; Caine, K.; Sen, S.; Halter, R.; Kotz, D. The case for a commodity hardware solution for stress detection. In Proceedings of the 2018 ACM International Joint Conference and 2018 International Symposium on Pervasive and Ubiquitous Computing and Wearable Computers, Singapore, 8–12 October 2018; pp. 1717–1728.
30. Boucsein, W. *Electrodermal Activity*; Springer Science & Business Media: Berlin/Heidelberg, Germany, 2012.
31. Gashi, S.; Min, C.; Montanari, A.; Santini, S.; Kawsar, F. A multidevice and multimodal dataset for human energy expenditure estimation using wearable devices. *Sci. Data* **2022**, *9*, 1–14. [CrossRef]
32. Van der Maaten, L.; Hinton, G. Visualizing data using t-SNE. *J. Mach. Learn. Res.* **2008**, *9*, 2579–2605.
33. Mishra, V.; Pope, G.; Lord, S.; Lewia, S.; Lowens, B.; Caine, K.; Sen, S.; Halter, R.; Kotz, D. Continuous detection of physiological stress with commodity hardware. *ACM Trans. Comput. Healthc.* **2020**, *1*, 1–30. [CrossRef]
34. Gjoreski, M.; Mahesh, B.; Kolenik, T.; Uwe-Garbas, J.; Seuss, D.; Gjoreski, H.; Luštrek, M.; Gams, M.; Pejović, V. Cognitive Load Monitoring with Wearables–Lessons Learned From a Machine Learning Challenge. *IEEE Access* **2021**, *9*, 103325–103336. [CrossRef]
35. Hochreiter, S.; Schmidhuber, J. Long short-term memory. *Neural Comput.* **1997**, *9*, 1735–1780. [CrossRef] [PubMed]
36. Bock, M.; Hölzemann, A.; Moeller, M.; Van Laerhoven, K. Improving Deep Learning for HAR with Shallow LSTMs. In Proceedings of the 2021 International Symposium on Wearable Computers, Virtual Event, 21–26 September 2021; pp. 7–12.
37. Freund, Y.; Schapire, R.; Abe, N. A short introduction to boosting. *J.-Jpn. Soc. Artif. Intell.* **1999**, *14*, 1612.
38. Breiman, L. Random forests. *Mach. Learn.* **2001**, *45*, 5–32. [CrossRef]
39. Chen, T.; Guestrin, C. Xgboost: A scalable tree boosting system. In Proceedings of the 22nd ACM Sigkdd International Conference on Knowledge Discovery and Data Mining, San Francisco, CA, USA, 13–17 August 2016; pp. 785–794.
40. Ke, G.; Meng, Q.; Finley, T.; Wang, T.; Chen, W.; Ma, W.; Ye, Q.; Liu, T.Y. Lightgbm: A highly efficient gradient boosting decision tree. *Adv. Neural Inf. Process. Syst.* **2017**, *30*, 52.
41. James, G.; Witten, D.; Hastie, T.; Tibshirani, R. *An Introduction to Statistical Learning*; Springer: Berlin/Heidelberg, Germany, 2013; Volume 112.
42. Ma, Y.; Liu, D.; Cai, L. Deep learning-based upper limb functional assessment using a single kinect v2 sensor. *Sensors* **2020**, *20*, 1903. [CrossRef]
43. Schmidt, P.; Reiss, A.; Dürichen, R.; Van Laerhoven, K. Labelling Affective States "in the Wild" Practical Guidelines and Lessons Learned. In Proceedings of the 2018 ACM International Joint Conference and 2018 International Symposium on Pervasive and Ubiquitous Computing and Wearable Computers, Singapore, 8–12 October 2018; pp. 654–659.
44. Siirtola, P.; Peltonen, E.; Koskimäki, H.; Mönttinen, H.; Röning, J.; Pirttikangas, S. Wrist-worn Wearable Sensors to Understand Insides of the Human Body: Data Quality and Quantity. In Proceedings of the 5th ACM Workshop on Wearable Systems and Applications, Seoul, Republic of Korea, 17–21 June 2019; pp. 17–21.
45. Gjoreski, M.; Luštrek, M.; Gams, M.; Gjoreski, H. Monitoring stress with a wrist device using context. *J. Biomed. Inform.* **2017**, *73*, 159–170. [CrossRef]
46. Attig, A.; Perner, P. The Problem of Normalization and a Normalized Similarity Measure by Online Data. *Trans. Case Based Reason.* **2011**, *4*, 3–17.
47. Siirtola, P.; Koskimäki, H.; Röning, J. Personalizing human activity recognition models using incremental learning. In Proceedings of the ESANN, Bruges, Belgium, 25–27 April 2018; pp. 627–632.

sensors

MDPI

Article

Technology Acceptance Model for Exoskeletons for Rehabilitation of the Upper Limbs from Therapists' Perspectives

Beatrice Luciani [1,2,*], Francesco Braghin [1], Alessandra Laura Giulia Pedrocchi [2,3] and Marta Gandolla [1,2,3]

1 Department of Mechanical Engineering, Politecnico di Milano, Via La Masa 1, 20156 Milano, Italy
2 NeuroEngineering And Medical Robotics Laboratory (NEARLab), Department of Electronics, Information and Bioengineering, Politecnico di Milano, Piazza Leonardo da Vinci, 32, 20133 Milano, Italy
3 WE-COBOT Lab, Politecnico di Milano, Polo Territoriale di Lecco, Via G. Previati, 1/c, 23900 Lecco, Italy
* Correspondence: beatrice.luciani@polimi.it

Abstract: Over the last few years, exoskeletons have been demonstrated to be useful tools for supporting the execution of neuromotor rehabilitation sessions. However, they are still not very present in hospitals. Therapists tend to be wary of this type of technology, thus reducing its acceptability and, therefore, its everyday use in clinical practice. The work presented in this paper investigates a novel point of view that is different from that of patients, which is normally what is considered for similar analyses. Through the realization of a technology acceptance model, we investigate the factors that influence the acceptability level of exoskeletons for rehabilitation of the upper limbs from therapists' perspectives. We analyzed the data collected from a pool of 55 physiotherapists and physiatrists through the distribution of a questionnaire. Pearson's correlation and multiple linear regression were used for the analysis. The relations between the variables of interest were also investigated depending on participants' age and experience with technology. The model built from these data demonstrated that the perceived usefulness of a robotic system, in terms of time and effort savings, was the first factor influencing therapists' willingness to use it. Physiotherapists' perception of the importance of interacting with an exoskeleton when carrying out an enhanced therapy session increased if survey participants already had experience with this type of rehabilitation technology, while their distrust and the consideration of others' opinions decreased. The conclusions drawn from our analyses show that we need to invest in making this technology better known to the public—in terms of education and training—if we aim to make exoskeletons genuinely accepted and usable by therapists. In addition, integrating exoskeletons with multi-sensor feedback systems would help provide comprehensive information about the patients' condition and progress. This can help overcome the gap that a robot creates between a therapist and the patient's human body, reducing the fear that specialists have of this technology, and this can demonstrate exoskeletons' utility, thus increasing their perceived level of usefulness.

Keywords: technology acceptance model; rehabilitation exoskeletons; therapists; neuro-rehabilitation; multiple linear regression; Pearson's correlation; integrated sensor systems

Citation: Luciani, B.; Braghin, F.; Pedrocchi, A.L.G.; Gandolla, M. Technology Acceptance Model for Exoskeletons for Rehabilitation of the Upper Limbs from Therapists' Perspectives. *Sensors* **2023**, *23*, 1721. https://doi.org/10.3390/s23031721

Academic Editor: Yvonne Tran

Received: 9 January 2023
Revised: 27 January 2023
Accepted: 2 February 2023
Published: 3 February 2023

1. Introduction

Upper-limb exoskeletons offer an innovative solution to support the rehabilitation pathway of patients in need of re-educational motor training. They are external structural mechanisms provided with joints and links that are intended to be coupled with those of the human body [1]. Such structures, which are provided with systems of actuators and sensors, are meant to substitute, support, and enhance the activities and movements of the arm when it has been impaired by paralytic effects related to pathologies such as spinal cord injury or stroke. Some examples of exoskeletons for upper-limb rehabilitation are shown in Figure 1.

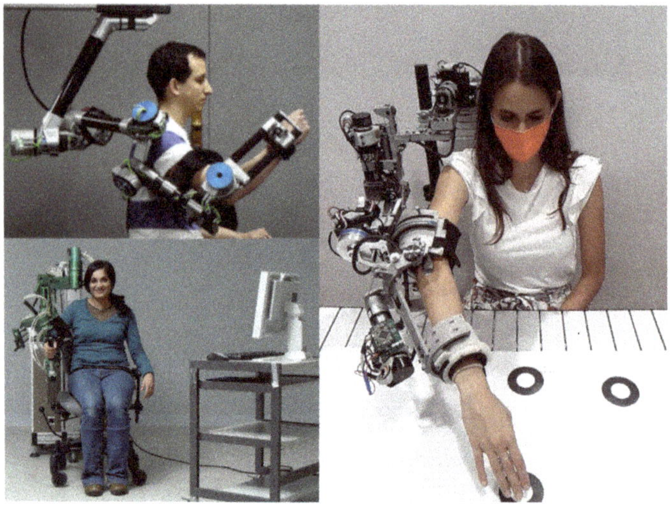

Figure 1. Some examples of upper-limb rehabilitation exoskeletons. The top-left one is ANYexo by ETH Zurich (©2019, Zimmerman et al. from Ref. [2]), the bottom-left one is ARMin (©2010, Nef et al., from Ref. [3]), and the one on the right is AGREE, the prototype from our research group at Politecnico di Milano [4].

Their application for rehabilitation purposes is at least comparable, in terms of efficacy, with conventional therapy, and it produces more functional benefits than other kinds of interventions [5]. The key elements for effective rehabilitation therapy include (i) a large amount of practice, (ii) goal-oriented training, (iii) feedback to the patients, (iv) rewarding and interactive exercises, and (v) individualized therapy [6]. The use of exoskeletons guarantees the fulfillment of all of these requirements, allowing the intensive training sessions with specific therapeutic purposes to be carried out while always adapting to the residual motor skills of the patients [7]. Nowadays, despite all of the advantages that we described, exoskeletons are poorly diffused in daily clinical practice [8]. Therapists tend to find them challenging to use and often do not think that robots can offer an actual improvement to the classical therapy that they perform every day. Moreover, they tend to perceive the presence of an exoskeleton as a barrier to their direct contact with the human limb, reducing the feedback on the patient's conditions. The technology acceptance model (TAM) is a theory that studies the various possible factors influencing users' acceptance of a certain technology [9]. Introduced by Davis in 1989 [10], the TAM was then expanded and applied in various fields to understand what affects human behavior toward a specific technology, and the acquired knowledge was applied to possibly modify the levels of users' acceptance or rejection. Other authors have applied the TAM to study users' intentions to use robotic systems for rehabilitation and assistance, but they always focused only on patients' points of view [11–13]. Therapists, however, are the counterparts of patients, and their opinions on this type of technology can strongly influence its diffusion and use. To the best of our knowledge, no previous studies have been carried out on the acceptability of upper-limb rehabilitation exoskeleton(s) or, in particular, considering therapists as target users. This paper, instead, applies the principles of the TAM to investigate the causes that, according to the therapists' perspectives, limit the acceptability and, consequently, the use of upper-limb exoskeletons in everyday clinical practice. Data to be fed to the model were collected from a questionnaire that we proposed to a pool of therapists, physiotherapists, and physiatrists. We believe that the investigation of this novel point of view can help identify new methods for improving the quality and usability of robotic systems for rehabilitation.

The rest of this paper is organized as follows. Section 2 describes the state of the art of TAM studies, especially those applied to healthcare technologies. The data collection and analysis process that we used for the construction of our TAM is presented in Section 3. Section 4 presents the results of the work, which are discussed in Section 5. Finally, Section 6 draws the conclusions of the work.

2. Related Works

2.1. Technology Acceptance Studies

When Davis proposed the TAM, he wanted to understand why people would choose to use a particular technology (such as emails and web processing systems) in the context of their work or daily life. The TAM's basis comes from physiological theories. The core model by Davis considered two main factors influencing the users' intentions: *perceived usefulness* (PU) and *perceived ease of use* (EOU) [10,14]. The aim was not to determine whether a technology is actually useful or easy to use, but to understand how potential customers perceive it. This perception is, of course, subject to variations due to age, gender, and experience, which are considered the control variables of the model. The TAM owes its success to the fact that it is an easily understandable and simple model. It is, in any case, subject to wide variations in the correlations among the analyzed variables depending on the users and the system under investigation. Furthermore, it starts from the assumption that human beings are rational in their decisions and behavior, which is not always true [15].

Since its introduction, the TAM has undergone several adaptations, such as extensions to include some "custom variables" in the model. These can be added by each author to better explain the main elements of their TAM [9]. The extensions to the model can be grouped into:

- External predictors or prior factors: These have a direct effect on the *perceived usefulness* and the *perceived ease of use* variables. They include self-confidence in technology, prior usage, and anxiety towards a technology.
- Factors coming from other theories: These should increase the reliability of the model. Subjective norms, risk, trust, expectations, and user participation belong to this category.
- Contextual factors: Gender, technological characteristics, and cultural diversity can influence the global effects of the model.
- Usage measures: These are related to attitudes toward technology and actual or expected usage of technology according to user's opinions [9,15].

2.2. The TAM Applied to Healthcare Technologies

Even though the TAM was developed for other contexts, it has become progressively more diffused in the healthcare technology field [12]. According to [16], at least 142 empirical studies were conducted on technology acceptance in healthcare by 2021. They mainly dealt with telemedicine, mobile applications, health websites, e-learning in medical education, and electronic health records, and they interviewed nurses, therapists, and patients—especially older people. Some of the most influential factors that they found in those studies were anxiety, computer self-efficacy, innovativeness, and trust. Studies about robotics for healthcare have included a variety of options: social robots, assistive robots, socially assistive robots, telerobots, and telepresence robots [17,18]. Table 1 summarizes works in the literature about the TAM for healthcare robotics. Especially for what concerns the use of rehabilitative and assistive exoskeletons, no study seems to have investigated therapists' perspectives.

Jankowski and colleagues [11] evaluated long-term changes in technology acceptance during patients' use of a robotic system for stroke rehabilitation and showed how experience could increase the intention to use the technology. Shore and colleagues [13] proposed a selection of possible TAMs to assess the acceptability level among the elderly with respect to the adoption of assistive exoskeletons in their daily lives. Onofrio and colleagues [12] specifically studied patients' opinions on the use of upper-limb exoskeletons for assistance in activities of daily living (ADLs). In particular, this study divided the variables influencing the model output into those related to emotional or functional perspectives and into individual or relational ones. PU and EOU, in this sense, were considered individual and connected to the functional perspective. The subjective norm was a relational variable that was connected to both emotional (if coming from relatives and beloved ones) and functional (if coming from clinicians) perspectives. Anxiety, aesthetics, and trust are factors that come from other studies related to individual emotional perspectives. They concluded that for an exoskeleton to be appreciated by patients, the most crucial aspect is that it must be perceived as useful and inspire confidence in the users.

Table 1. Summary of relevant works from the literature investigating applications of the TAM for robotics in healthcare. Types of technologies and interviewed users are indicated in the last two columns.

Study	Author and Date	Technology	Point of View
[19]	He et al., 2022	Social robots in elderly care facilities	Elderly people
[20]	Turja et al., 2020	Service robots	Healthcare professionals (nurses and doctors)
[21]	Nertinger et al., 2022	Remote assistive robots (Humanoids)	Adult patients
[11]	Jankowski et al., 2020	Rehabilitation end-effector (Bi-Manu-Interact robot)	Stroke patients
[22]	Hall et al., 2019	Assistive robots for activities of daily living	Patients
[13]	Shore et al., 2018	Assistive exoskeletons	Elderly people
[12]	Onofrio et al., 2020	Assistive technologies for neurological motor impairments	Neurological patients

3. Methods

3.1. A Novel Point of View

Despite the existence of multiple studies dedicated to the acceptability of robotic systems (including those introduced in Section 2.2), we could not find any from the literature that considered physiotherapists as the users to be interviewed in relation to this topic. Our study aims to investigate therapists' and physiatrists' perspectives, with the awareness that they, too, are the end users who are asked to interface with exoskeletal technology. Their perception is crucial for guaranteeing the integration of rehabilitation robots into classical therapy sessions.

3.2. Data Collection

Data were collected through the distribution of an anonymous questionnaire (see Appendix A.1). It was distributed both online and in paper form to therapists working in different hospitals in Italy. At the beginning of the survey, we asked the participants to confirm that they belonged to one of the following professional groups: occupational therapists, physiotherapists, or physiatrists. No other eligibility criteria were considered. The data that were collected were anonymized, and the survey was developed according to the law of data protection, according to Art. 13 of the UE 2016/679 norm (General Data Protection Regulation). Its distribution was approved by the Ethical Committee of our university (approval no. 8/2022—16 February 2022).

The questionnaire was composed of twenty-five questions related to the topic of the study. The questions belonged to eight different categories, representing the variables of interest of our TAM:

- *Intention to use*—ITO: Independent variable and output of the model;
- *Perceived usefulness*—PU: How useful the therapists perceive an exoskeleton to be in supporting part of their rehabilitation sessions;

- *Perceived ease of use*—EOU: Level of ease of use of the robotic system in terms of both setting and application during the therapy;
- *Subjective norm*: The extent to which the opinions and suggestions received from other people (e.g., patients, doctors, and other people who the compiler of the application deems reliable) are favorable to the use of exoskeletons;
- *Willingness to interact*: How much do the therapists that are interviewed consider it desirable to interact with the system and to be personally involved in the robotic therapy?
- *Anxiety*: How much do the participants fear that the use of exoskeletons is a source of risk for patients or has negative effects on the therapy?
- *Time saving*: level of perception of an exoskeleton as helpful in saving time and working with more patients;
- *Effort saving*: Level of perception of an exoskeleton as helpful in reducing the physical burden on therapists during the execution of rehabilitation exercises.

According to what was introduced in Section 2, the variables representing the core of the TAM are the EOU, PU, and the output, ITO. The other variables that we included belong to the "prior factors" group (*time saving* and *effort saving*) and to the "factors from other studies" group (*anxiety*, *subjective norm*, and *willingness to interact*). Figure 2 shows the structure of the model and the relations among the variables that we proposed.

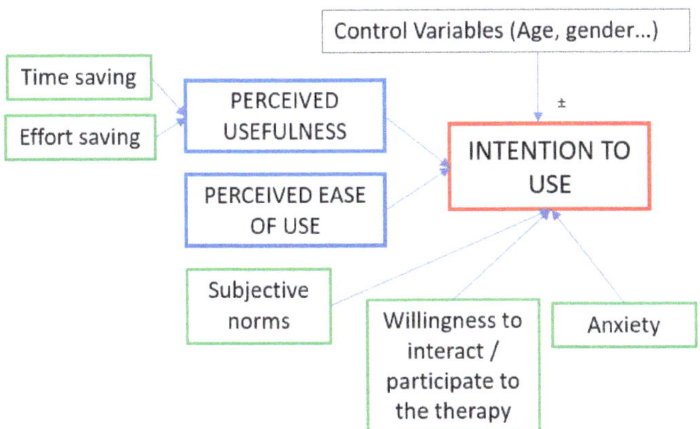

Figure 2. Structure of the TAM. Light blue variables are those of the core of the model, and gray variables are those that we added for our specific study. The dark-blue box represents the output (i.e., the predicted variable).

The following table (Table 2) reports the number of questions referred to each category in the trade-off between the need for a proper number of questions in view of data analysis (i.e., the more, the better) and the total time required to complete the questionnaire (i.e., the less, the better).

The order of presentation of the questions was random and was not related to the categories in order to avoid any possible bias.

Answers were expressed as five levels of agreement with the information provided in each question. They were then converted into numerical values. Scores going from one to five corresponded to the scale of answers from "strongly disagree" to "strongly agree".

At the beginning of the survey, some additional questions were also proposed with the aim of gathering some personal (age, sex, occupation) and attitude (relationship with the technology, previous experiences with rehabilitation exoskeletons) information from the participants.

Table 2. Number of questions in each category of the TAM.

Category (Variable)	No. of Questions
Intention to use	2
Perceived usefulness	3
Anxiety	4
Time saving	2
Effort saving	2
Subjective norm	6
Perceived ease of use	2
Willingness to interact with the system	4

3.3. Data Analysis

Once we collected all of the answers and we built their correspondence to the numerical scores, we grouped them according to the categories. At this point, we analyzed data as follows according to the process proposed by the literature [23–25].

- **Cronbach's alpha**: We evaluated Cronbach's alpha for each variable of the model. Cronbach's alpha is a measure of reliability used to assess the internal consistency of the answers given to questions belonging to the same category. Acceptable reliability is represented by values of alpha ranging from around 0.7 to 0.95 [26].
- **Consistency adjustments**: If some categories obtained alphas lower than 0.7, we further investigated them. We removed the questions that, from an inner correlation study, were revealed to be uncorrelated with the other questions belonging to the same group (under the acceptability threshold of $\rho = 0.3$ for Pearson's coefficient [27]). If the correlation values were acceptable, we kept the questions in the dataset. We concluded by checking whether defections actually improved the alphas of the various categories.
- **Pearson's pairwise correlation**: For every category, we evaluated the mean score from the answers provided by each participant. The literature is unclear about the use of the mean value rather than the median when studying Likert-type categories data [28]. Given that no agreement seems to have been reached, we tried to be consistent with Davis's work, which carried out TAM analyses by using mean scores, Pearson's correlation, and multiple linear regression [14,23]. Using the mean data, we built a correlation matrix to highlight the correlations between the variables involved in the TAM. We studied Pearson's coefficients and their statistical significance through an evaluation of their p-values [23].
- **Multiple regression model**: We created a multiple regression model with the variables of the TAM. ITO was our output variable, while the other categories were the regressors [23].
- **Effects of control variables**: Given that the control variables could influence the results of our model, we decided to divide the data coming from people who had a previous experience with rehabilitation exoskeletons (for both upper and lower limbs) from the data coming from those who did not. In both cases, we analyzed the correlations between the variables and studied the differences.

4. Results

4.1. Participants and Answers

Fifty-five people completed the questionnaire. Table 3 shows a summary of the answers that we collected for some questions that we made to characterize the population.

Table 3. Summary of the answers to the general questions.

Information	Answer
Age	Mean: 37.4 ± 10.1 y.o. Range: 23–59 y.o. Median: 35 y.o
Gender	• 23 men • 31 women • 1 other
Occupation	• 3 occupational therapists • 3 physiatrists • 1 clinical researcher in physiotheraphy • 48 physiotherapists
Already knew what an exoskeleton is?	• 54 yes • 1 no
Already used an exoskeleton?	• 31 yes • 24 no

Figure 3 reports a summary of the statistics of the scores attributed to the twenty-five questions of the survey, divided into the eight aforementioned classes.

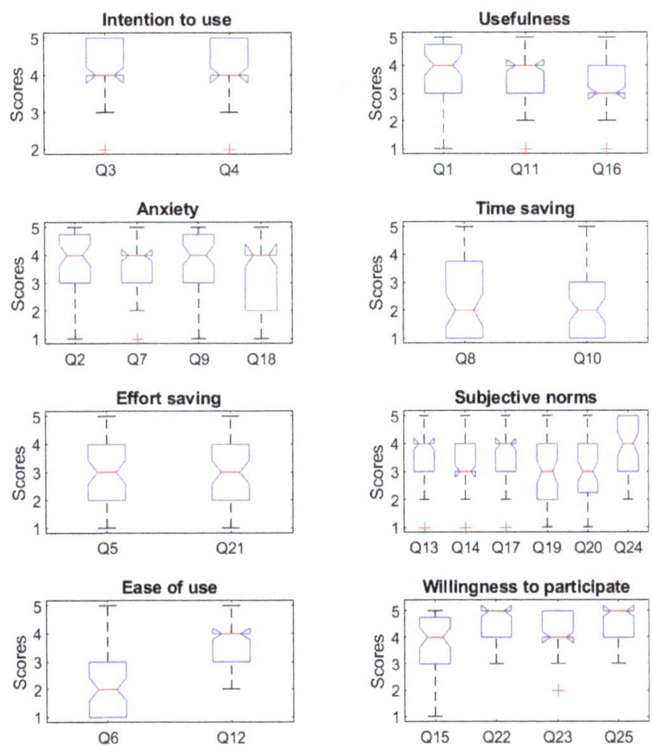

Figure 3. Statistical analysis of the scores given by the 55 users to the questions. Question numbers correspond to those indicated in Appendix A.1. We gathered the questions by category. In each box plot, the central mark indicates the median, and the bottom and top edges of the box indicate the 25th and 75th percentiles, respectively.

4.2. Results of the Data Analysis

4.2.1. Cronbach's Alpha and Consistency Adjustments

The analysis of Cronbach's alpha gave the following results (see Table 4).

Table 4. Cronbach's alpha for the categories of the TAM. ANX: anxiety, ES: effort saving, TS: time saving, SUBJN: subjective norm, WTI: willingness to interact.

	ITO	PU	ANX	TS	ES	SUBN	EOU	WTI
α	0.865	0.829	0.795	0.674	0.464	0.705	−0.123	0.424

The alpha values were acceptable for the ITO, PU, *anxiety*, and *subjective norm*, and they were slightly under the threshold for *time saving*. For all of the categories whose alpha was considered unacceptable, we tested the inner correlations of the answers that we collected. The correlations were evaluated through Pearson's coefficient. From the evaluation of the acceptability of such correlations, we had to eliminate one of the four questions (and its results) related to the variable *willingness to interact*. As a consequence, Cronbach's alpha passed from $\alpha = 0.424$ to $\alpha = 0.759$. The alpha could be under the threshold when a category had too few items (i.e., questions). We presented just two items for *effort saving* and *ease of use*, and this could be the cause of their low alpha values. After the correlation analysis, we decided to keep both of the questions for *effort saving* ($\rho = 0.303 > 0.3$) and to remove the question of the two whose answers had a greater variance for *ease of use* ($\rho = -0.0586 < 0.3$).

4.2.2. Pearson's Correlation

Table 5 reports the results of the analysis of correlation. We have reported in green the correlation coefficients of the couples of variables whose relations are relevant to our TAM.

Table 5. Results of Pearson's correlation analysis on the dataset. If ** is indicated, the correlation is significant at the 0.01 level. If * is indicated, the correlation is significant at the 0.1 level.

	Int. to Use	Perc. Useful.	Anxiety	Time Saving	Effort Saving	Subj. Norm	Ease of Use	Will. to Interact
Intention to use	1	0.765 **	−0.153 *	0.221	0.359	0.440 **	0.347 **	0.230 *
Perceived Usefulness	0.765	1	−0.273	0.408 **	0.582 **	0.312	0.367	0.168

4.2.3. Effects of Control Variables

As anticipated in Section 2.1, categorical variables can have a strong influence on the results of the analysis.

4.2.4. Experience

We decided to consider the *experience* variable, and we split the dataset into two groups. We divided the results coming from participants who had already used exoskeletons from the data coming from those who had not (see Table 6). As indicated in Table 3, 31 out of 55 therapists and physiatrists declared that they had previously used exoskeletons for their therapy sessions.

Table 6. Comparison of the correlations between the variables of the model. Global values refer to the analysis of the whole dataset. "Already used" refers to data coming from participants who had already used exoskeletons, and "never used" refers to data coming from those who did.

Independent Variable	Dependent Variable	Pearsons's Coefficient		
		Global	Already Used	Never Used
Perc. usefulness	ITO	0.765	0.796	0.719
Anxiety	ITO	−0.153	−0.105	−0.221
Subj. norm	ITO	0.440	0.411	0.589
Ease of use	ITO	0.347	0.352	0.414
Will. to interact	ITO	0.230	0.347	0.089
Time saving	PU	0.408	0.522	0.249
Effort saving	PU	0582	0.708	0.322

The scheme of the TAM with the results of the correlation analysis coming from the two groups is presented in Figure 4.

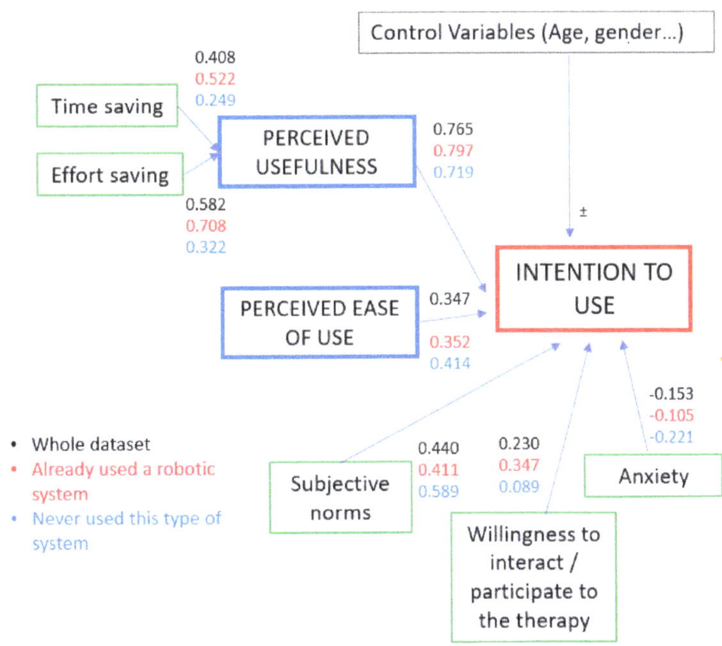

Figure 4. Structure of the TAM with references of the correlations between the various variables involved in the study.

As can be observed, apart from values related to the correlation between PU and ITO, all of the others significantly changed when isolating data coming from already experienced therapists from data coming from those who had never used an exoskeleton before the questionnaire.

4.2.5. Age

Given the relatively wide range of participants' ages, we decided to study how the correlations between our variables of interest changed when passing from younger therapists to older ones. From their answers to the general questions, younger therapists seemed more used to the technology (the statistics of the scores that they attributed to questions related to attitude towards technology, presented in Figure 5, confirmed this). They were also those who were more likely to have come into contact with exoskeletons for rehabilitation during their study path. The global age range was 36 years (from 23 to 59 years old). We divided this range into three equivalent sub-ranges (23 ÷ 34 years old, 35 ÷ 46 years old, and 47 ÷ 59 years old) and built a correlation model for the participants belonging to each age group.

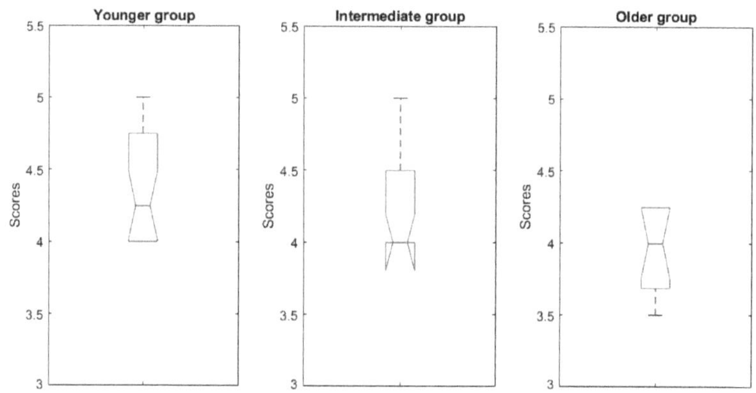

Figure 5. Statistics of the scores given to questions related to participants' attitudes towards technology, divided according to the three age ranges that we identified.

When comparing the three models, the values showing an appreciable monotone age-related trend were related to the correlations between *perceived usefulness* and *subjective norms* with the variable *intention to use*. Table 7 shows Pearson's coefficient values for these two relations.

Table 7. Pearson's coefficients for the correlations between *perceived usefulness* and ITO and between *subjective norms* and ITO for the three age-range groups.

	Younger Group	Intermediate Group	Older Group
Perceived usefulness	0.881	0.722	0.545
Subjective norms	0.290	0.370	0.551

4.2.6. Multiple Regression Model

When trying to infer cause–consequence relationships, the results of a TAM can also be explained with a regression model. Table 8 reports the beta coefficients of the multiple regression that we modeled on the whole dataset, the standard error, and the results of the F-test, which checked whether the model fit significantly better than a degenerated model consisting of only a constant term. The values of the coefficient of determination and adjusted coefficient of determination of the regression were, respectively, $R^2 = 0.649$ and $R^2adj = 0.613$. This meant that the model explained approximately 65% of the variability of the response variable *intention to use*. The results were statistically significant, given that $p_{value} = 3.89 \times 10^{-10}$ (which was under the acceptability threshold). Conversely, we obtained high p_{values} for the pairwise relations of *ease of use*, *willingness to interact*, and *anxiety* with the output variable. These values were, in any case, due to the relatively small sample of participants, and in a future expansion of the study, we can expect to see them be reduced below the acceptability threshold (other works that obtained statistically significant regressions included around 110 participants in their TAMs; see [24]).

We built a second regression model to find the beta coefficients linking *time saving* and *effort saving* (i.e., the prior factors) to *perceived usefulness*. We found that the path coefficients indicating the influences of *time saving* and *effort saving* on PU were, respectively, equal to

$\beta_{TS} = 0.001606$ and $\beta_{ES} = 0.32972$. This second model was also statistically significant ($p_{value} = 0.02$; under the threshold), but it suffered from the limited dataset.

Table 8. Results of the construction of a multiple regression model for our data. Significant values are written in bold.

Independent Variable	Dependent Variable	β	SE	tStat	p Values
Perc. Usefulness	**ITO**	**0.7090**	0.0981	7.2302	2.9203×10^{-9}
Ease of Use	ITO	0.0228	0.0891	0.2556	0.7993
Subj. Norm	**ITO**	**0.2794**	0.1249	2.2423	0.0295
Will. to Interact	**ITO**	**0.2098**	0.1341	1.5643	0.0422
Anxiety	ITO	−0.0154	0.0893	−0.1720	0.8642

5. Discussion

The correlation analysis provided information on the percentages of the variance of the latent variables that were explained by the other variables in the model. The correlations that were relevant to our model were all found to be significant. The regression model that we constructed, on the other hand, was statistically reliable overall, but the causal effects of variables such as *willingness to interact*, *anxiety*, and *ease of use* need to be further investigated with additional data to increase the consistency of the results. We hypothesize the following interpretation of the obtained results:

- As we can see from the global correlation model, the *perceived usefulness* of the exoskeleton explained the majority of the variance of the output (around 77%). This correlation value did not change much when splitting the dataset and comparing the results for the two *experience* subgroups. This point is in agreement with the results obtained from the analysis of patients' opinions about exoskeletons [12]. The regression coefficient for *perceived usefulness* was $\beta = 0.7090$, and it had two-tailed significance. We conclude that, from therapists' perspectives, the first requirement to be considered for application during everyday rehabilitation sessions is to perceive exoskeletons as useful instruments. Nowadays, various benchmarking frameworks are used to evaluate the efficacy of rehabilitation for the motor abilities of neurological patients [29]. They include the use of multiple sensors: from EMG sensors for muscular activation to optoelectronic systems and inertial measurement units for kinematic performance. Applying these systems to the measurement of the improvement of patients who used a rehabilitation robot for their treatment could increase the level of usefulness perceived by therapists.
- The correlations of *time saving* and *effort saving* with PU showed Pearson's coefficients that are, respectively, equal to $\rho = 0.408$ and $\rho = 0.582$, and both correlations were significant. The beta coefficient of *effort saving* was statistically significant, while that related to *time saving* was lower than the acceptability threshold of 0.05. Overall, we can infer that therapists tend to find a robotic system that is able to reduce their physical effort in the execution of the rehabilitation exercises slightly more useful than one that makes them save time (i.e., allowing them to treat a patient while a second one uses the exoskeleton for his/her therapy session).
- The effect of *ease of use* on the output variable in our model proved to be lower than that of *perceived usefulness*. The correlation between *ease of use* and ITO was, in any case, higher for inexperienced therapists, who may have been held back by the prospect of a system that was too complex to learn to use (especially if they did not know about its advantages).
- The correlation between anxiety towards the technological system and ITO, as can be guessed, was negative (and so was the β coefficient). It is interesting to notice that for participants who had already experienced the use of an exoskeleton for their sessions, the negative effect of the *anxiety* variable on ITO was reduced by about 12%. This information suggests that the use of robotic systems for rehabilitation

could be encouraged if therapists have the chance of getting in contact with this kind of technology. Raising the public's level of knowledge, at least in hospitals and rehabilitation centers, could be a good way to increase the level of confidence in this technology and reduce apprehension in those who do not know how it works. In general, it is important to find methods for reducing the negative impact that the fear of not being able to control the therapy has on the willingness to use robotic systems. As we can understand from the answers collected for **Q7** and **Q9** (see Appendix A.1), therapists' *anxiety* was caused by the fact that they felt that they would have no information about how a session conducted by a robot was proceeding if they did not continuously observe the patient. This leads to us losing the advantage in terms of time represented by making one patient use the robot while we work on another patient. An efficient solution to this problem could be investing in complete systems of sensors to be coupled with the exoskeletons and provide reliable and remote feedback to therapists. Other studies proved that feedback is crucial for therapists; rehabilitation experts think that having information about muscular activation and joint positions could be very useful in assessing a patient's conditions [30]. In this sense, surface electromyography sensors can be integrated into the structure of the robot to record the amount of muscular participation of the patients [31]. Precise position sensors can provide real-time information on the 3D configuration of the arm of the patient. Compact force sensors at the interface with the robot [32] can be used to tune the level of assistance provided by the exoskeleton and assure the therapist that the patient is not harmed. The work described in [33] already moves in this direction; it presented a telerehabilitation system that collected haptic data from the interaction between a patient and a robot and provided them to therapists, who felt confident about being distant from the user while they performed rehabilitation with the device.

- When studying the results of the relation between the *subjective norm* and the output variable, we could observe that participants' intention to use exoskeletons had a significant positive correlation with the opinions and suggestions received by doctors and patients (as indicated in the questions that we proposed). The effect of others' opinions on the use of robots for rehabilitation was reduced by almost 18% for the respondents to the survey who had already used such systems. It also seemed to be reduced when studying the answers of younger therapists, who were more experienced with technologies such as exoskeletons.
- The questions that we proposed that were related to the *willingness to interact* category aimed to understand if the therapists preferred dealing with robotic systems that gave them many chances for interaction and personalization of the therapy or leaving the exoskeletons in charge of the organization of the entire therapy. We wanted to understand whether it is better to invest in autonomous devices or if it is preferable to find new ways to make therapists cooperate and exchange information with robots. The correlation analysis between *willingness to interact* and *intention to use* produced a Pearson's coefficient that was statistically irrelevant for inexperienced participants ($\rho = 0.089 < 0.1$). In any case, the correlation increased by 25.8% when studying the answers provided by therapists who had already used an exoskeleton before compiling the questionnaire ($\rho = 0.347$). We can infer that if inexperienced clinicians prefer the advantages offered by a higher level of automatization of the therapy, therapists who have already come into contact with exoskeletal technology consider collaborating with the system more relevant.

6. Conclusions

The study that we conducted aimed, for the first time, to understand the factors that influence the acceptability level of exoskeletons for rehabilitation of upper limbs from therapist's perspectives. Other works from the literature showed that understanding which factors influence users' trust and approval towards a certain technology is crucial for improving the quality of human–robot interaction [16,34]. Such studies focused only on patients' perspectives. With our work, we investigated a new point of view that we believe adds fundamental information for increasing the acceptability and use of rehabilitation robots in clinical environments.

From the analysis of the collected data, we concluded that the perceived level of usefulness was the most relevant aspect influencing users' willingness to use the technology. The usefulness perception and the level of satisfaction towards the functionalities of rehabilitation technology were demonstrated to increase patients' trust in robots [35]. Our work confirms that these aspects are also relevant according to therapists and physiatrists. According to our model, the fact that an exoskeleton can reduce the physical effort required of therapists is an element in favor of their perceived utility. In a potential future version of the model, we could look for other possible factors that increase this perception. Both the anxiety produced by the technology and the importance that is given to what other people (even if they are relevant ones) think decrease when analyzing data from people with previous experience with exoskeletons. This is why we see the need to invest in the diffusion of technology and train rehabilitation professionals on the potential that exoskeletons offer. This conclusion is also supported when comparing answers collected by younger therapists with those from older ones. New generations of physiotherapists who have more experience with exoskeletons and often come into contact with them during their studies seem to be less influenced by others' opinions about this new technology. Our model also supports the conclusion that integrating multi-sensor systems into rehabilitation robots can have an impact on reductions in the effects of *anxiety*, thus increasing therapists' trust in this technology and augmenting the level of *perceived usefulness*. Coupling joint positions with data coming from electromyography, electroencephalography, and force sensors at the interface between the arm and the exoskeleton can tell a therapist about the user's level of participation and performance and allow the therapist to monitor their safety. We should invest in new methods for integrating information coming from all of these sensors and make it easily interpretable by therapists. Evaluating it at the end of the therapeutic path can prove the usefulness of the system, while monitoring it in real-time would reassure the therapist about the progress of the robotic sessions while they are busy with other patients, thus possibly increasing the perceived relevance of the *time saving* variable. This study can be improved by introducing new questions into the survey, which can be formulated as clearly as possible to increase the level of inner consistency of the data that we collect. It can also be expanded by finding new therapists and physiatrists to participate in the study. Increasing the number of answers that we gather would also increase the statistical reliability of the model.

Author Contributions: Conceptualization, B.L. and M.G.; methodology, B.L.; validation, B.L., M.G., F.B. and A.P.; formal analysis, B.L.; investigation, B.L. and M.G.; data curation, B.L.; writing—original draft preparation, B.L.; writing—review and editing, M.G., F.B. and A.P.; supervision, M.G.; project administration, F.B. and A.P. All authors have read and agreed to the published version of the manuscript.

Funding: This research received no external funding.

Institutional Review Board Statement: The study was conducted in accordance with the Declaration of Helsinki and approved by the Ethics Committee of Politecnico di Milano (protocol code 8/2022 - 16/02/2022).

Informed Consent Statement: Informed consent was obtained from all subjects involved in the study.

Data Availability Statement: The mean answer scores given to all of the questions—as summaries of the data that we collected—are reported in Figure 3. The population-characterizing information is reported in Table 3. If needed, the raw data analyzed in this study are available on request from the corresponding author. The raw data have not been published online to be consistent with the privacy statement that was declared at the beginning of the questionnaire and was approved by the Ethical Committee of Politecnico di Milano.

Acknowledgments: The authors thank all of the participants of the survey for their time and precious contributions.

Conflicts of Interest: The authors declare no conflicts of interest.

Abbreviations

The following abbreviations are used in this manuscript:

TAM	Technology acceptance model
PU	Perceived usefulness
EOU	Perceived ease of use
ADL	Activity of daily living
ITO	Intention to use

Appendix A. Research Methods

Appendix A.1. Questionnaire

EXOSKELETONS FOR THE REHABILITATION OF THE UPPER LIMB: QUESTIONNAIRE FOR THE THERAPISTS

Welcome to the questionnaire "Exoskeletons for the rehabilitation of the upper limb: therapists' usability". This survey is part of a research project that is being conducted by our university as part of the realization of exoskeletons for the neuro-rehabilitation of the upper limbs. The questionnaire is anonymous and aimed at healthcare personnel working in the rehabilitation field.

By answering the following questions, you will help us understand what the most critical needs that we should satisfy to realize a robotic system for rehabilitation that is useful and appreciated by therapists are.

GENERAL QUESTIONS

- Have you ever worked/are you working in an environment in which this kind of technology for rehabilitation is used?
- Have you ever used this kind of technology with your patients?
- Have you ever used similar technologies, such as exoskeletons, for the rehabilitation of a hand or a lower limb?

RESEARCH QUESTIONS: We kindly ask you to express how much you agree with the following sentences, ranging from "strongly disagree" to "strongly agree".

- **Q1**: I think that performing some sessions with the help of an exoskeleton for rehabilitation could make the treatment provided to the patients more effective.
- **Q2**: I am afraid that my patient could be hurt if I leave him/her alone during the session with the exoskeleton.
- **Q3**: I'd like to use/I'd love to continue using exoskeletal systems during some of the therapy sessions that I perform with my patients.
- **Q4**: I'd like to integrate the use of an exoskeleton for rehabilitation into the exercises that I propose to my patients daily.
- **Q5**: I think that the use of an exoskeleton makes the therapy sessions less tiring for me.
- **Q6**: I think that using an exoskeleton does not require too much concentration effort on my side.
- **Q7**: I am afraid that the therapy may not go as planned if I do something else while the patient is using the exoskeleton.
- **Q8**: I believe it is positive to be able to save time by dedicating myself to one patient while another performs the therapy while wearing the exoskeleton.
- **Q9**: I am afraid that, if I leave my patient alone while using the exoskeleton, the therapy session may not have the effects that I hope for.
- **Q10**: I appreciate the fact that the exoskeleton gives me the chance to work on more than one patient at the same time.

- **Q11**: I think that introducing some therapy sessions that use an exoskeleton would improve the overall effect of my physiotherapy treatments.
- **Q12**: I believe that I could understand how to approach the technology proposed by an exoskeleton without any problem.
- **Q13**: My patients would like me to offer physiotherapy sessions carried out with the aid of exoskeletons for rehabilitation.
- **Q14**: The doctors who treat my patients expect me to offer physiotherapy sessions carried out with the aid of exoskeletons for rehabilitation.
- **Q15**: I appreciate that a robotic system has many parameters and functionalities to be tuned, so the therapy is sure to be personalized according to my needs.
- **Q16**: I think that the use of an exoskeleton for rehabilitation would make my job easier.
- **Q17**: Clinical studies indicate that I should use exoskeletons for rehabilitation in my physiotherapy sessions, and therefore, I am inclined to use them.
- **Q18**: I am afraid that the exoskeleton could cause damage to the patient if I do not monitor it carefully.
- **Q19**: My patients expect me to offer them physiotherapy sessions carried out with the aid of exoskeletons for rehabilitation.
- **Q20**: The doctors who treat my patients would like me to offer physiotherapy sessions carried out with the aid of exoskeletons for rehabilitation.
- **Q21**: I appreciate that the exoskeletal structure supports and moves the patient's limbs in my place.
- **Q22**: I would like to be the one in charge of deciding how to regulate the behavior of the exoskeleton based on my perception of the conditions of the patient.
- **Q23**: I prefer to have direct control over the course of therapy, without letting a device make decisions for me.
- **Q24**: I am aware of scientific studies that highlight the benefits of using exoskeletons for rehabilitation, and therefore, I am inclined to use them in my work.
- **Q25**: I appreciate the possibility that an exoskeleton has many parameters and functions to adjust so that I can be sure that the therapy is tailored to my needs.

PERSONAL QUESTIONS:

- Age;
- Gender;
- Educational qualification;
- Job;
- Relationship with technology:
  - I like using electronic devices (smartphone, computer, tablet, etc.);
  - I always do my best to learn how to use a new technology that I am not familiar with;
  - I think technology is really important in our everyday life;
  - I am familiar with technological devices (computer, mobile telephone, etc.).

References

1. Perry, J.; Rosen, J.; Burns, S. Upper-Limb Powered Exoskeleton Design. *Mechatronics IEEE/ASME Trans.* **2007**, *12*, 408–417. [CrossRef]
2. Zimmermann, Y.; Forino, A.; Riener, R.; Hutter, M. ANYexo: A Versatile and Dynamic Upper-Limb Rehabilitation Robot. *IEEE Robot. Autom. Lett.* **2019**, *4*, 3649–3656. [CrossRef]
3. Nef, T.; Klamroth-Marganska, V.; Riener, R. *ARMin—Exoskeleton Robot for Stroke Rehabilitation*; Springer: Berlin/Heidelberg, Germany, 2010; Volume 25, pp. 127–130. [CrossRef]
4. Gasperina, S.D.; Longatelli, V.; Panzenbeck, M.; Luciani, B.; Morosini, A.; Piantoni, A.; Tropea, P.; Braghin, F.; Pedrocchi, A.; Gandolla, M. AGREE: An upper-limb robotic platform for personalized rehabilitation, concept and clinical study design. In Proceedings of the 2022 International Conference on Rehabilitation Robotics (ICORR), Rotterdam, The Netherlands, 2022; pp. 1–6. [CrossRef]
5. Mehrholz, J.; Pohl, M.; Platz, T.; Kugler, J.; Elsner, B. Electromechanical and robot-assisted arm training for improving activities of daily living, arm function, and arm muscle strength after stroke. *Cochrane Database Syst. Rev.* **2015**. [CrossRef] [PubMed]
6. Maier, M.; Ballester, B.; Verschure, P. Principles of Neurorehabilitation After Stroke Based on Motor Learning and Brain Plasticity Mechanisms. *Front. Syst. Neurosci.* **2019**, *13*, 74. [CrossRef] [PubMed]

7. Mehrholz, J.; Todhunter-Brown, A.; Pohl, M.; Kugler, J.; Elsner, B. Systematic review with network meta-analysis of randomized controlled trials of robotic-assisted arm training for improving activities of daily living and upper limb function after stroke. *J. Neuroeng. Rehabil.* **2020**, *17*, 83. [CrossRef] [PubMed]
8. Rose, C.G.; Deshpande, A.D.; Carducci, J.; Brown, J.D. The road forward for upper-extremity rehabilitation robotics. *Curr. Opin. Biomed. Eng.* **2021**, *19*, 100291. [CrossRef]
9. Marangunić, N.; Granić, A. Universal Access in the Information Society International Journal Technology acceptance model: A 521 literature review from 1986 to 2013. *Univers. Access Inf. Soc.* **2014**, *14*, 1–15. [CrossRef]
10. Davis, F.D. Perceived Usefulness, Perceived Ease of Use, and User Acceptance of Information Technology. *MIS Q.* **1989**, *13*, 319–340. [CrossRef]
11. Jankowski, N.; Ivanova, E.; Wiehe, L.; Wahl, M. Long-term changes in technology acceptance of a robotic system in stroke treatment: A pilot study. *Curr. Dir. Biomed. Eng.* **2020**, *6*, 20202012. [CrossRef]
12. Onofrio, R.; Gandolla, M.; Lettieri, E.; Pedrocchi, A. *Acceptance Model of an Innovative Assistive Technology by Neurological Patients with a Motor Disability of Their Upper Limb*; Springer: Cham, Switzerland, 2020; pp. 907–913. [CrossRef]
13. Shore, L.; Power, V.; de Eyto, A.; O'Sullivan, L. Technology Acceptance and User-Centred Design of Assistive Exoskeletons for Older Adults: A Commentary. *Robotics* **2018**, *7*, 3. [CrossRef]
14. Davis, F.D. User Acceptance of Information Technology: System Characteristics, User Perceptions and Behavioral Impacts. *Int. J. Man Mach. Stud.* **1993**, *38*, 475–487. [CrossRef]
15. King, W.; He, J. A meta-analysis of the Technology Acceptance Model. *Inf. Manag.* **2006**, *43*, 740–755. [CrossRef]
16. AlQudah, A.A.; Al-Emran, M.; Shaalan, K. Technology Acceptance in Healthcare: A Systematic Review. *Appl. Sci.* **2021**, *11*, 10537. [CrossRef]
17. Tam, L.; Khosla, R. Using assistive social robots in health settings: Implications of personalization for human-machine communication. *Communication* **2016**, *5*, 9.
18. Mois, G.; Beer, J. The Role of Healthcare Robotics in Providing Support to Older Adults: A Socio-ecological Perspective. *Curr. Geriatr. Rep.* **2020**, *9*, 82–89. [CrossRef]
19. He, Y.; He, Q.; Liu, Q. Technology Acceptance in Socially Assistive Robots: Scoping Review of Models, Measurement, and Influencing Factors. *J. Healthc. Eng.* **2022**, *2022*, 6334732. [CrossRef]
20. Turja, T.; Aaltonen, I.; Taipale, S.; Oksanen, A. Robot acceptance model for care (RAM-care): A principled approach to the intention to use care robots. *Inf. Manag.* **2020**, *57*, 103220. [CrossRef]
21. Nertinger, S.; Kirschner, R.; Naceri, D.; Haddadin, S. Acceptance of Remote Assistive Robots with and without Human-in-the-Loop for Healthcare Applications. *Int. J. Soc. Robot.* **2022**. [CrossRef]
22. Hall, A.K.; Backonja, U.; Painter, I.; Cakmak, M.; Sung, M.; Lau, T.; Thompson, H.J.; Demiris, G. Acceptance and perceived usefulness of robots to assist with activities of daily living and healthcare tasks. *Assist. Technol.* **2019**, *31*, 133–140. PMID: 29125804. [CrossRef]
23. Yousafzai, S.; Foxall, G.; Pallister, J. Technology acceptance: A meta-analysis of the TAM: Part 2. *J. Model. Manag.* **2007**, *2*, 281–304. [CrossRef]
24. Paluri, R.; Mehra, S. Exploring the acceptance for e-learning using technology acceptance model among university students in India. *Int. J. Process Manag. Benchmarking* **2015**, *5*, 194–210. [CrossRef]
25. Zhao, J.; Fang, S.; Jin, P. Modeling and Quantifying User Acceptance of Personalized Business Modes Based on TAM, Trust and Attitude. *Sustainability* **2018**, *10*, 356. [CrossRef]
26. Tavakol, M.; Dennick, R. Making Sense of Cronbach's Alpha. *Int. J. Med. Educ.* **2011**, *2*, 53–55. [CrossRef]
27. Ratner, B. The correlation coefficient: Its values range between $+1/-1$, or do they? *J. Target. Meas. Anal. Mark.* **2009**, *17*. [CrossRef]
28. Jamieson, S. Likert Scales: How to (ab) Use Them. *Med. Educ.* **2005**, *38*, 1217–1218. [CrossRef]
29. Longatelli, V.; Torricelli, D.; Tornero, J.; Pedrocchi, A.; Molteni, F.; Pons, J.; Gandolla, M. A unified scheme for the benchmarking of upper limb functions in neurological disorders. *J. Neuroeng. Rehabil.* **2022**, *19*, 102. [CrossRef]
30. Lu, E.C.; Wang, R.H.; Hebert, D.; Boger, J.; Galea, M.P.; Mihailidis, A. The development of an upper limb stroke rehabilitation robot: identification of clinical practices and design requirements through a survey of therapists. *Disabil. Rehabil. Assist. Technol.* **2011**, *6*, 420–431. PMID: 21184626. [CrossRef] [PubMed]
31. Steele, K.; Papazian, C.; Feldner, H. Muscle Activity After Stroke: Perspectives on Deploying Surface Electromyography in Acute Care. *Front. Neurol.* **2020**, *11*, 576757. [CrossRef]
32. Choi, H.; Seo, K.; Hyung, S.; Shim, Y.; Lim, S.C. Compact Hip-Force Sensor for a Gait-Assistance Exoskeleton System. *Sensors* **2018**, *18*, 566. [CrossRef]
33. Baur, K.; Rohrbach, N.; Hermsdörfer, J.; Riener, R.; Klamroth-Marganska, V. The "Beam-Me-In Strategy"—Remote haptic therapist-patient interaction with two exoskeletons for stroke therapy. *J. Neuroeng. Rehabil.* **2019**, *16*, 85. [CrossRef]
34. Hancock, P.; Kessler, T.; Kaplan, A.; Brill, J.; Szalma, J. Evolving Trust in Robots: Specification through Sequential and Comparative Meta-Analyses. *Hum. Factors J. Hum. Factors Ergon. Soc.* **2020**, *63*, 001872082092208. [CrossRef] [PubMed]
35. Koren, Y.; Feingold Polak, R.; Levy-Tzedek, S. Extended Interviews with Stroke Patients Over a Long-Term Rehabilitation Using Human–Robot or Human–Computer Interactions. *Int. J. Soc. Robot.* **2022**, *14*, 1–19. [CrossRef] [PubMed]

sensors

MDPI

Article

Toward Early and Objective Hand Osteoarthritis Detection by Using EMG during Grasps

Néstor J. Jarque-Bou *, Verónica Gracia-Ibáñez, Alba Roda-Sales, Vicente Bayarri-Porcar Joaquín L. Sancho-Bru and Margarita Vergara

Department of Mechanical Engineering and Construction, Universitat Jaume I, E12071 Castellón, Spain
* Correspondence: jarque@uji.es; Tel.: +34-964-728125

Abstract: The early and objective detection of hand pathologies is a field that still requires more research. One of the main signs of hand osteoarthritis (HOA) is joint degeneration, which causes loss of strength, among other symptoms. HOA is usually diagnosed with imaging and radiography, but the disease is in an advanced stage when HOA is observable by these methods. Some authors suggest that muscle tissue changes seem to occur before joint degeneration. We propose recording muscular activity to look for indicators of these changes that might help in early diagnosis. Muscular activity is often measured using electromyography (EMG), which consists of recording electrical muscle activity. The aim of this study is to study whether different EMG characteristics (zero crossing, wavelength, mean absolute value, muscle activity) via collection of forearm and hand EMG signals are feasible alternatives to the existing methods of detecting HOA patients' hand function. We used surface EMG to measure the electrical activity of the dominant hand's forearm muscles with 22 healthy subjects and 20 HOA patients performing maximum force during six representative grasp types (the most commonly used in ADLs). The EMG characteristics were used to identify discriminant functions to detect HOA. The results show that forearm muscles are significantly affected by HOA in EMG terms, with very high success rates (between 93.3% and 100%) in the discriminant analyses, which suggest that EMG can be used as a preliminary step towards confirmation with current HOA diagnostic techniques. Digit flexors during cylindrical grasp, thumb muscles during oblique palmar grasp, and wrist extensors and radial deviators during the intermediate power–precision grasp are good candidates to help detect HOA.

Keywords: hand function; hand osteoarthritis; electromyography; diagnosis; discriminant analysis

Citation: Jarque-Bou, N.J.; Gracia-Ibáñez, V.; Roda-Sales, A.; Bayarri-Porcar, V.; Sancho-Bru, J.L.; Vergara, M. Toward Early and Objective Hand Osteoarthritis Detection by Using EMG during Grasps. *Sensors* **2023**, *23*, 2413. https://doi.org/10.3390/s23052413

Academic Editors: Alessandro Scano, Alfonso Mastropietro and Massimo W. Rivolta

Received: 19 January 2023
Revised: 17 February 2023
Accepted: 20 February 2023
Published: 22 February 2023

1. Introduction

Hand osteoarthritis (HOA) is a chronic disease that may affect hand function. HOA can be found at different degrees in 81% of the elderly population [1,2], with a high prevalence especially in females aged over 50 years. HOA consequences are pain, joint deformity, and reduced hand mobility, strength and function [3,4]. Despite its high prevalence, HOA is a silent degenerating disorder that is clinically treated only in very severe situations. However, applying adequate treatments in early stages would benefit patient quality of life and could prevent disease progression [5].

HOA is usually diagnosed with a combination of different approaches, such as looking at risk factors, clinical presentations (e.g., nodes), radiographic images, laboratory results and subjective questionnaires [6]. Radiographic HOA is often diagnosed with the presence of osteophytes, loss of joint space, juxta articular sclerosis, local erosion and geodes, whereas clinical HOA is defined as the experience of joint pain, stiffness and discomfort [7]. However, symptoms often persist before HOA is observed via these methods [8]. Similarly, disability assessment in HOA is frequently performed using subjective questionnaires based on pain, satisfaction or physical hand function [9]. Therefore, patients' diagnosis and follow-up very much depend on their willingness to recognize their functional limitations [10]. Very little attention has been paid to study forearm and hand muscles in individuals with HOA, perhaps because HOA is considered a problem of the joints. However, periarticular structures such as muscles, ligaments and synovial membranes may also be affected. Some studies have highlighted reduced muscle strength in patients with HOA [11,12]. Subjects diagnosed with HOA usually face increasing difficulty in performing simple handling tasks, reduced strength in lifting a ten pound weight and 10% less hand grasp strength [8]. Nunes et al. [4] found that HOA affects hand function and leads to functional deficits. However, none have studied whether the forearm muscles are significantly affected by HOA or differently used as a result of joint deterioration. Electromyographic (EMG) studies performed with knees have shown that strength deficits in the knee extensors of persons with osteoarthritis are partly due to the decreased recruitment of muscle fibers [13]. If muscle activation differs in the muscles around an osteoarthritic knee, then perhaps there are similar problems in the osteoarthritic hand's forearm muscles. Surface EMG (sEMG) is a noninvasive technique that provides information on both the neural drive (amplitude) and temporal/phasic (shape) activation characteristics of muscles. In patients with osteoarthritis, Aspden [14] found that changes in muscle tissue seem to occur before joint degeneration and negatively affect joint stabilization. Brorsson et al. [15] studied the electromyography activity of extensor digitorum communis (EDC) and flexor carpi radialis (FCR) while female subjects with HOA performed functional activities to compare the results to a group of healthy subjects. They found statistically significant differences between the groups, finding that the HOA group used higher levels of muscle activation in daily tasks than the healthy group, and wrist extensors and flexors appeared to be equally affected. On the contrary, a recent work [16] compared the EMG signals of healthy individuals' forearm muscles to those of HOA patients, and found an activation deficit of the wrist's flexor and extensor muscles, even in initial HOA stages.

The merging of technology and medical science plays an essential role in the prevention, diagnosis and treatment of illnesses and diseases, including patient diagnostic data [17]. Health technology helps clinicians screen abnormalities and contributes to detecting clinical signs [18]. Thus, studying the forearm's muscle signals while performing the most relevant grasps in daily life can lead to the finding of indicators that help detect HOA before the main symptoms appear. Given the large number of muscles that overlap in the forearm [19], it is practically impossible to isolate the surface EMG signal from each one. Therefore, in a previous work [20], we identified seven forearm areas with similar muscle activation patterns that can be used to characterize the forearm's muscle activity while performing ADLs. However, such a study is hindered by the many EMG characteristics and their combinations that can be used to study muscle function. Selecting optimal EMG characteristics and the best combination between features and channels are challenging problems for accomplishing satisfactory classification performance [21,22]. In addition, an increment in EMG characteristics not only introduces redundancy into the function vector, but also increases complexity [21,23]. Of the existing characteristics, and besides muscle activation, new zero crossing (NZC), enhanced wavelength (EWL) and enhanced mean absolute value (EMAV) are those most frequently used in the literature for their efficiency and simplicity [24–26]. To date, no study has examined these EMG characteristics in an attempt to diagnose functional diseases such as HOA. Therefore, a study into the electromyography of forearm muscles (by considering the cited characteristics) would allow researchers to investigate whether subjects with HOA use different neuromuscular control compared to healthy subjects, especially in early disease stages.

One way to characterize the hand is studying hand grasp execution, which is composed mainly of two stages: the reach-to-object and grasp. The force needed to close a hand around and grasp an object is determined by several parameters, such as grasp stability (ability to resist external forces), and grasp security (resistance to slippery objects). Both depend on the grasp configuration [27,28], among other factors. Grasp configuration is determined by the type of applied grasp, and several grasp taxonomies have been reported in the literature in accordance with their purpose [29,30], such as the nine-type classification proposed in [30] for the commonest grasps used in activities of daily living (ADLs). This paper presents a study of the surface electromyography of forearm muscles (considering muscle activation, NZC, EWL and EMAV characteristics from seven representative forearm areas) while performing the commonest grasps used in day-to-day life with a twofold objective: (i) look for muscular forearm areas that are significantly affected or differently used by HOA in EMG terms; (ii) study if the affected EMG characteristics can be used as predictors to detect HOA in an early stage by using different combinations of them in discriminant analyses.

2. Materials and Methods

2.1. Experimental Study

Twenty HOA patients, all right-handed females (72 ± 9 years of age), and 22 right-handed healthy subjects (10 females and 12 males aged 32 ± 9 and 37 ± 11 years, respectively) were recruited for the experiment. All the subjects gave their written informed consent before participating in this study, which was approved by both the University and Hospital Ethics Committees (reference numbers CD/31/2019 and CD/27/2022). HOA patients were recruited by clinicians from among hospital patients showing different disease stages and levels of compromise, and none had undergone surgery. The recruitment was managed by our collaborator P. Granell in the framework of the collaboration agreement signed with the hospital. Healthy subjects were recruited among members of the research team, staff of the university and their relatives, and students, and inclusion criteria included subjects without a history of neuromuscular problems or injuries in the upper arm.

In a comfortable sitting posture, all the participants were asked to exert maximum effort without the help of other muscles other than those of the forearm and hand while performing six representative ADL grasps (Figure 1) based on the grasp taxonomy used in Vergara et al. [30], while recording muscular activity by means of sEMG: two-finger pad-to-pad pinch (P2D); cylindrical grasp (Cyl); lumbrical grasp (Lum); lateral pinch (LatP); oblique palmar grasp (Obl); and intermediate power–precision grasp (IntPP).

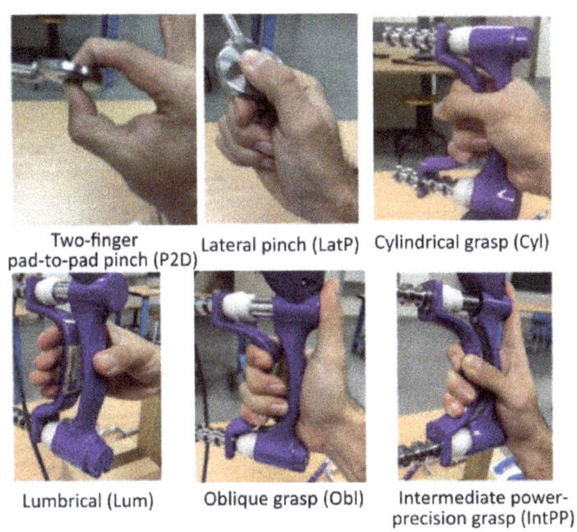

Figure 1. Six grasp types whose maximum grasping effort (MGE) was recorded. Grasp type definitions according to [30].

All the participants performed each grasp following an operator's instructions: with their arm aligned with their trunk and an arm–forearm angle of 90°, the subject held a dynamometer by simulating the grasp to be analyzed without exerting force on it, and then exerted MGE for 2 s while maintaining the posture. Each MGE grasp was performed in a random order, with a 3-min break between each grasp to avoid muscle fatigue. For the normalization of sEMG signals, seven maximum voluntary contraction (MVC) records were measured with each subject (Figure 2): flexion and extension of the wrist, flexion and extension of fingers, ulnar and radial deviation of the wrist, and pronation of the forearm.

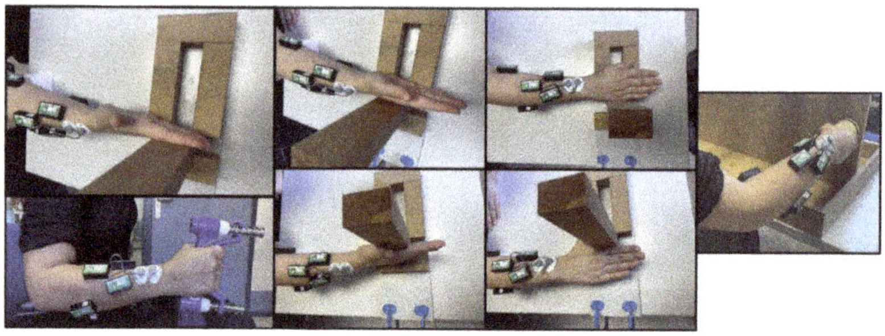

Figure 2. Seven MVC records for the normalization of the muscle activity signal. From left to right: flexion and extension of the fingers, flexion and extension of the wrist, ulnar and radial deviation of the wrist, and pronation of the forearm.

EMG signals were recorded with an 8-channel sEMG Biometrics Ltd. device at a sampling frequency of 1000 Hz. sEMG electrodes and dynamometer signals were synchronized by using the software provided by Biometrics. To place the sEMG electrodes, a grid was drawn on the forearm by using five easily identifiable anatomical landmarks, while the subject sat comfortably with their elbow resting on a table at an arm–forearm angle of 90° and the palm of their hand facing the subject. The grid defined 30 different spots covering the entire forearm surface (Figure 2). Following SENIAM recommendations [31], electrodes were placed longitudinally in the center of seven of these spots based on the spot groups obtained in a previous work [20] (Figure 3). Before placing electrodes, hair was removed by shaving and the skin was cleaned with alcohol.

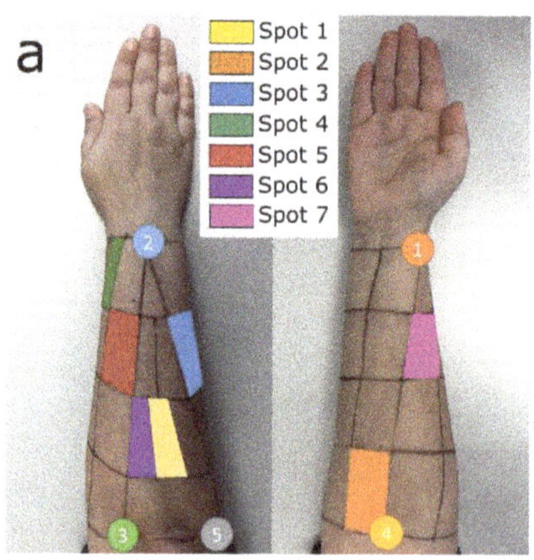

b

	Landmark
1	Styloid process of the radius
2	Ulna head
3	Medial epicondyle of the humerus
4	Centrepoint of the elbow
5	Humeral lateral epicondyle

Figure 3. (**a**) Grid and spot areas selected for the sEMG recordings. (**b**) Five anatomical landmarks used to draw the grid. The signals from these seven spots are related to seven different movements according to [20]. Spot 1: wrist flexion and ulnar deviation (WF_UD); spot 2: wrist flexion and radial deviation (WF_RD); spot 3: digit flexion (DF); spot 4: thumb extension and abduction/adduction (TM); spot 5: finger extension (FE); spot 6: wrist extension and ulnar deviation (WE_UD); spot 7: wrist extension and radial deviation (WE_RD).

2.2. Data Analysis

2.2.1. Computed Parameters

Figure 4 shows the flowchart followed the data analysis. For efficiency and simplicity, those waveform characteristics most frequently used in the literature [21,23] were extracted (muscle activity, NZC, EWL and EMAV).

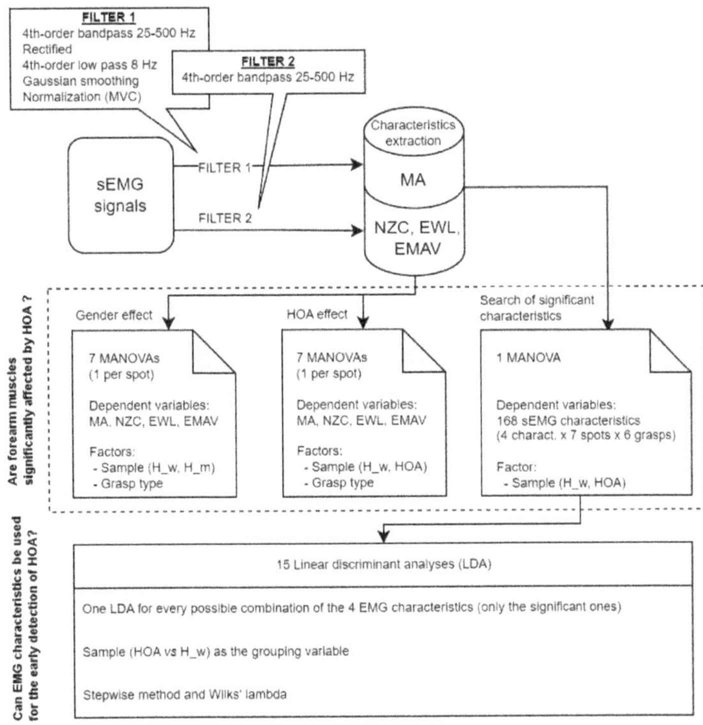

Figure 4. Flowchart of the methodology followed in this paper.

First of all, in order to define NZC, EWL and EMAV, the sEMG signals from the MGE records were filtered with a fourth-order bandpass filter between 25–500 Hz. Waveform characteristics (NZC, EWL, EMAV) were extracted from each record by considering the two seconds during which the maximum effort was made (according to the force signal recorded by the dynamometer). The proposed EMG characteristics were formulated according to [24,32], where x is the sEMG signal (mV), L is signal length and T is the selected threshold:

$$EWL = \sum_{i=2}^{L} \left|(x_i - x_{i-1})^p\right| \tag{1}$$

$$EMAV = \frac{1}{L}\sum_{i=1}^{L} \left|(x_i)^p\right| \tag{2}$$

$$where\ p = \begin{cases} 0.75, & if\ i \geq 0.2L\ and\ i \leq 0.8L \\ 0.50, & otherwise \end{cases}$$

$$NZC = \begin{cases} 1, & if\ x_i > T\ and\ x_{i+1} < T \\ & or\ x_i < T\ and\ x_{i+1} > T\ ; T = 0 \\ 0, & otherwise \end{cases} \tag{3}$$

To determine muscle activity, the sEMG signals from the MGE records were filtered with a fourth-order bandpass filter between 25–500 Hz, rectified, filtered by a fourth-order low pass filter at 8 Hz and smoothed using Gaussian smoothing [33]. Later, they were normalized with the maximal values obtained in any of the seven MVC records measured with each subject. Finally, for each record, the average muscle activity recorded during the 2 s while performing maximum effort was computed for each spot (MA from this point onward).

2.2.2. Global Description

First, as the HOA patients were all female, the gender effect was assessed among the four characteristics in the healthy subjects. For this purpose, the control group subjects were segregated by gender: subsample H_w (10 females) and subsample H_m (12 males). Then, a set of MANOVAs (one for each spot) was applied with the four characteristics as dependent variables, and with subsample and grasp type as factors. The MANOVAs compared subsamples H_w and H_m to assess the gender effect. For an overview of the results, the descriptive statistics (box-and-whisker plot) of all the characteristics (EWL, EMAV, NZC and MA values) per spot and grasp were computed for both subsamples H_w and H_m.

After checking the gender effect, a second set of MANOVAs (one per spot) was applied with the four characteristics as the dependent variable, and with sample (H_w and HOA) and grasp type as factors, as well as their interactions. For an overview of the results, the descriptive statistics (box-and-whisker plot) of all the characteristics (EWL, EMAV, NZC and MA values) per spot and grasp were computed for both samples H_w and HOA patients.

Finally, the four EMG characteristics were converted into 168 variables (4 EMG characteristics x 7 spots x 6 grasps). A MANOVA was performed with the EMG characteristics (168 variables) as dependent variables and sample (H_w and HOA) as the factor to identify which EMG characteristics, spots and grasps presented differences and which of them were, therefore, hindered by HOA.

2.2.3. Can EMG Characteristics Be Used for Early HOA Diagnosis?

As a classification's accuracy depends on the number and type of variables introduced into the model, 15 linear discriminant analyses (LDA) were performed (one for every possible combination of the four EMG characteristics; see Table 1) to locate a small set of predictive parameters to detect HOA. For each LDA, the EMG characteristics of spots and grasps that presented significant differences in the previous MANOVAs were taken as independent variables, and sample (HOA patient vs. H_w) was considered to be the grouping variable. Table 1 shows all the possible combinations of the EMG characteristics proposed in each LDA.

For LDAs, the stepwise method was used (predictors were entered sequentially), which searches for the highest correlated predictors. In particular, Wilks' lambda was employed, which checks how well each independent variable (potential predictor) contributes to the model: 0 means total discrimination, and 1 denotes no discrimination. Each independent variable was tested by placing it in the model and then taking it out to generate a Λ statistic. The significance of change in Λ was measured using an F-test. The variable was entered in the model if the significance level of its F value was lower than the entry value (0.05), and it was removed if the significance level was higher than the removal value (0.1). Classification ability goodness was checked by a leave-one-out cross-validation, which repeats the analysis by taking one case out in each repetition. In addition, the percentage of correctly and incorrectly classified patients was checked.

Table 1. All the different combinations of the EMG characteristics of all the performed LDAs.

	NZC	EWL	EMAV	MA
LDA1	X			
LDA2		X		
LDA3			X	
LDA4				X
LDA5	X	X		
LDA6	X		X	
LDA7	X			X
LDA8		X	X	
LDA9		X		X
LDA10			X	X
LDA11	X	X	X	
LDA12	X		X	X
LDA13	X	X		X
LDA14		X	X	X
LDA15	X	X	X	X

3. Results

3.1. *Are Forearm Muscles Significantly Affected or Differently Used by HOA in Terms of EMG Characteristics?*

Table S1 of supplementary material presents the statistics (average and SD) of all the EMG characteristics for each spot, grasp and group. The next sections present that data in terms of box-and-whisker plots.

3.1.1. Gender Effect in the Control Group Subjects

Figures 5 and 6 show the box-and-whisker plots of the EMG characteristics segregated by gender and calculated for every grasp in each sample. As expected, the statistics shown in the box-and-whisker plots and the results of the first set of MANOVAs (Table 2) when comparing H_w and H_m found that gender significantly affected most of the EMG characteristics ($p < 0.05$), except for the ulnar deviators of the wrist (WR_UD and WE_UD). NZC was less affected by gender, and was affected only in FE and WE_RD. As gender affected the EMG characteristics, and to compare both target populations, from this point onward we only considered subsample H_w for the subsequent analyses as being representative of the control group.

Table 2. Results in columns of the set of MANOVAs. The EMG characteristics that significantly differed between H_w and H_m are indicated. Abbreviations are defined the Figure 5 caption.

	Spot						
Factor	**WF_UD**	**WF_RD**	**DF**	**TM**	**FE**	**WE_UD**	**WE_RD**
Gender		EWL EMAV MA	EWL EMAV MA	EWL EMAV MA	NZC EWL EMAV MA		NZC
Grasp type	NZC EWL EMAV MA	NZC EWL EMAV MA	NZC EWL EMAV MA	NZC EWL EMAV MA	NZC EWL EMAV MA	EWL EMAV MA	NZC EWL EMAV MA
Interaction			EWL EMAV MA				

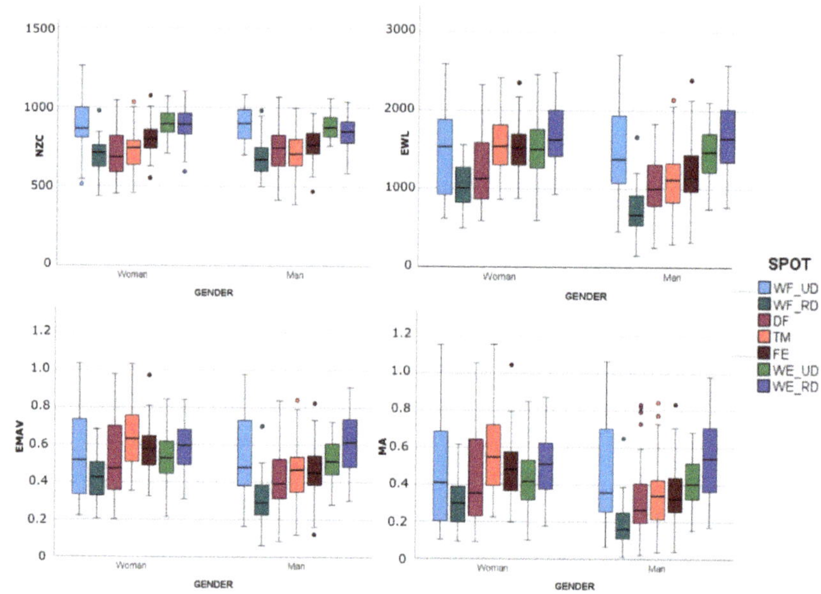

Figure 5. Box–and-whisker plots (horizontal central mark in the boxes is the median; the edges of the boxes are the 25th and 75th percentiles; whiskers extend to 1.5 times the interquartile range and outliers are marked as color circles) of the EMG characteristics segregated by gender and calculated per spot in each sample. Wrist flexion and ulnar deviation (WF_UD); wrist flexion and radial deviation (WF_RD); digit flexion (DF); thumb extension and abduction/adduction (TM); finger extension (FE); wrist extension and ulnar deviation (WE_UD); wrist extension and radial deviation (WE_RD).

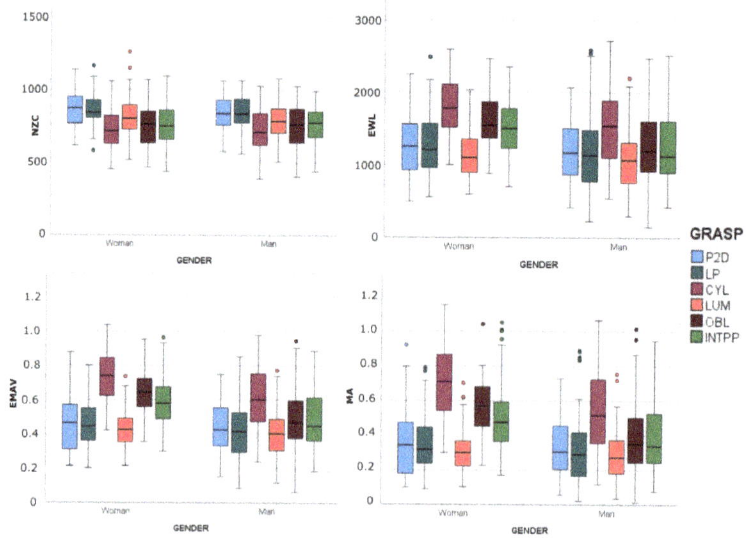

Figure 6. Box-and-whisker plots (horizontal central mark in the boxes is the median; the edges of the boxes are the 25th and 75th percentiles; whiskers extend to 1.5 times the interquartile range and outliers are marked as color circles) of the EMG characteristics segregated by gender and calculated per grasp in each sample.

3.1.2. HOA Effect

Figures 7 and 8 show the statistics of the EMG characteristics segregated by sample (HOA and H_w) and calculated per grasp in each sample by means of box-and-whisker plots. The results of the MANOVAs (Table 3) for comparing samples H_w and HOA show that group and grasp significantly affected most of the EMG characteristics ($p < 0.05$). Once again, NZC was that less affected by sample and its interaction with grasp.

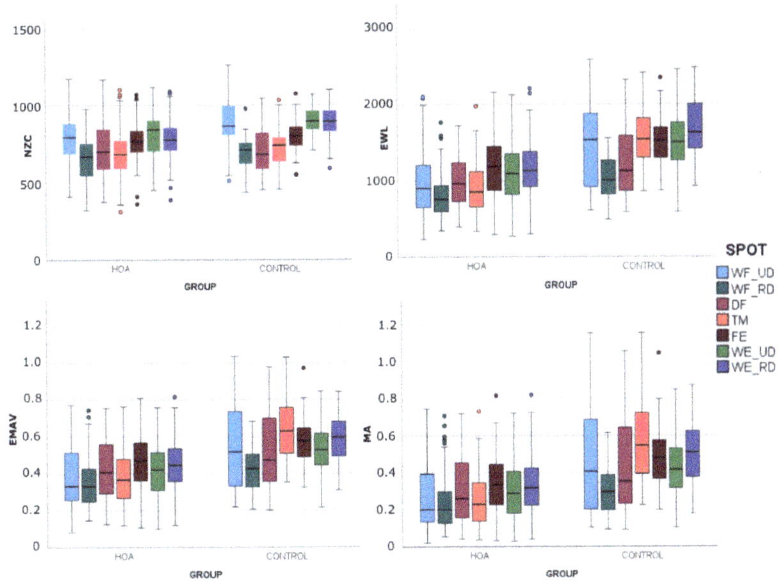

Figure 7. Box-and-whisker plots (horizontal central mark in the boxes is the median; the edges of the boxes are the 25th and 75th percentiles; whiskers extend to 1.5 times the interquartile range and outliers are marked as color circles) of the EMG characteristics segregated by group and calculated per spot in each sample. Abbreviations are defined the Figure 5 caption.

Table 3. Results in columns of the set of MANOVAs. The EMG characteristics that significantly differ between the H_w and HOA patients samples are indicated. Abbreviations are defined in the Figure 5 caption.

	Spot						
Factor	**WF_UD**	**WF_RD**	**DF**	**TM**	**FE**	**WE_UD**	**WE_RD**
Sample	NZC EWL EMAV MA	EWL EMAV MA	EWL EMAV MA	EWL EMAV MA	NZC EWL EMAV MA	NZC EWL EMAV MA	NZC EWL EMAV MA
Grasp type	NZC EWL EMAV MA	NZC EWL EMAV MA	NZC EWL EMAV MA	NZC EWL EMAV MA	NZC EMAV MA	EWL EMAV MA	NZC EWL EMAV MA
Interaction	EWL EMAV MA	EWL EMAV MA	EWL EMAV MA	EWL EMAV MA	MA		EMAV MA

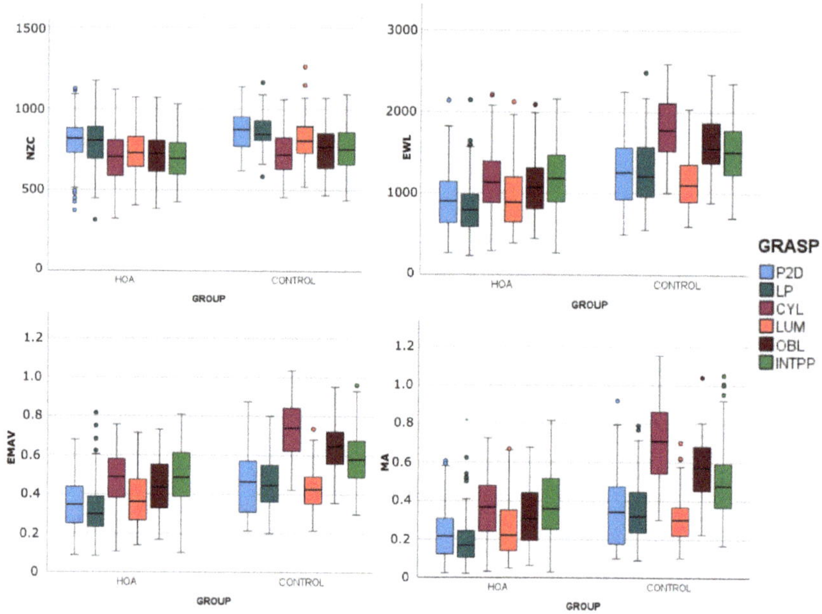

Figure 8. Box-and-whisker plots (horizontal central marks in the boxes correspond to the median; the edges of the boxes are the 25th and 75th percentiles; whiskers extend to 1.5 times the interquartile range and outliers are marked as color circles) of the EMG characteristics segregated by group and calculated per grasp in each sample.

Table 4 shows the results of the MANOVA ($p < 0.05$) performed to look for the EMG characteristics with significant differences between H_w and HOA. Regarding grasp types, Lum and IntPP were the grasps with the fewest significant variables in the different spots. On the contrary, Cyl and Obl were the grasps with the most significant variables. WF_UD, TM and WE_RD were the spots with the most significant variables, while WF_RD and DF were those with the least significant variables. Of the initial 168 variables (4 EMG characteristics $\times$ 7 spots $\times$ 6 grasps), 100 presented significant differences between samples. These 100 variables were used in the next LDAs.

3.2. Can EMG Characteristics Be Used for the Early Detection of HOA?

Table 5 shows the results of the discriminant analyses. The models in the table can be used to calculate discriminant scores F for each subject in such a way that when F is positive, the prediction is a healthy subject, and if F is negative, the subject has HOA. Superscripts ^{i,j} correspond to spot i, grasp j. The success ratio of the prediction using these discriminant scores ranged from 73.3% to 100%.

LDA1 had the worst success ratio, which was composed of only the NZC values. LDA2, LDA4, LAD5, LDA9, LDA10 and LDA14 had the highest success ratios (100%), with LDA4, LDA9, LDA10 and LDA14 requiring fewer characteristics and grasps with similar resulting models. Some LDAs obtained the same model, as can be observed in Table 5. LDA3 and LDA8 were the models with the fewest characteristics and required grasps (thumb muscles and Cyl grasp) and had a high success ratio (93.3%).

Table 4. Results of the MANOVA with the combined variable grasp x spot x EMG characteristic as input. Variables that significantly differ between H_w and HOA patients depend on the spot and grasp type. Abbreviations are defined in the Figure 5 caption.

	Spot						
Grasp Type	**WF_UD**	**WF_RD**	**DF**	**TM**	**FE**	**WE_UD**	**WE_RD**
P2D	EWL	EWL EMAV MA		EWL EMAV MA	EWL EMAV MA	EWL EMAV MA	NZC EWL EMAV MA
LatP	EWL EMAV MA	EWL EMAV MA		NZC EWL EMAV MA	NZC EWL EMAV MA	EWL EMAV MA	NZC EWL EMAV MA
CyL	EWL EMAV MA	EWL EMAV MA	EWL EMAV MA	EWL EMAV MA	EWL EMAV MA	EWL EMAV MA	EWL EMAV MA
Lum	EWL			EWL EMAV MA		NZC	NZC EWL
Obl	EWL EMAV MA	EWL EMAV	EWL EMAV MA	EWL EMAV MA	EWL EMAV MA	NZC EWL EMAV MA	NZC EWL EMAV MA
IntPP	EWL EMAV MA			EWL EMAV MA	NZC	NZC	NZC EWL EMAV MA

Table 5. The success ratios and models obtained from the different performed LDAs.

	Success Ratio	Model
LDA1	73.3%	$0.013 \cdot NZC^{WE\_RD,Lum}$ -10.383
LDA2 & LDA5	100%	$0.002 \cdot EWL^{WF\_UD,Obl} + 0.003 \cdot EWL^{TM,LatP} + 0.004 \cdot EWL^{TM,Cyl} - 0.004 \cdot EWL^{TM,Lum} - 0.002 \cdot EWL^{WE\_RD,P2D} - 4.198$
LDA3 & LDA8	93.3%	$8.065 \cdot EMAV^{TM,Cyl} - 4.399$
LDA4	100%	$3.163 \cdot MA^{DF,Obl} + 8.121 \cdot MA^{TM,Cyl} - 4.986 \cdot MA^{WE\_RD,IntPP} - 3.232$
LDA6 & LDA11	93.3%	$7.902 \cdot EMAV^{TM,Cyl} + 0.005 \cdot NZC^{WE\_RD,IntPP} - 8.483$
LDA7	93.3%	$6.514 \cdot MA^{TM,Cyl} + 0.006 \cdot NZC^{WE\_RD,IntPP} - 7.512$
LDA9	100%	$0.002 \cdot EWL^{DF,Obl} + 8.542 \cdot MA^{TM,Cyl} - 5.566 \cdot MA^{WE\_RD,IntPP} - 4.140$
LDA10	100%	$3.277 \cdot MA^{DF,Obl} + 8.215\ MA^{TM,Cyl} - 6.313 \cdot EMAV^{WE\_RD,IntPP} - 2.127$
LDA12, LDA13 & LDA15	93.3%	$6.514 \cdot MA^{TM,Cyl} + 0.006 \cdot NZC^{WE\_RD,IntPP} - 7.512$
LDA14	100%	$0.002 \cdot EWL^{DF,Obl} + 8.681 \cdot MA^{TM,Cyl} - 7.112 \cdot EMAV^{WE\_RD,IntPP} - 2.938$

4. Discussion

In this work, an EMG study of forearm muscles (considering muscle activation, NZC, EWL and EMAV characteristics from seven representative forearm areas) while performing the commonest grasps of ADL was carried out with a twofold objective: (1) check if the EMG characteristics obtained from different forearm areas during grasps presented significant differences in HOA patients; (2) if these significant EMG characteristics can be used to help diagnose HOA.

4.1. Are Forearm Muscles Significantly Affected or Differently Used by HOA in EMG Terms?

First, and as expected, gender significantly affected the EMAV, MA and EWL characteristics, except for the ulnar deviators of the wrist (WR_UD and WE_UD). NZC was the least affected by gender, and was only affected in finger/wrist extensors and radial deviators (FE and WE_RD). Similarly, the vast majority of the EMG characteristics were also affected by condition (healthy women and HOA patients) and grasp type. In addition, NZC was once again the least affected. This seems reasonable because NZC is meant to approximate signal frequency, unlike EMAV, MA and EWL, which are related to signal amplitude and are, consequently, related more to grasping force, the decrease in which is a HOA symptom [34].

4.2. Can EMG Characteristics Be Used to Detect HOA Early?

From the LDA results, we observed that the EMG characteristics could help in detecting HOA. From all the tested combinations, six models presented the highest success ratio (100%), some of which presented similarities:

- LDA2 and LDA5 were composed only of EWL values and required recording EMG signals from wrist flexors, ulnar deviators, thumb muscles, wrist extensors and radial deviators while performing all the grasps except the intermediate power–precision grasp. Not requiring MA characteristics would prevent MVC recordings and simplify the diagnosis method;
- LDA4, LDA9, LD10 and LDA14 required different combinations of EMG characteristics, but always from the same muscular forearm spots and grasps: digit flexors, thumb muscles, wrist extensors and radial deviators while performing the cylindrical, oblique-palmar grasp and intermediate power–precision grasp.

From the other models, LDA3 and LDA8 were seen to be the models with the fewest characteristics and required grasps (thumb muscles and Cyl grasp), and they also had a high success ratio (93.3%). This means that recording only one muscle spot while performing the cylindrical grasp could suffice to detect 93.3% of cases. Furthermore, not requiring MA characteristics would prevent MVC recordings and simplify the diagnosis method.

However, there were also differences in these models regarding the employed EMG characteristics:

- NZC values did not well-discriminate HOA patients;
- Muscle activity (MA) did not require any other characteristic to discriminate HOA patients, but required MVC recordings;
- EWL could very accurately discriminate, but needed information of more grasps;
- EMAV could very accurately discriminate, but always had to be accompanied by other EMG characteristics (MA or EWL).

There are few previous works that study different muscle activations in HOA, and comparisons with them must be made with caution, since the measurement protocols and the analyses performed are not the same. In [35], intrinsic muscles were considered in a fine manipulation activity, analyzing integrated activation as the only indicator and reaching the conclusion that although there were differences, when considering the longer execution time required by HOA patients, these differences disappeared. Despite [15] concluding that HOA patients require greater muscle activation for activities such as writing or cutting with scissors, ref. [16] indicates that this activation is lower. However, although in both studies the signal was normalized, in [15], they do not indicate the application of any filtering. The novelty and importance of our work lies in considering different grasp types representative of ADLs; muscles whose activity is also representative of these ADLs; and indicators based not only on the amplitude of muscle activity, but also on the frequency domain of the signal. Therefore, ours is a broadening study in the pursuit of checking for differences in muscle activity due to HOA. The equations provided in this work show that the digit flexors during the cylindrical grasp, thumb muscles during the oblique palmar grasp and wrist extensors and radial deviators during the intermediate power–precision grasp were much more significant for detecting HOA than the other muscles and grasps. The recent work [16] found that early-stage HOA may contribute to the activation deficit of the flexor and extensor muscles of the wrist. The results herein reiterate wrist extensors, along with thumb muscles and digit flexors, as possible muscle indicators for detecting HOA.

5. Conclusions

This paper proposes using EMG characteristics to identify discriminant functions for the early detection of HOA. The discriminant results show very high success rates (between 93.3% and 100%), which suggests that EMG can be used as a preliminary step to confirm current HOA diagnostic techniques. In particular, digit flexors during the cylindrical grasp, thumb muscles during the oblique palmar grasp, and wrist extensors and radial deviators during the intermediate power–precision grasp are good candidates to help detect HOA. These results highlight the possibilities of merging technology and medical science as an essential role in the prevention, diagnosis and treatment of illnesses and diseases such as HOA. Furthermore, the results presented herein may help improve the control of hand prostheses and assistive exoskeletons, especially those intended for HOA patients. As limitations, note that there are other EMG parameters that have not been considered, that the sample of participants is limited both in number and degree of HOA compromise, and that we do not know what would happen with other pathologies (that could give similar indicators and be mislabeled as HOA). More studies are needed to check if these differences in EMG characteristics between healthy and HOA patients are present before strength loss in HOA.

Supplementary Materials: The following supporting information can be downloaded at: https://www.mdpi.com/article/10.3390/s23052413/s1.

Author Contributions: J.L.S.-B. and M.V. conceived the study, designed the experiments and contributed to the writing phase. N.J.J.-B., A.R.-S., V.G.-I. and V.B.-P. performed the experiments and data analysis and helped with the writing phase. All authors have read and agreed to the published version of the manuscript.

Funding: This research was part of the project PGC2018-095606-B-C21, funded by MCIN/AEI/10.13039/501100011033 and "ERDF A way of making Europe" and also of projects CIGE/2021/024 and UJI-A2021-03. These authors wish to thank our collaborator, Pablo Granell, who actively helped us during this study thanks to the collaboration agreement signed with the Consorci Hospitalari Provincial of Castellón de la Plana (Spain).

Institutional Review Board Statement: The study was conducted according to the guidelines of the Declaration of Helsinki and was approved by the Universitat Jaume I Ethics Committee.

Informed Consent Statement: Informed consent was obtained from all the subjects involved in the study.

Data Availability Statement: The data presented in this study are available upon request from the corresponding author.

Conflicts of Interest: The authors declare no conflict of interest.

Abbreviations

ADLs	Activities of daily living
Cyl	Cylindrical grasp
DF	Digit flexion
EMAV	Enhanced mean absolute value
EMG	Electromyography
EWL	Enhanced wavelength
FE	Finger extension
HOA	Hand osteoarthritis
IntPP	Intermediate power–precision grasp
LatP	Lateral pinch

Lum	Lumbrical grasp
MGE	Maximum grasping effort
MVC	Maximum voluntary contraction
NZC	New zero crossing
Olb	Oblique palmar grasp
P2D	Two-finger pad-to-pad pinch
TM	Thumb extension and abduction/adduction
WE_UD	Wrist extension and ulnar deviation
WE_RD	Wrist extension and radial deviation
WF_RD	Wrist flexion and radial deviation
WF_UD	Wrist flexion and ulnar deviation

References

1. Blanco, F.J.; Silva-Díaz, M.; Quevedo Vila, V.; Seoane-Mato, D.; Pérez Ruiz, F.; Juan-Mas, A.; Pego-Reigosa, J.M.; Narváez, J.; Quilis, N.; Cortés, R.; et al. Prevalence of symptomatic osteoarthritis in Spain: EPISER2016 study. *Reumatol. Clin.* **2020**, *17*, 461–470. [CrossRef]
2. Kloppenburg, M.; Kwok, W.Y. Hand osteoarthritis a heterogeneous disorder. *Nat. Rev. Rheumatol.* **2012**, *8*, 22–31. [CrossRef] [PubMed]
3. Fautrel, B.; Hilliquin, P.; Rozenberg, S.; Allaert, F.A.; Coste, P.; Leclerc, A.; Rossignol, M. Impact of osteoarthritis: Results of a nationwide survey of 10,000 patients consulting for OA. *Jt. Bone Spine* **2005**, *72*, 235–240. [CrossRef] [PubMed]
4. Nunes, P.M.; de Oliveira, D.G.; Aruin, A.S.; dos Santos, M.J. Relationship between hand function and grip force control in women with hand osteoarthritis. *J. Rehabil. Res. Dev.* **2012**, *49*, 855. [PubMed]
5. Hermann, W.; Lambova, S.; Müller-Ladner, U. Current Treatment Options for Osteoarthritis. *Curr. Rheumatol. Rev.* **2018**, *14*, 108–116. [CrossRef] [PubMed]
6. Leung, G.J.; Rainsford, K.D.; Kean, W.F. Osteoarthritis of the hand I: Aetiology and pathogenesis, risk factors, investigation and diagnosis. *J. Pharm. Pharmacol.* **2014**, *66*, 339–346. [CrossRef]
7. Zhang, W.; Doherty, M.; Leeb, B.F.; Alekseeva, L.; Arden, N.K.; Bijlsma, J.W.; Dincer, F.; Dziedzic, K.; Hauselmann, H.J.; Kaklamanis, P.; et al. EULAR evidence-based recommendations for the diagnosis of hand osteoarthritis: Report of a task force of ESCISIT. *Ann. Rheum. Dis.* **2008**, *68*, 8–17. [CrossRef]
8. Zhang, Y. Prevalence of Symptomatic Hand Osteoarthritis and Its Impact on Functional Status among the Elderly: The Framingham Study. *Am. J. Epidemiol.* **2002**, *156*, 1021–1027. [CrossRef]
9. Kroon, F.P.B.; Boersma, A.; Boonen, A.; van Beest, S.; Damman, W.; van der Heijde, D.; Rosendaal, F.R.; Kloppenburg, M. Performance of the Michigan Hand Outcomes Questionnaire in hand osteoarthritis. *Osteoarthr. Cartil.* **2018**, *26*, 1627–1635. [CrossRef]
10. Sierakowska, M.; Wysocka-Skurska, I.; Kułak, W. Identification of demographic factors and health problems that affect the acceptance of disease and health behaviors of patients with osteoarthritis. *PeerJ* **2017**, *2017*, e3276. [CrossRef]
11. Bagis, S.; Gunsah, A.E.; Ae, S.; Yapici, Y.; Bolgen, O.; Ae, C.; Erdogan, C. The effect of hand osteoarthritis on grip and pinch strength and hand function in postmenopausal women. *Clin. Rheumatol.* **2003**, *22*, 420–424. [CrossRef] [PubMed]
12. Kalichman, L.; Hernández-Molina, G. Hand Osteoarthritis: An Epidemiological Perspective. *Semin. Arthritis Rheum.* **2010**, *39*, 465–476. [CrossRef] [PubMed]
13. Lewek, M.D.; Rudolph, K.S.; Snyder-Mackler, L. Quadriceps Femoris Muscle Weakness and Activation Failure in Patients with Knee Osteoarthritis. *J. Orthop. Res.* **2011**, *22*, 110–115. [CrossRef]
14. Aspden, R.M. Osteoarthritis: A problem of growth not decay? *Rheumatology* **2008**, *47*, 1452–1460. [CrossRef]
15. Brorsson, S.; Nilsdotter, A.; Thorstensson, C.; Bremander, A. Differences in muscle activity during hand-dexterity tasks between women with arthritis and a healthy reference group. *BMC Musculoskelet. Disord.* **2014**, *15*, 154. [CrossRef]
16. Tossini, N.B.; Lessi, G.C.; Simões Zacharias, A.L.; Corrêa e Silva, G.R.; Seraphim Abrantes, L.S.; Mendes da Silva Serrão, P.R. Impairment of electrical activation of wrist flexor and extensor muscles during gripping and functional activities in the early stage of hand osteoarthritis: A cross-sectional study. *J. Hand Ther.* **2021**, *34*, 109–115. [CrossRef]
17. Harada, T.; Miyagami, T.; Kunitomo, K.; Shimizu, T. Clinical decision support systems for diagnosis in primary care: A scoping review. *Int. J. Environ. Res. Public Health* **2021**, *18*, 8435. [CrossRef] [PubMed]
18. Rizzi, M.; Guaragnella, C. A Decision Support System for Melanoma Diagnosis from Dermoscopic Images. *Appl. Sci.* **2022**, *12*, 7007. [CrossRef]
19. Yu, H.-L.; Chase, R.A.; Robert, A.; Strauch, B. *Atlas of Hand Anatomy and Clinical Implications*; Mosby: St. Louis, MO, USA, 2004; ISBN 0815179278.
20. Jarque-Bou, N.J.; Vergara, M.; Sancho-Bru, J.L.; Alba, R.-S.; Gracia-Ibáñez, V. Identification of forearm skin zones with similar muscle activation patterns during activities of daily living. *J. NeuroEng. Rehabil. Under Rev.* **2018**, *15*, 91. [CrossRef]
21. Too, J.; Abdullah, A.R.; Saad, N.M.; Tee, W. EMG Feature Selection and Classification Using a Pbest-Guide Binary Particle Swarm Optimization. *Computation* **2019**, *7*, 12. [CrossRef]

22. Tepe, C.; Demir, M.C. Real-Time Classification of EMG Myo Armband Data Using Support Vector Machine. *IRBM* **2022**, *43*, 300–308. [CrossRef]
23. Phinyomark, A.; Quaine, F.; Charbonnier, S.; Serviere, C.; Tarpin-Bernard, F.; Laurillau, Y. EMG feature evaluation for improving myoelectric pattern recognition robustness. *Expert Syst. Appl.* **2013**, *40*, 4832–4840. [CrossRef]
24. Too, J.; Abdullah, A.R.; Saad, N.M. Classification of Hand movements based on discrete wavelet transform and enhanced feature extraction. *Int. J. Adv. Comput. Sci. Appl.* **2019**, *10*, 83–89. [CrossRef]
25. Arunraj, M.; Srinivasan, A.; Arjunan, S.P. A Real-Time Capable Linear Time Classifier Scheme for Anticipated Hand Movements Recognition from Amputee Subjects Using Surface EMG Signals. *IRBM* **2021**, *42*, 277–293. [CrossRef]
26. Sravani, C.; Bajaj, V.; Taran, S.; Sengur, A. Flexible Analytic Wavelet Transform Based Features for Physical Action Identification Using sEMG Signals. *IRBM* **2020**, *41*, 18–22. [CrossRef]
27. Cutkosky, M.R. On grasp choice, grasp models, and the design of hands for manufacturing tasks. *IEEE Trans. Robot. Autom.* **1989**, *5*, 269–279. [CrossRef]
28. Cipriani, C.; Zaccone, F.; Micera, S.; Carrozza, M.C. On the shared control of an EMG-controlled prosthetic hand: Analysis of user-prosthesis interaction. *IEEE Trans. Robot.* **2008**, *24*, 170–184. [CrossRef]
29. Feix, T.; Romero, J.; Schmiedmayer, H.-B.; Dollar, A.M.; Kragic, D. The GRASP Taxonomy of Human Grasp Types. *IEEE Trans. Hum.-Mach. Syst.* **2016**, *46*, 66–77. [CrossRef]
30. Vergara, M.; Sancho-Bru, J.L.; Gracia-Ibáñez, V.; Pérez-González, A. An introductory study of common grasps used by adults during performance of activities of daily living. *J. Hand Ther.* **2014**, *27*, 1–28. [CrossRef]
31. Hermens, H.J.; Freriks, B.; Disselhorst-Klug, C.; Rau, G. Development of recommendations for SEMG sensors and sensor placement procedures. *J. Electromyogr. Kinesiol.* **2000**, *10*, 361–374. [CrossRef]
32. Toledo-Perez, D.C.; Rodriguez-Resendiz, J.; Gomez-Loenzo, R.A. A study of computing zero crossing methods and an improved proposal for EMG signals. *IEEE Access* **2020**, *8*, 8783–8790. [CrossRef]
33. Konrad, P. *The ABC of EMG A Practical Introduction to Kinesiological Electromyography*; Noraxon: Scottsdale, AZ, USA, 2005.
34. Haugen, I.K.; Aaserud, J.; Kvien, T.K. Get a Grip on Factors Related to Grip Strength in Persons With Hand Osteoarthritis: Results From an Observational Cohort Study. *Arthritis Care Res.* **2021**, *73*, 794–800. [CrossRef]
35. Calder, K.M.; Galea, V.; Wessel, J.; MacDermid, J.C.; MacIntyre, N.J. Muscle activation during hand dexterity tasks in women with hand osteoarthritis and control subjects. *J. Hand Ther.* **2011**, *24*, 207–215. [CrossRef]

sensors

Article

COVID-19 Detection Using Photoplethysmography and Neural Networks

Sara Lombardi [1,*], Piergiorgio Francia [1], Rossella Deodati [2], Italo Calamai [2], Marco Luchini [2], Rosario Spina [2] and Leonardo Bocchi [1]

1 Department of Information Engineering, University of Florence, 50139 Florence, Italy
2 Ospedale San Giuseppe, 50053 Empoli, Italy
* Correspondence: sara.lombardi2@unifi.it

Abstract: The early identification of microvascular changes in patients with Coronavirus Disease 2019 (COVID-19) may offer an important clinical opportunity. This study aimed to define a method, based on deep learning approaches, for the identification of COVID-19 patients from the analysis of the raw PPG signal, acquired with a pulse oximeter. To develop the method, we acquired the PPG signal of 93 COVID-19 patients and 90 healthy control subjects using a finger pulse oximeter. To select the good quality portions of the signal, we developed a template-matching method that excludes samples corrupted by noise or motion artefacts. These samples were subsequently used to develop a custom convolutional neural network model. The model accepts PPG signal segments as input and performs a binary classification between COVID-19 and control samples. The proposed model showed good performance in identifying COVID-19 patients, achieving 83.86% accuracy and 84.30% sensitivity (hold-out validation) on test data. The obtained results indicate that photoplethysmography may be a useful tool for microcirculation assessment and early recognition of SARS-CoV-2-induced microvascular changes. In addition, such a noninvasive and low-cost method is well suited for the development of a user-friendly system, potentially applicable even in resource-limited healthcare settings.

Keywords: photoplethysmogram; microcirculation; deep learning; convolutional neural network; modelling; classification

Citation: Lombardi, S.; Francia, P.; Deodati, R.; Calamai, I.; Luchini, M.; Spina, R.; Bocchi, L. COVID-19 Detection Using Photoplethysmography and Neural Networks. *Sensors* **2023**, *23*, 2561. https://doi.org/10.3390/s23052561

Academic Editors: Alessandro Scano, Alfonso Mastropietro and Massimo W. Rivolta

Received: 1 February 2023
Revised: 23 February 2023
Accepted: 23 February 2023
Published: 25 February 2023

1. Introduction

COVID-19 is an infectious respiratory disease caused by SARS-CoV-2, a coronavirus discovered in the city of Wuhan, China, in 2019 [1]. Since then, the virus has spread rapidly to other countries around the world, causing a global health and economic crisis. According to data from the World Health Organization (WHO), more than 664 million cases and more than 6.6 million deaths have been recorded as of January 2023 [2]. The rapid spread and the difficulties of treating patients with SARS-CoV-2 infection have led to the development of several diagnostic methods for the early recognition and treatment of patients with COVID-19. Except for the molecular test, based on reverse transcription-polymerase chain reaction (RT-PCR), which remains the reference diagnostic tool, low-cost and easy-to-perform procedures, and tests have also been proposed. Among these, the analysis of the photoplethysmogram (PPG) signal as a means for the early recognition of patients with COVID-19 in the hospital setting has been suggested.

COVID-19 infection typically presents with symptoms such as weakness or fatigue with fever, dry cough and shortness of breath. In severe infection, the symptomatology may progress to serious complications such as pneumonia, acute respiratory distress syndrome (ARDS), requiring intubation and emergency treatment. The virus binds to upper respiratory tract epithelial cells primarily through the ACE-2 receptor, which is highly expressed in adult nasal epithelial cells. The virus then undergoes replication and propagation within the upper respiratory tract, triggering the immune response responsible for the onset of typical symptomatology. If the immune response is not sufficient to contain the spread of the infection, lower respiratory tract (pulmonary alveoli) involvement and progression to acute respiratory distress syndrome (ARDS) occurs in severe cases [3]. Infected lung cells release a storm of cytokines (CS) that triggers an exaggerated host immune system response that can culminate in widespread cellular damage. As previously observed in other clinical conditions such as sepsis [4,5], the body's immune response results in endothelial dysfunction that can induce microvascular damage, coagulation alterations, and consequently contribute to organ dysfunction [6]. In this regard, it has been reported that in COVID-19 patients, systemic microcirculatory changes accompanied by endothelial dysfunction correlate with the severity of ARDS [7]. The role of endothelial dysfunction is important considering that it has been associated with poor prognosis in the acute phase and with persistent symptoms, such as chest pain and fatigue, during the long COVID-19 period (4 weeks or more after onset infection) [8]. Therefore, an analysis of microcirculation and endothelial damage may play a key role in both the clinical course of COVID-19 and the evaluation of the long-term effects of this clinical condition. This evaluation could allow the development of new tools for monitoring patients to reduce the number of severe cases requiring intensive care units. In this context, the use of devices such as the pulse oximeter may be a valuable solution. The pulse oximeter is a non-invasive optical device based on the technique of photoplethysmography that allows the measurement of blood volume changes in a peripheral district, usually the fingertip or earlobe. The definition of the anatomical site where measurement is performed is a key point in the acquisition protocol, since perfusion characteristics vary according to the measurement location [9,10]. This device is commonly used for the estimation of heart rate and for the measurement of blood oxygenation (SpO2). In addition to these commonly monitored parameters, it is known that the characteristic components of the pulse oximeter waveform (PPG) are associated with specific circulatory functions [11,12]. In this perspective, a detailed analysis of the PPG waveform could provide important information on the microcirculatory function abnormalities and enable early recognition of patients with SARS-CoV-2 infection.

In our previous work, Rossi et. al [13], we investigated the feasibility of using the photoplethysmographic signal through a multi-exponential model to recognise patients hospitalised with COVID-19 and the severity of the disease itself. The photoplethysmographic signal was evaluated in 93 subjects with the aim of discriminating between healthy controls and COVID-19 patients of different severity. Using the parameters of the mathematical model, three different classifiers (Bayesian, SVM and KNN) were trained and tested, validating the results obtained by the leave-one-subject-out method. In this work, we will use the same dataset used in that study by proposing a different method for the analysis of the PPG signal. In particular, this article presents a new method for PPG signal pre-processing and a custom deep learning model that, starting from PPG signal analysis only, performs classification between COVID-19 patients and control subjects. Regarding the pre-processing phase, we developed a method that analyzes waveform morphology. Specifically, we adopted a Template Matching approach that performs a pulse-by-pulse comparison with a reference signal. With regard to the deep learning model, we developed a convolutional neural network architecture, a type of model that is finding increasing application in the field of biosignal analysis [14]. The method proposed in this paper, based only on the pulse oximeter signal, could be applied as a first assessment tool for the identification of COVID-19-induced microcirculatory alterations. Moreover, since the pulse oximeter is a low-cost device, as well as already widely used in hospital settings, the introduction of such a method would not imply any additional costs and could also be applied in healthcare settings with limited resources, such as those in underdeveloped countries and territorial emergency. Furthermore, due to the easy usage and the non-invasiveness of the device, this method could be useful in the development of a clinician-friendly system that could potentially be applied to other clinical conditions that have an impact on peripheral circulation, such as hypertension or sepsis.

The purpose of this study was to define a method, based on deep learning approaches, for the identification of COVID-19 subjects from the analysis of the raw PPG signal only. In addition, comparison with the results obtained with the different procedure [13] applied on the same sample of patients is a further objective of this study. In the Section 2, we reported the main studies that, similar to ours, have adopted a template-matching method for analysing the PPG signal or have used an artificial intelligence method applied to the photoplethysmogram. Our method is described in detail in Section 3. In particular, we described the data acquisition aspects, the implementation details of the pre-processing algorithm and the architecture of the neural network, together with the strategy adopted for the model training. The obtained results are reported in Section 4, while a discussion of them and a comparison with other work is given in Section 5, where limitations and future developments of the present study are highlighted.

2. Related Methods

There are many techniques that can be used to analyze the PPG signal. In this sense, knowing the performance and characteristics of different methods can contribute to optimising the treatment of patients. In this section, we report the main studies resulting from the literature review which, similarly to our method, adopted a template-matching approach for processing or an artificial intelligence approach for analysing the PPG signal.

Proposed template matching techniques differ from each other both in the method of reference signal (template) generation and in the metric used in the pulse-by-pulse comparison. Sukor et al. [15] derived a reference template by averaging the individual pulses of a PPG segment. The authors compared all pulses with the template by evaluating the Euclidean distance and the ratio between the amplitudes of the two signals. Acceptability thresholds for the two metrics were determined heuristically. Orphanidou et al. [16] and Karlen et al. [17] used Pearson's correlation coefficient as a metric for the Template Matching. Orphanidou et al. derived the reference signal as the average of pulses in a PPG segment and then evaluated the correlation of each pulse with the template. The average correlation coefficient over the segment was then used as a metric for selecting good samples by imposing heuristic thresholds obtained from applying the method to different PPG sensors. Karlen et al., conversely, assessed the quality of each pulse by calculating the correlation between consecutive pulses, imposing a threshold for the correlation coefficient of 0.99. Considering a maximum number of consecutive pulses equal to 10 the assumption is that clean pulses taken from a short time interval are more or less equal to each other, unless they are corrupted with artefacts. Li et al. [18] used dynamic time warping (DTW) to match each beat to a template. By calculating the correlation and by using a signal clipping algorithm, the authors derived 4 features, which were used to train a multilayer perceptron with the goal of identifying good and bad-quality pulses. The DTW technique was also used in the study of Papini et al. [19]. The authors compared the morphology of PPG pulses with an adaptive template obtained by DTW barycenter averaging several beats, to consider physiological differences among individual pulses. The quality index of each sample was evaluated by taking into account the mean square error of dissimilarities between pulse and template. We recently proposed a PPG pre-processing method that required the generation of an ideal synthetic signal [20]. In particular, 3-s windows of the signal were compared with the reference signal by calculating the correlation coefficient. The use of a synthetic signal allows for very selective sample selection but it has the limitation of not accounting for the morphological variability of the waveform among the subjects.

Several studies used PPG waveform analysis applied to the study of cardiovascular disease. Nayan et al. [21] analyzed a set of 20 features extracted from the PPG signal using machine learning approaches for classification between healthy and COVID-19 subjects. The considered characteristics included amplitudes and time intervals of the main morphological features of the waveform: pulse onset, systolic peak, diastolic peak and dichrotic notch. The authors evaluated the performance of different classifiers, such as discriminant analysis (DA), k-nearest neighbour (KNN), decision tree (DT), support vector machine (SVM) and artificial neural network (ANN). The results obtained showed that ANN performed best in discriminating the two classes, achieving 95.45% of accuracy on the test set and 84.62% of accuracy on the validation set. Praveen et al. [22] used a feature vector extracted from the PPG signal to train three machine learning models (Random forest, Gradient boost, Xgboost) to classify blood pressure into 4 different stages of hypertension. Other approaches involve the use of deep learning methods to analyze the raw PPG signal, without requiring the process of feature selection and extraction from the data. Paviglianiti et al. [23] trained several neural networks to infer arterial blood pressure starting from photoplethysmogram (PPG) and electrocardiogram waveforms, obtaining good results on the estimation of diastolic and systolic pressure. Mahmud et al. [24] proposed a new approach for predicting the severity of hypoxia using deep learning applied to the PPG signal. This method is an alternative to the traditional application of the pulse oximeter, which, having a high sensitivity in detecting oxygen degradation, often has a high rate of false positives that could lead to desensitization of healthcare operators.

3. Materials and Methods

3.1. Data Acquisition

Data acquisition was carried out at S. Giuseppe Hospital in Empoli, Italy. A total of 183 subjects were recruited for the study, including 93 subjects affected by COVID-19 and 90 control, healthy subjects not affected by the target disease. The COVID-19 group included RT-PCR-positive subjects admitted to the hospital with medium to high disease severity, identified by the need for treatment with a high-flow nasal cannula (HFNC) or noninvasive ventilation (NIV). Subjects in the control group were recruited from the hospital's healthcare staff including healthy subjects not affected by COVID-19 or by other cardiovascular diseases. Only subjects older than 18 years and of white Caucasian ethnicity were included. The inclusion criterion on ethnicity resulted from the fact that as shown in recent studies, many factors can influence the PPG waveform, and among them one of the most significant is the skin color [25,26]. All participants accepted informed consent before being enrolled into the study.

The patient cohort recruited for the study was the same as the one used in the work of Rossi et al. [13] except for the number of control patients, which was increased to balance the number of subjects with COVID-19. Among the covid group, 64% of subjects were men and 36% were women, while in the control group, men accounted for 37% of subjects and women for 63%. The mean and standard deviation of age were (65.93 ± 17.75) for septic subjects and (43.99 ± 11.16) for control subjects.

For each subject, the protocol consisted of the acquisition of the photoplethysmographic trace using a finger pulse oximeter. The acquisition took place under resting conditions and for a duration of at least 5 min. The measurement site was the index finger of the right hand for all subjects involved. The acquisition system consisted of a finger pulse oximeter connected to the Mindray ePM-10 monitor, commonly used in the hospital for continuous monitoring of patients' vital parameters. A Raspberry Pi 3 device, connected to the monitor using a network connection and an HL7 (Health level seven) protocol, was used to store the waveforms. Data were acquired with a 60 Hz sampling frequency and stored as standard HL7 messages. In the first decoding step, PPG waveform values were extracted from the HL7 message for each subject. Then the signals were stored with a progressive numerical code so as to eliminate any identifying data that could trace back to the patient.

3.2. PPG Quality Assessment

The PPG waveform is susceptible to various forms of noise. Among these, one of the most common is the presence of motion artefacts that distort the shape of the signal. In this study, we developed an algorithm for the evaluation of PPG signal quality based on waveform morphology. A PPG pulse is characterised by a rising phase (anacrotic phase), which represents the systolic phase of the heart, and a falling phase (catacrotic phase), which represents the diastolic phase. A valley, called a dichrotic notch is often present in the catacrotic phase of the waveform, and is associated with aortic valve closure and good arterial function [27]. The use of morphological features to assess PPG signal quality has been widely used in the literature. One of the most common methods is a pulse-by-pulse comparison with a reference signal, which is called Template Matching. In this work, we implemented a Template Matching method by deriving, from each acquisition, a patient-specific reference pulse. This pulse was then compared with the entire PPG signal through the calculation of the Pearson correlation coefficient. The good quality portions of the signal were selected by imposing a threshold for the correlation coefficient.

3.2.1. Template Calculation

In our study, each patient-acquired signal was processed to obtain a specific reference pulse. Each PPG acquisition was normalised to have values between −1 and 1, and then the signal was filtered with a Butterworth bandpass filter with cutoff frequencies of 0.5 and 5 Hz [15]. The filtering allowed the preservation of spectral components related to cardiac activity, thus, facilitating subsequent identification of systolic peaks. From the filtered PPG, the lower and upper envelope of the signal were calculated to identify the position of the pulse onset and systolic peaks (Figure 1a).

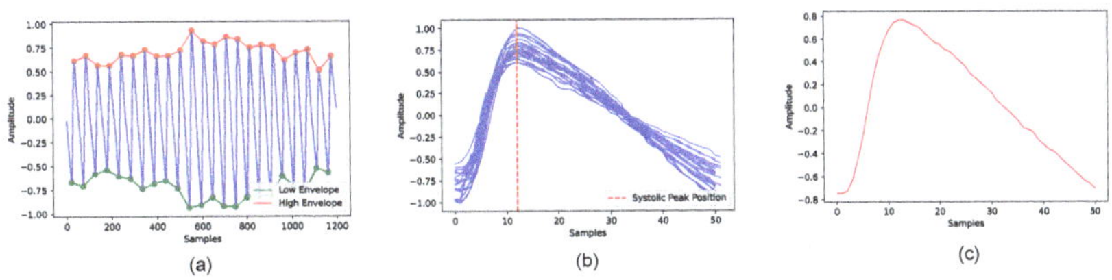

Figure 1. Main steps for calculating the reference template for each patient. (**a**) shows the lower and upper envelope of the signal. (**b**) shows the alignment of segmented pulses on the systolic peak. The calculated template is presented in the (**c**).

This allowed the segmentation of each individual pulse of the signal, identified as the waveform between two consecutive onsets. At this stage we provided limits to the pulse duration imposed by natural cardiovascular physiology so that only those peaks that met the physiological limits were considered for template calculation. Specifically, the limits imposed include minimum and maximum values for the systolic phase (SP), that is the rising wave between the pulse onset and the systolic peak, and limits for the pulse wave duration (PWD). The acceptable values for the duration of the systolic phase were in the range of 0.08 to 0.49 s, as described in the study of Fisher et al. [28]. The constraints for PWD were calculated, as described in Equation (1), by imposing a minimum mean heart rate of 40 bpm and a maximum mean heart rate of 180 bpm, considering that the subject was in a resting state during acquisition.

$$PWD_{min} = \frac{60 \times F_s}{HR_{max}}; \quad PWD_{max} = \frac{60 \times F_s}{HR_{min}} \tag{1}$$

With regard to the pulse duration, we also derived a PWD reference value by calculating the median of the width of the pulses. Samples with PWD that differed from the median value of more than 30% were not considered for template calculation. The selected pulses were then aligned on the systolic peak, as shown in Figure 1b. The reference point for alignment was calculated as the mean of the position of the systolic peak of all pulses. To obtain pulses of the same length, we performed truncation of the longer samples and constant-value padding at the beginning or end of the signal for the shorter samples. Once the samples were aligned, we obtained the template by calculating the median of the pulse waveforms (Figure 1c). The implemented algorithm for template calculation is summarised in Figure 2

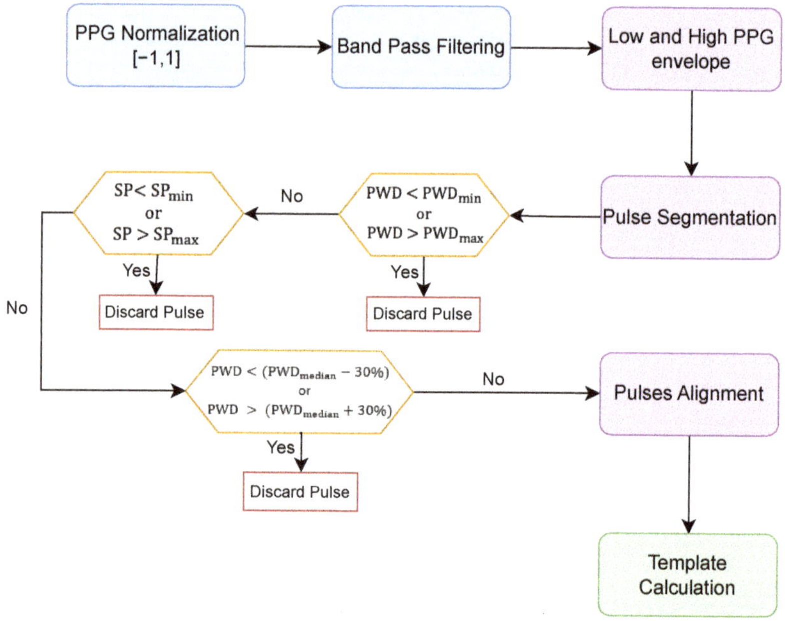

Figure 2. Flowchart of algorithm for template generation.

3.2.2. Quality Assessment

Once the reference template was obtained for each patient, signal quality was assessed by calculating the Person's correlation between the template and each segmented pulse of the patient acquisition. Each pulse was rated of acceptable quality if it correlated with the template equal to or greater than 0.8. The threshold for the correlation coefficient was determined empirically by visual inspection of the waveforms. Therefore, we stored all portions of the signal that contained consecutive pulses labelled as being of good quality. As a result, we obtained PPG samples of varying lengths associated with the same subject. Among these, we only selected for further analysis those samples with a minimum duration of 30 s. The minimum duration of 30 s was chosen experimentally, considering the need to select a waveform of the longest possible duration and the need to have as much data as possible available for training the neural network. As a result of the preprocessing algorithm, we obtained 336 PPG samples. Specifically, 186 samples from 81 patients of the control group, while a total of 150 samples from 84 patients from the covid group.

3.3. *Dataset Construction*

The selected PPG samples were then divided into a training set and test set. As a result of the pre-processing algorithm, multiple PPG samples could be associated with each patient. The division of the available data was done to ensure that data from a specific patient was present in only one of the two sets. The assignment of subjects to the training or test set was done completely randomly. The training set was used for neural network model development, while the test set was used for model performance evaluation. Since the PPG samples can have variable durations, we segmented the test samples to obtain a fixed set of PPG segments on which to perform neural network performance evaluation. Segmentation was performed by deriving for each PPG sample all possible 30-s duration windows from the onset points of individual pulses. The number of samples and the number of subjects in each set of data are reported in Table 1.

Table 1. Description of training and test sets data.

	Training Set		Test Set		Test Set Segmented	
	COVID	Control	COVID	Control	COVID	Control
N subjects	68	65	16	16	16	16
N samples	119	143	29	36	2159	2339
Ratio [%]	45.4%	54.6%	44.6%	55.4%	48.0%	52.0%

3.4. Neural Network Architecture

The design, training and testing of the neural network were implemented in Python using the Tensorflow and Keras frameworks. All experiments were conducted on a computer with an Intel i9-11900 2.5 GHz processor and 48 GB RAM within the Microsoft Windows 10 Pro operating system (Lenovo Italy S.R.L., 20054 Milano, Italy). The model structure used in this study is an architecture based on a convolutional neural network (CNN). CNN architectures are made of 3 main layers: the convolution layer, the pooling layer and the fully connected dense layer. The convolution layers and pooling layers compose the first block of the model, which is devoted to featuring extraction from the input data. The last block of the architecture consists of a fully connected network formed by dense layers, and is responsible for associating the extracted features with the desired output. Our custom model consists of 4 feature extraction blocks (CONV Block), each comprising a 1D Convolution layer, ReLu activation and a Max Pooling layer. The first two CONV Blocks have a number of filters equal to 64, while in the last two, the number of filters is 128. All convolution layers have a kernel size of 11 and all Max Pooling layers have a filter width of 4 and stride size of 2. The fully connected network includes a first dense layer with 100 units, followed by a layer with 50 units. For both layers, we included the dropout method with a rate of 0.2 as a regularization strategy to prevent model overfitting. The output layer contains two nodes with softmax activation, as we want to discriminate between two classes. As input, the model takes 30-s PPG segments normalised to have values in the range $[-1, +1]$. The detailed description of the proposed architecture is shown in Figure 3.

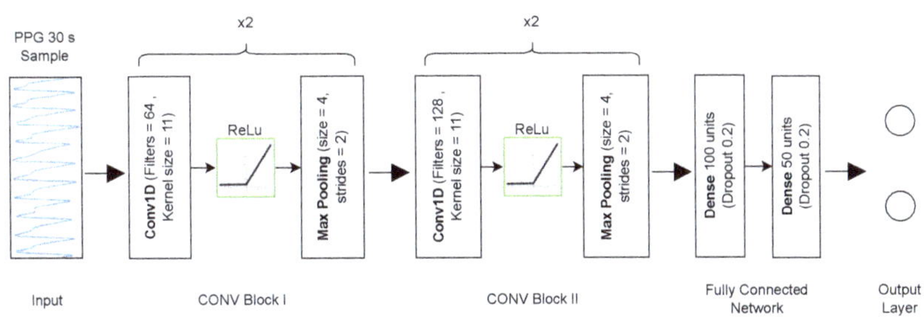

Figure 3. Description of our custom CNN architecture.

Regarding the complexity of the proposed model, we analyzed some of the most commonly used metrics to assess the complexity of artificial neural networks: the number of trainable parameters, the number of Floating Point Operations (FLOP) and the inference time. As for the first metric, our model has 1,614,532 trainable parameters. With regard to the number of FLOP, this metric represents the total number of calculations (for example, additions or multiplications) that the model has to perform to process an input sample. Each layer of the model involves performing a number of operations that depend on the structure of the layer itself, e.g., the number of FLOP for a one-dimensional convolutional layer depends on the number of filters, the kernel size, the number of input features and the output size. For our architecture, we estimated the number of floating-point operations equal to 236.74 MFLOP using the TensorFlow Python API. Finally, the inference time represents how long it takes to process an input and produce the output. This parameter depends on the available hardware and, in particular, on the number of Floating Point Operations per Second (FLOPS). This measure can be obtained from the CPU specification and, in our case, is 3.2×10^5 MFLOPS. The inference time was then calculated by dividing the number of FLOP required from the model by the number of operations per second supported by the CPU, yielding an inference time of 0.74 ms.

3.5. Model Training

When working with neural networks, three sets of data are usually used for training, validation and testing of the model, respectively. At the same time, to evaluate the generalization ability of the model, cross-validation is typically adopted. There are several ways to validate a model, in this case, we adopted 5-fold cross-validation. Validation data were derived from the training set by performing stratified group sampling, where each group contains PPG samples related to a specific patient. In this way, we obtained 5 sets of PPG samples containing data from different subjects, thus, permitting evaluation of the robustness of the method with respect to data variation. In each iteration, 1 of the 5 groups constituted the validation set, and the other 4 were used to train the model. The same architecture, previously described in Figure 3, was used for each cross-validation iteration.

Our architecture takes 30-s PPG segments as input examples. Therefore, a segment of the desired duration was derived from each sample in the training set. The selection of that segment was made during the training process by considering a 30-s window that had as its starting point the onset of one of the individual pulses that constitute the waveform. At each iteration, the selected window was different; thus, the model was trained with many different portions of a signal from the same patient. The range of values assumed by each input data was between -1 and 1.

Regarding the selection of the training hyperparameters, a trial-and-error approach was used, evaluating the trend of the learning curves and the performance obtained by the model on the validation and test set. The investigated parameters were batch size, learning rate, optimiser, loss function and the number of epochs. The chosen parameters for the final version of the model are summarised in Table 2.

Table 2. Chosen hyperparameters for model training.

Hyperparameter	Value
Learning rate	1×10^{-6}
Number of Epochs	800
Batch size	8
Optimiser	Adam
Loss Function	Mean Squared Error (MSE)

Given the limited amount of data available, after validating our method in cross-validation, we re-trained the model using all the data in the training set, assuming to improve the performance due to the utilization of more data.

4. Evaluation Results

The model was evaluated in the training phase by considering the average performance obtained on the cross-validation sets and then on the data selected for the test set. In both cases, the evaluated metrics were: accuracy, sensitivity, specificity and precision. Areas under the curve (AUC) of the Receiver Operating Characteristic (ROC) curve and the Precision-Recall (PR) curve were also measured.

The cross-validation process produced 5 different models. Each model differed from the other in the subjects used in training and validation, thus, permitting assessment of the robustness of the method with respect to the physiological variability of the subjects. The average performance of our architecture on the validation sets resulted in an accuracy of 79.01%, a sensitivity of 80.02%, a specificity of 76.57% and a precision of 74.95%.

Then the performance of each model was evaluated on the test set data. All models showed consistent performance on test data, as shown in Table 3. In addition to the average performance of the models, we evaluated an "ensemble" approach, previously used in our other work [29], in which all models were combined in the prediction process. In this method, for each test sample, all models were consulted and the class obtaining the majority of votes was considered as the final prediction.

Table 3. Performances of 5-fold cross-validation sets on test data.

	Accuracy [%]	Sensitivity [%]	Specificity [%]	Precision [%]	AUC ROC	AUC PR
Fold I	83.95	87.45	80.72	80.72	0.85	0.88
Fold II	82.33	83.97	80.80	80.15	0.84	0.84
Fold III	78.21	81.15	75.50	75.35	0.82	0.84
Fold IV	80.50	81.24	79.82	78.80	0.81	0.80
Fold V	83.06	83.00	83.11	81.94	0.83	0.81
Average Performances	81.61	83.36	79.99	79.39	0.83	0.83
Model Ensemble	82.88	85.27	80.68	80.29	0.84	0.86

Each fold of data differs in the subjects used in training and validation. Each fold produced a different trained model, whose evaluation on the test set data is shown in the table. The table reports both the performance of the individual models, the average performance and the performance obtained using the ensemble of models.

As we expected, the combined use of the 5 models resulted in an increase in performance over that achieved by a single one. Similarly, we evaluated the performance on the test set after using all the training set data to train the neural network (hold-out validation). The obtained results are summarised by the ROC Curve and the PR Curve shown in Figure 4.

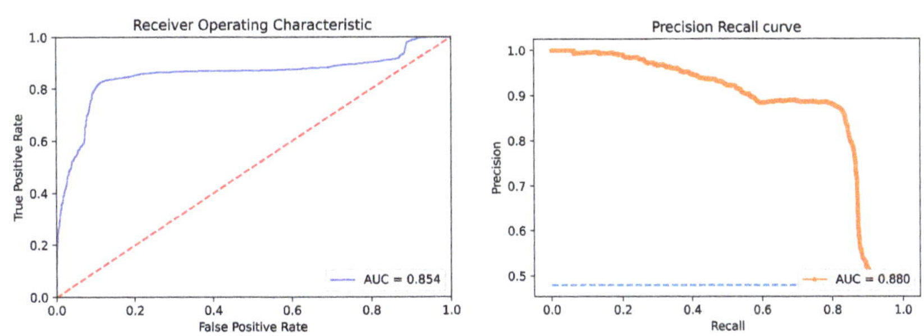

Figure 4. Model performance on the test set. (**a**) shows the ROC Curve and the correspondent AUC. (**b**) illustrates the PR and the associated AUC. Each point, on both curves, is derived from the values of the confusion matrix associated with the application of a specific cutoff to the predictions of the classifier.

These curves show the ability of a model to classify binary outcomes for each possible cutoff value applied to the classifier's predictions. Specifically, the ROC curve is generated by plotting a model's false positive rate against the true positive rate, while the PR curve plots the true positive rate (recall or sensitivity) against the positive predictive value (precision). With a threshold equal to 0.5, our model achieved an accuracy of 83.86%, a sensitivity of 84.30%, a specificity of 83.45% and a precision of 82.46%. The total number of predictions for each class is described in the confusion matrix, shown in Figure 5.

		Predicted label		
		Covid	**Control**	**Total**
True Label	**Covid**	**1820**	**339**	2159
	Control	**357**	**1952**	2307
	Total	2177	2291	

Figure 5. Confusion matrix on the test set using the hold-out validation.

The results obtained in the hold-out validation confirm our hypothesis that more data available for model implementation could lead to improved performance.

The most significant parameter for our study is sensitivity, which identifies the percentage of COVID-19 samples correctly identified. This parameter reached a good value of 84.30%. However, since multiple 30-s PPG windows are associated with each subject, to assess the true percentage of correctly classified subjects, we performed the test on the individual patient. In this case, we evaluated the number of correctly identified PPG samples for each patient. Therefore, each subject was considered to be correctly classified if most of his/her signal samples were associated with the right class. In this testing modality, our method correctly classified 25 of the 32 patients assigned to the test set, corresponding to an accuracy of 78%, sensitivity of 75% and specificity of 81%.

5. Discussion

In this study, we evaluated the possibility of using the PPG signal to identify patients infected with COVID-19. Specifically, we presented a new template matching method for PPG signal pre-processing and we developed a CNN deep learning-based model for analyzing the photoplethysmographic signal acquired with a common pulse oximeter. Data acquisition was carried out at S. Giuseppe Hospital in Empoli, using the Mindray multiparameter monitor commonly used in the Intensive Care Unit, thus, simulating a real application of the developed method. The collected data were divided into training set and test set, which were, respectively, used for classifier training and performance evaluation. To assess the robustness of the classifier with respect to variation in the subjects used for performance evaluation, we initially implemented cross-validation and then performed hold-out validation. In the hold-out validation, our model showed good performance in the classification between COVID-19 patients and control subjects by achieving an accuracy of 83.86% and a sensitivity of 84.30% on the test data. Observing the ROC curve related to our classifier (Figure 4), it can be seen that the curve reaches a plateau. This means that as the threshold applied to the model's predictions increases, there is no corresponding improvement in the performance of the classifier. Therefore, it can be deduced that there are some patients whom the model fails to classify. We can hypothesise that this performance may be due to unimpaired microcirculation in these subjects.

Overall, the obtained results confirmed the presence of microvascular changes due to SARS-CoV-2 infection and the potential of photoplethysmography as a tool for microcirculation assessment. This method, based only on samples of the PPG signal, seems to be suitable for a rapid screening procedure, which could provide the clinician with an early warning signal and allow, for example, the use of specific diagnostic procedures. Moreover, given the noninvasiveness and wide use of this device, especially in the hospital setting, this method could be a tool for the evaluation of microcirculatory changes that does not introduce additional costs. The results obtained allow us to compare the usefulness of using deep learning approaches versus other methods based on feature extraction from the photoplethysmogram. In the study of Nayan et al. [21] a set of features extracted from the

PPG signal was used to classify COVID-19 patients using Machine learning approaches. Similar to our study, the classifiers were trained by implementing a 5-fold cross-validation on the training data and then evaluated on the test set. In particular, the best performing classifier was a feed-forward multilayer perceptron network, which achieved consistent performance on both the validation set and the test set, in contrast to other classifiers that had significantly lower performance on the validation set. The authors obtained excellent results achieving more than 90% accuracy on the test set and 84.62% accuracy on the validation set. Although our work yielded lower performance than the method described by Nayan et al. we believe it still has advantages. Differently from that study, our method does not require the process of extracting and selecting morphological features from the PPG signal, but processes 30-s windows of the raw PPG signal. This could be particularly advantageous in the case of signals acquired under uncontrolled conditions, such as those acquired from multi-parameter monitors in Intensive Care Units, for which the extraction of PPG features could be challenging. In our previous study, Rossi et al. [13], the PPG features were derived by fitting the waveform with a 3-exponential model. The model parameters were then used in ML approaches to identify COVID-19 patients with different severity. Since this study is based on the same data set, the aim of this study is to compare the obtained results with those achieved using the exponential photoplethysmogram model. Given the limited number of subjects enrolled in the study and, consequently, the limited amount of data available for training the neural network, our work focused on the classification between control healthy subjects, indicated as group 0, and covid subjects, regardless of severity, identified as a group (1, 2) in our previous study. In that study, the comparison between group 0 and group (1, 2) was performed in three different ways utilising the Bayesian Classifier with the Leave-One-Subject-Out (LOSO) validation method. The classifier was trained both with features extracted from a single beat and with features averaged over two consecutive beats. The classification of the patient was then obtained based on the majority of the classifications of the single or pairs of cycles. Furthermore, performances were evaluated by considering a single feature vector per patient, obtained by averaging the characteristics over the entire acquisition. The best performance was obtained using the average feature vector, resulting in an accuracy of 70%, sensitivity of 68% and specificity of 74% in the classification of subjects. Although the methods are not directly comparable, as they were validated using different methods, we are interested in comparing the performance of the two approaches in terms of correctly classified subjects. In this respect, we can observe that the method proposed in this work, based on a deep learning model, performed better in classifying individual subjects, achieving an accuracy of 78%, a sensitivity of 75% and a specificity of 81%.

Study Limitations and Future Developments

Overall, the findings confirm the potential of the proposed method for the early assessment of microcirculation alterations in COVID-19 patients. However, it is necessary to consider some limitations of this work that open the way for further investigation and development of the implemented method. The main limitations are related to the dataset used. The data for training and evaluation of the model were all acquired in the same hospital. To assess the generalization ability of the model, we plan to evaluate its performance on at least one other database. In addition, we hypothesise that a greater number of data available may improve the performance of the model, as well as allow for an evaluation of performance on a larger population. Finally, the two groups of subjects enrolled in the study, although balanced in number, are biased in terms of gender and age. Investigating the influence of these two parameters on the performance of our method will be our further goal. In this regard, we are aware that interpretability of the model is a fundamental requirement for applying this method in the medical field. For this reason, further validations will be necessary to make the model explainable and consequently improve the clinician confidence in using this method. Finally, given the potential shown in this work by the photoplethysmographic technique in the evaluation of microcirculatory alterations, in our future work, we are interested in exploring the use of PPG imaging since optical imaging techniques may also allow a description of the spatial distribution of peripheral blood flow [30,31].

6. Conclusions

In this study, we evaluated the possibility of using the PPG signal for the screening and classification of patients with COVID-19. Specifically, we developed a custom convolutional neural network model that discriminates between Covid patients and control subjects by analyzing only the PPG signal. The proposed method achieved interesting results in terms of accuracy (78%), sensitivity (75%) and specificity (81%) on the test set data. Overall this study confirms that PPG signal may be used for the screening of patients with COVID-19 and the assessment of microcirculatory alterations. Moreover, these results are important because acquiring the photoplethysmographic trace is simple, noninvasive and inexpensive. In this regard, this method could be used to develop a user-friendly system that could represent an initial assessment tool for the clinician, applicable even in clinical settings with limited resources. Further studies with a larger sample size of patients and with data from other databases, as well as an evaluation of the interpretability of the model, will be needed to assess the effectiveness of the proposed method.

Author Contributions: S.L.: Data elaboration, formal analysis, methodology, software and results extraction, writing: original draft—review—editing; P.F.: methodology, writing: original draft—review—editing, formal analysis, reviewing; R.D.: data acquisition, reviewing, validation. I.C.: data acquisition, reviewing, validation. M.L.: data acquisition and curation, writing: original draft. R.S.: data acquisition, supervision, validation, funding acquisition. L.B.: conceptualization, data modelling, writing: original draft—review, editing, validation, funding acquisition. All authors have read and agreed to the published version of the manuscript.

Funding: This research was funded by Tuscany Region (Italy), Bando Ricerca COVID-19 Toscana, Covid Research Grant D75F21000980002.

Institutional Review Board Statement: The study was conducted in accordance with the Declaration of Helsinki, and approved by the Institutional Ethics Committee of Area Vasta Centro Ethics Committee (CEAVC) (protocol code CEAVC19059).

Informed Consent Statement: Informed consent was obtained from all subjects involved in the study.

Data Availability Statement: Not applicable.

Conflicts of Interest: The authors declare no conflict of interest. The funders had no role in the design of the study; in the collection, analyses, or interpretation of data; in the writing of the manuscript; or in the decision to publish the results.

References

1. Huang, C.; Wang, Y.; Li, X.; Ren, L.; Zhao, J.; Hu, Y.; Zhang, L.; Fan, G.; Xu, J.; Gu, X.; et al. Clinical features of patients infected with 2019 novel coronavirus in Wuhan, China. *Lancet* **2020**, *395*, 497–506. [CrossRef] [PubMed]
2. World Health Organization (WHO): Weekly Epidemiological Update on COVID-19. Available online: https://www.who.int/publications/m/item/weekly-epidemiological-update-on-COVID-19--27-january-2023 (accessed on 27 January 2023).
3. Parasher, A. COVID-19: Current understanding of its Pathophysiology, Clinical presentation and Treatment. *Postgrad. Med. J.* **2020**, *97*, 312–320. [CrossRef] [PubMed]
4. De Backer, D.; Donadello, K.; Sakr, Y.; Ospina-Tascon, G.; Salgado, D.; Scolletta, S.; Vincent, J.L. Microcirculatory alterations in patients with severe sepsis: Impact of time of assessment and relationship with outcome. *Crit. Care Med.* **2013**, *41*, 791–799. [CrossRef] [PubMed]
5. Charlton, M.; Sims, M.; Coats, T.; Thompson, J. The microcirculation and its measurement in sepsis. *J. Intensive Care Soc.* **2016**, *18*, 221–227. [CrossRef]
6. Colantuoni, A.; Martini, R.; Caprari, P.; Ballestri, M.; Capecchi, P.L.; Gnasso, A.; Lo Presti, R.; Marcoccia, A.; Rossi, M.; Caimi, G. COVID-19 Sepsis and Microcirculation Dysfunction. *Front. Physiol.* **2020**, *11*, 747. [CrossRef]
7. Mesquida, J.; Caballer, A.; Cortese, L.; Vila, C.; Karadeniz, U.; Pagliazzi, M.; Zanoletti, M.; Pacheco, A.; Castro, P.; García-de Acilu, M.; et al. Peripheral microcirculatory alterations are associated with the severity of acute respiratory distress syndrome in COVID-19 patients admitted to intermediate respiratory and intensive care units. *Crit. Care* **2021**, *25*, 381. [CrossRef]

8. Charfeddine, S.; Ibn Hadj Amor, H.; Jdidi, J.; Torjmen, S.; Kraiem, S.; Hammami, R.; Bahloul, A.; Kallel, N.; Moussa, N.; Imen, T.; et al. Long COVID 19 Syndrome: Is It Related to Microcirculation and Endothelial Dysfunction? Insights From TUN-EndCOV Study. *Front. Cardiovasc. Med.* **2021**, *8*, 1702. [CrossRef]
9. Sorelli, M.; Stoyneva, Z.; Mizeva, I.; Bocchi, L. Spatial heterogeneity in the time and frequency properties of skin perfusion. *Physiol. Meas.* **2017**, *38*, 860–876. [CrossRef]
10. Bandini, A.; Orlandi, S.; Manfredi, C.; Evangelisti, A.; Barrella, M.; Bevilacqua, M.; Bocchi, L. Effect of local blood flow in thermal regulation in diabetic patient. *Microvasc. Res.* **2013**, *88*, 42–47. [CrossRef]
11. Elgendi, M. On the Analysis of Fingertip Photoplethysmogram Signals. *Curr. Cardiol. Rev.* **2012**, *8*, 14–25. [CrossRef]
12. Castaneda, D.; Esparza, A.; Ghamari, M.; Soltanpur, C.; Nazeran, H. A review on wearable photoplethysmography sensors and their potential future applications in health care. *Int. J. Biosens. Bioelectron.* **2018**, *4*, 195. [CrossRef]
13. Rossi, E.; Aliani, C.; Francia, P.; Deodati, R.; Calamai, I.; Luchini, M.; Spina, R.; Bocchi, L. COVID-19 detection using a model of photoplethysmography (PPG) signals. *Med. Eng. Phys.* **2022**, *109*, 103904. [CrossRef]
14. Alaskar, H. Convolutional neural network application in biomedical signals. *J. Comput. Sci. Inform. Technol.* **2018**, *6*, 45–59. [CrossRef]
15. Sukor, J.; Lovell, N. Signal quality measures for pulse oximetry through waveform morphology analysis. *Physiol. Meas.* **2011**, *32*, 369–84. [CrossRef]
16. Orphanidou, C.; Bonnici, T.; Charlton, P.; Clifton, D.; Vallance, D.; Tarassenko, L. Signal-Quality Indices for the Electrocardiogram and Photoplethysmogram: Derivation and Applications to Wireless Monitoring. *IEEE J. Biomed. Health Inform.* **2014**, *19*, 832–838. [CrossRef]
17. Karlen, W.; Kobayashi, K.; Ansermino, M.; Dumont, G. Photoplethysmogram signal quality estimation using repeated Gaussian filters and cross-correlation. *Physiol. Meas.* **2012**, *33*, 1617–1629. [CrossRef]
18. Li, Q.; Clifford, G.D. Dynamic time warping and machine learning for signal quality assessment of pulsatile signals. *Physiol. Meas.* **2012**, *33*, 1491–1501. [CrossRef] [PubMed]
19. Papini, G.; Fonseca, P.; Eerikäinen, L.; Overeem, S.; Bergmans, J.; Vullings, R. Sinus or not: A new beat detection algorithm based on a pulse morphology quality index to extract normal sinus rhythm beats from wrist-worn photoplethysmography recordings. *Physiol. Meas.* **2018**, *39*, 1491–1501. [CrossRef]
20. Lombardi, S.; Partanen, P.; Bocchi, L. Detecting sepsis from photoplethysmography: Strategies for dataset preparation. In Proceedings of the 2022 44th Annual International Conference of the IEEE Engineering in Medicine & Biology Society (EMBC), Glasgow, Scotland, UK, 11–15 July 2022; pp. 2286–2289. [CrossRef]
21. Nayan, N.A.; Jie Yi, C.; Suboh, M.Z.; Mazlan, N.F.; Periyasamy, P.; Abdul Rahim, M.Y.Z.; Shah, S.A. COVID-19 Prediction With Machine Learning Technique From Extracted Features of Photoplethysmogram Morphology. *Front. Public Health* **2022**, *10*, 920849. [CrossRef]
22. Durga Siva Praveen, B.; Sandeep, D.; Raghavendra, I.; Yuvaraj, M.; Sarath, S. Non-invasive Machine Learning approach for classifying Blood Pressure using PPG Signals in COVID situation. In Proceedings of the 2021 12th International Conference on Computing Communication and Networking Technologies (ICCCNT), Kharagpur, India, 6–8 July 2021; pp. 1–7 . [CrossRef]
23. Paviglianiti, A.; Randazzo, V.; Villata, S.; Cirrincione, G.; Pasero, E. A comparison of deep learning techniques for arterial blood pressure prediction. *Cogn. Comput.* **2022**, *14*, 1689–1710. [CrossRef]
24. Mahmud, T.I.; Imran, S.A.; Shahnaz, C. Res-SE-ConvNet: A Deep Neural Network for Hypoxemia Severity Prediction for Hospital In-Patients Using Photoplethysmograph Signal. *IEEE J. Transl. Eng. Health Med.* **2022**, *10*, 4901409. [CrossRef] [PubMed]
25. Sinaki, F.Y.; Ward, R.; Abbott, D.; Allen, J.; Fletcher, R.R.; Menon, C.; Elgendi, M. Ethnic disparities in publicly-available pulse oximetry databases. *Commun. Med.* **2022**, *2*, 59. [CrossRef] [PubMed]
26. Fine, J.; Branan, K.L.; Rodriguez, A.J.; Boonya-Ananta, T.; Ramella-Roman, J.C.; McShane, M.J.; Coté, G.L. Sources of inaccuracy in photoplethysmography for continuous cardiovascular monitoring. *Biosensors* **2021**, *11*, 126. [CrossRef]
27. Allen, J. Photoplethysmography and its application in clinical physiological measurement. *Physiol. Meas.* **2007**, *28*, R1–R39. [CrossRef]
28. Fischer, C.; Domer, B.; Wibmer, T.; Penzel, T. An Algorithm for Real-Time Pulse Waveform Segmentation and Artifact Detection in Photoplethysmograms. *IEEE J. Biomed. Health Inform.* **2017**, *21*, 372–381. [CrossRef]
29. Lombardi, S.; Partanen, P.; Francia, P.; Calamai, I.; Deodati, R.; Luchini, M.; Spina, R.; Bocchi, L. Classifying sepsis from photoplethysmography. *Health Inf. Sci. Syst.* **2022**, *10*, 30. [CrossRef]
30. Frassineti, L.; Giardini, F.; Perrella, A.; Sorelli, M.; Sacconi, L.; Bocchi, L. Evaluation of spatial distribution of skin blood flow using optical imaging. In Proceedings of the CMBEBIH 2017: Proceedings of the International Conference on Medical and Biological Engineering 2017, Sarajevo, Bosnia and Herzegovina, 16–18 March 2017; pp. 74–80. [CrossRef]
31. Allan, D.; Chockalingam, N.; Naemi, R. Validation of a non-invasive imaging photoplethysmography device to assess plantar skin perfusion, a comparison with laser speckle contrast analysis. *J. Med. Eng. Technol.* **2021**, *45*, 170–176. [CrossRef]

MDPI
St. Alban-Anlage 66
4052 Basel
Switzerland
www.mdpi.com

Sensors Editorial Office
E-mail: sensors@mdpi.com
www.mdpi.com/journal/sensors

www.ingramcontent.com/pod-product-compliance
Lightning Source LLC
Chambersburg PA
CBHW040802250926
51850CB00007BA/1063
9783036598239